Inorganic Chemistry Concepts
Volume 13

Editors

Christian K. Jørgensen, Geneva · Michael F. Lappert, Brighton
Stephen J. Lippard, Cambridge, MA · John L. Margrave, Houston
Kurt Niedenzu, Lexington · Heinrich Nöth, Munich
Robert W. Parry, Salt Lake City · Hideo Yamatera, Nagoya

Charles W. Struck · William H. Fonger

Understanding Luminescence Spectra and Efficiency Using W_p and Related Functions

With 70 Figures

Springer-Verlag Berlin Heidelberg
New York London Paris Tokyo
Hong Kong Barcelona Budapest

Dr. Charles W. Struck
GTE Laboratories Inc.
40 Sylvan Road
Waltham, MA 02254
USA

Dr. William H. Fonger
174 Guyot Avenue
Princeton, NJ 08540
USA

ISBN 3-540-52766-4 Springer-Verlag Berlin Heidelberg New York
ISBN 0-387-52766-4 Springer-Verlag New York Heidelberg Berlin

Library of Congress Cataloging-in-Publication Data
Struck, Charles W., 1927 –
Understanding luminescence spectra and efficiency using W_p and related functions / Charles W. Struck, William H. Fonger.
p. – cm. – (Inorganic chemistry concepts ; v. 13)
Includes bibliographical references and index.
ISBN 3-540-52766-4 (Berlin). – ISBN 0-387-52766-4 (New York)
1. Luminescence spectroscopy. I. Fonger, William H., 1925 –
II. Title. III. Series.
QD96.L85S77   1991   543'.0852–dc20   90-10290

This work is subject to copyright. All rights are reserved, whether the whole or part of the material is concerned, specifically the rights of translation, reprinting, re-use of illustrations, recitation, broadcasting, reproduction on microfilms or in other ways, and storage in data banks. Duplication of this publication or parts thereof is only permitted under the provisions of the German Copyright Law of September 9, 1965, in its current version, and a copyright fee must always be paid. Violations fall under the prosecution act of the German Copyright Law.

© Springer-Verlag Berlin Heidelberg 1991
Printed in Germany

The use of registered names, trademarks, etc. in this publication does not imply, even in the absence of a specific statement, that such names are exempt from the relevant protective laws and regulations and therefore free for general use.

Typesetting: ASCO Trade Typesetting Ltd., Hongkong
Offsetprinting: Color-Druck, Berlin
Binding: Lüderitz & Bauer, Berlin
2151/3020-543210 – Printed on acid-free paper

Preface

There are both a remote and a proximate history in the development of this book. We would like to acknowledge first the perceptiveness of the technical administrators at RCA Laboratories, Inc. during the 1970s, and in particular Dr. P. N. Yocom. Buoyed up by the financial importance of yttrium oxysulfide: europium as the red phosphor of color television tubes, they allowed us almost a decade of close cooperation aimed at understanding the performance of this phosphor.

It is significant that we shared an approach to research in an industrial laboratory which allowed us to avoid the lure of "first-principles" approaches (which would have been severely premature) and freed us to formulate and to study the important issues directly.

We searched for a semiquantitative understanding of the properties observed in luminescence, i.e., where energy absorption occurs, where emission occurs, and with what efficiency this conversion process takes place. We were aware that the nonradiative transition rates found in practice vary enormously with temperature and, for a given activator, with small changes in its environment. We traced the source of this enormous variation to the magnitude of the vibrational overlap integrals, which have strong dependences on the rearrangements occurring during optical transitions and on the vibrational number of the initial electronic state. We were willing to excise from the problem the electronic aspects – the electronic wavefunctions and their transition integrals – by treating them as parameters to be obtained from the experimental data.

We learned to describe the vibrational components of the transition rates quantum-mechanically, using the single-configurational-coordinate diagram and the Condon operator, i.e., using the squared overlap integrals thermally weighted for the initial-state population. We learned the recursion algebra which proved fruitful for obtaining moments, for developing the W_p recursion formula, and then for developing the z and derivative operator expressions.

It was somewhat accidental that we recognized that the functions we were led to, namely W_p, are identical to the Huang-Rhys-Pekar functions derived in the 1950s for multiple coordinate models. We found a simple proof of the identity of these two model expressions, by using the W_p recursion formula and then performing some simple sums.

We then gave attention to the multiple-coordinate single-frequency model, for the z and derivative operators. Somewhat tedious algebra led to somewhat simple expressions. We wanted contact with the literature in this field, which was rich in approximate forms and, indeed, for Huang and Rhys, with the exact expression in a cumbersome form.

Finally, we gave attention to the multiple-frequency model and to its literature.

To us it is important to note the difficulty of distinguishing in practical experimental data among the various models: between one and many coordinates, between the Condon (unity) operator and the z or derivative operators, between one and many frequencies. Our conclusion is that experimental data will seldom demand adoption of one model over another and the simple W_p function emerges having the most practical value. Perhaps this emphasis, this downplaying the importance of the demand for the derivative operator which physics rigorously imposes, exhibits in yet another way our comfort within an industrial research environment.

Two classes of applications were considered and are presented here. First, an exposition of the broad classes of luminescence behavior is given and the W_p function and related functions are applied to each. One gets thereby an appreciation of the scope of the applicability of these functions and of their practical indistinguishability. Examples of each type of behavior are cited from the literature. Then we give an extensive discussion of the detailed application of these functions to specific experimental data, namely, to Eu^{3+}, Tm^{3+}, and Yb^{3+} in oxysulfides and oxychlorides and to Cr^{3+} in ruby.

We discuss certain effects seen in experimental data, due to anharmonicity dominantly, which are contained in this theory by ad-hoc embellishments. Here again, our industrial orientation might be apparent, since to us these embellishments seem an effective incorporation of the essential manifestation of anharmonicity: the expansion of the lattice and the consequent temperature variation in the energy levels.

We also discuss the surprising energy storage and loss processes encountered in oxysulfides. These are effects which are totally beyond the model of a luminescence center developed here. They therefore serve as a strong statement that not all effects are understandable within the framework of the model.

This is the remote history of the book and an exposition of the book's contents.

The more modern history of the evolution of this book leads to acknowledging the support of the technical administrators at GTE Laboratories Incorporated, particularly Dr. J. Gustafson, and the enthusiastic persistence of Professor B. DiBartolo of Boston College. Professor DiBartolo convinced one of us (CWS) to give a set of lectures on this work at his NATO school in Erice, Sicily, in the summer of 1989. It was in preparation for these lectures that this book was written. The word-processing capabilities at GTE Laboratory proved invaluable and we express our appreciation. All of the figures were prepared at the Technical Publications Center at GTE Laboratories, and we thank Shirley Blanchard, Ralph Ferrini, and Ellen Connor for their careful work.

We thank Elsevier Science Publishers, Inc. for permission to use Figures 1–13, 18, 22, 26, 27, 69, and 70, and Tables 1–4 and 12–15, and 42, all of which first appeared in The Journal of Luminescence under our authorship. We also thank them and Prof. J. C. Gâcon for the use of Figures 44 and 46, and them and Dr. E. Nakazawa for the use of Figure 42. We thank the American Institute of Physics for permission to use Figures 25–38, 45, and 47–49, and Tables 19, 22, 26, 27, and 32, which first appeared in our articles in the Journal of Chemical Physics. We thank The American Physical

Society for the use of Figures 50–56 and 60–68 and Tables 35, 37–39, and 41, which all appeared first in our articles in The Physical Review. We thank The Electrochemical Society for the use of Figures 57–59, which were first in our article in The Journal of the Electrochemical Society. We thank Plenum Press for the use of Figures 40, 41, and 43, which first appeared in an article by us in a book under Prof. B. DiBartolo's editorship entitled "Radiationless Processes". Finally, we thank Pergamon Press for permission to quote a statement of the method of steepest descents from a book by B. A. Fuchs and V. I. Levin. All of the above publishers deserve thanks also because much of the text of this book derives at least by paraphrasing from text in our articles in the various journals cited.

Finally we salute our wives, Marie Struck and, posthumously and in fond memory, Carol Fonger. They provided loving support which cost them many hours of loneliness.

Waltham, October 1990   Charles W. Struck
Princeton, October 1990   William H. Fonger

Table of Contents

1	**Introduction**	1
1.1	Luminescence Centers and Models of Them	1
1.2	The Simplest Model: One Coordinate and Equal Force Constants	2
1.2.1	The Optical Band Shapes	2
1.2.2	The Nonradiative Rate	4
1.2.3	Six Typical Thermal Quenching Behaviors	5
1.2.3.1	Fast Bottom Crossover	6
1.2.3.2	Outside Crossover	6
1.2.3.3	Small-Offset Multiphonon Emission	8
1.2.3.4	Two-Step Quenching With a Fast Second Step	10
1.2.3.5	Two-Step Quenching with a Slow Second Step	13
1.2.3.6	Low-temperature Tunnelling Crossover	14
1.3	The Franck-Condon Principle for Nonradiative Rates	14
2	**Harmonic Oscillator Wavefunctions**	19
2.1	Hermite Polynomials	19
2.2	Generating Function for the Harmonic Oscillator Wavefunctions	20
3	**The Manneback Recursion Formulas**	24
3.1	Introduction	24
3.2	The Overlap Integral	24
3.3	The Generating Function for the Overlap Integral	25
3.4	The Recursion Formulas for the Overlap Integrals	27
3.5	Familiarity	28
3.6	The Orthonormality of the A_{NM} Matrix	29
3.7	Additional Equal-Force-Constants Recursion Relations	30
4	**The Luminescence Center: the Single-Configurational-Coordinate Model**	32
4.1	The Model for the Radiative Rate	32
4.2	The Equal-Force-Constants Radiative Rate	34
4.3	The Unequal-Force-Constants Radiative Rate	37
4.4	The Model for the Nonradiative Rate	38

4.5	The W_p Recursion Formula	39
4.6	Explicit Series Expression for the W_p Function	40
4.7	I_p Modified Bessel Function Form for W_p	41
4.8	Limiting and Approximate Forms of W_p	42
4.9	The 5-W_p Formula for $W_{p,z}$	44
4.10	The $\langle m \rangle_p$ Formula	45
4.11	The $W_{p,d/dz}$ Expression	45
4.12	The W_{-p}/W_p and Related Ratios	48
4.13	Equal-Force-Constants Moments	49
4.14	Unequal-Force-Constants Moments	51
4.14.1	The Moments	51
4.14.2	Preliminaries I: The (α, m, β)	52
4.14.3	Preliminaries II: The Thermal Averages $\langle m^\alpha \rangle_v$	55
4.14.4	Preliminaries III: The $\langle n^\alpha m^\gamma \rangle_{uv}$	56
4.14.5	The Derivation of the Moment Expressions (4.109)	58
5	**Multiple Coordinate Models of a Luminescence Center**	60
5.1	The Einstein-Huang-Rhys-Pekar Single-Frequency Multiple-Coordinate Model	60
5.2	The z and d/dz Multiple-Coordinate Nuclear Factors	61
5.2.1	Preliminaries I: the Y_p Function	62
5.2.2	Preliminaries II: $X_{p,O\pm}$	63
5.2.3	Preliminaries III: XX Sums	64
5.2.4	Preliminaries IV: $W_{p,O\pm} W_p$ Sums	65
5.2.5	Proof of Eq. (5.6) for Two Coordinates	65
5.3	Multiple-Frequency Models of a Luminescence Center	66
5.3.1	The Selected Model	66
5.3.2	Definition of the 1, z, and d/dz Operator Rates	67
5.3.3	The Condon-Operator Distribution	68
5.3.4	The Recursion Algebra for the z and d/dz Operators	69
5.3.4.1	The Ψ Functions	70
5.3.4.2	The Ψ Recursion Algebras	70
5.3.4.3	The Φ Functions	71
5.3.4.4	The Φ Recursion Algebras in terms of Θ Functions	72
5.3.5	The Discretized Debye Equal S and A Model	72
6	**Energy Transfer**	75
6.1	The Model	75
7	**Compendium of Useful Equations**	77
7.1	The Wavefunctions	77
7.2	The Manneback Recursion Formulas	78
7.3	The Equal-Force-Constants W_p and Related Functions in One Dimension	79
7.4	The Unequal-Force-Constants Expressions	80

7.5	The Moments	81
7.6	Multiple Coordinate Models of a Luminescence Center	82
7.7	Energy Transfer	84
8	**Contact with the Theoretical Literature**	**86**
8.1	Unequal-Force-Constants A_{nm}	86
8.2	Equal-Force-Constants A_{nm}	86
8.2.1	Explicit Formulas	86
8.2.2	Laguerre Polynomial Expressions	87
8.2.3	Citations	88
8.3	The W_p Formula	88
8.4	The $W_{p,d/dz}$ Formula	91
8.5	The Equal-Force-Constants Moments	92
8.6	The Unequal-Force-Constants Moments	92
8.7	The Single-Frequency-Multiple-Coordinate Derivative Operator Expressions	94
8.7.1	Huang and Rhys	94
8.7.2	Perlin	95
8.7.3	Miyakawa and Dexter	95
8.8	Multiple-Frequency Rates	96
8.8.1	Perlin's Condon-Operator Distribution	96
8.8.1.1	The Distribution	96
8.8.1.2	Cauchy's Integral Theorem and its Consequences	97
8.8.1.3	The Saddle Point Approximation	98
8.8.1.4	The Use of the Saddle Point Approximation Here	99
8.8.1.5	The integral of dz/z	101
8.8.1.6	Expansion of v_m for Small Offset	102
8.8.1.7	Perlin's Derivation	104
8.8.2	The Correspondence between Perlin's and Our Multiple Frequency Expression	104
8.8.3	Perlin's Multiple-Coordinate Derivative Operator Expression	105
8.8.4	The Correspondence between Perlin's and Our Multiple Frequency Derivative Expression	107
8.8.5	Mostoller, Ganguly, and Wood	107
8.9	Energy Transfer	108
8.9.1	Förster and Dexter	108
8.9.2	Miyakawa and Dexter	109
9	**Representative Luminescence Centers**	**110**
9.1	Equal- and Unequal-Force-Constants Bandshapes and Nonradiative Transitions	110
9.1.1	Bandshapes	110
9.1.2	Nonradiative Rates	112
9.2	One- and N_{Av}-Dimensional Bandshapes	113
9.3	Vibrationally-Enhanced Radiative Transitions	119
9.4	Comparisons of Nonradiative Rate Expressions	121

10	**Experimental Studies**	129
10.1	Eu in Oxysulfides and in Oxyhalides	129
10.1.1	The Energy Level Diagram and Qualitative Behavior	129
10.1.2	Feeding Fractions	133
10.1.3	Efficiencies under Quenching Conditions	134
10.1.4	Absorption Spectra at $T > 0K$	135
10.1.5	The Fit of the Absorption Data	136
10.1.5.1	LaOCl	136
10.1.5.2	Oxysulfides	138
10.1.6	Fitting the Quenching Data	141
10.1.6.1	LaOCl	141
10.1.6.2	Oxysulfides	141
10.1.6.3	Fitting the $(\eta_i)_i$	143
10.1.7	Fitting The Feeding Fractions	145
10.1.7.1	LaOCl	145
10.1.7.2	Oxysulfides	145
10.2	Oxysulfides: Other Rare Earths	147
10.2.1	$La_2O_2S: Tm^{3+}$	147
10.2.1.1	The Experimental Behavior	147
10.2.1.2	Fitting with W_p Functions	150
10.2.2	$Y_2O_2S: Yb^{3+}$	153
10.2.2.1	The Experimental Behavior	153
10.2.2.2	Fitting the Optical Spectra with SCC Model Functions	155
10.2.2.3	Fitting the Quenchings with SCC Model Functions	155
10.3	Alkali and Alkaline Earth Halides: Sm	158
10.3.1	The Model and Expectations	158
10.3.2	The Fitting Parameters	162
10.4	Ruby	165

11	**Effects Beyond the Model: Oxysulfide: Eu Storage and Loss Processes**	173
11.1	The Need for Enhancement of the Model	173
11.2	Synopsis of the Experiments to Probe the Model	175
11.3	The Model Equations: Notation	176
11.4	CTS Dissociation: The B_0/G Behavior	178
11.5	The SCC Model for Understanding Storage-Loss Processes in Oxysulfide: Eu Phosphors	181
11.6	The Steady-State Efficiency and its Dependence on Excitation Intensity: B_∞/G	182
11.6.1	The Observed Behavior	182
11.6.2	The Equation for B_∞/G	184
11.6.3	Derivation of Nonlinear Efficiency Expression	186
11.7	The $n_{0\infty}$ Achieved	189
11.8	The Rise Time	189

11.9	The Assymetry Between Phosphorescence and Build-Up	190
11.10	An Expression for Phosphorescence	191
12	**The Exponential Energy-Gap "Law" for Small-Offset Cases**	194
13	**Conclusions**	198
14	**References**	201

Source Code 204

Subject Index 249

Source of Illustrations 255

1 Introduction

1.1 Luminescence Centers and Models of Them

A luminescence center is a construct aimed at understanding similarities in the absorption and emission performances found among luminescent materials. The word "center" implies that the most important observed properties can be understood while restricting one's attention to a particular lattice site and its immediate surroundings. Free excitonic absorption and emission, for example, are beyond the scope of a model of a luminescence "center", because the nature of these optical processes involves the whole lattice. But for the large majority of observed luminescent materials, the "center" is a useful concept.

There is found a wavelength dependence to the absorption of light and to its emission. There is often found a non-unit quantum efficiency, and one that depends strongly on the temperature. There is often found a strong temperature dependence to the emission lifetime. Understanding all these properties is part of the function of a model of a luminescence center.

Certain of these properties require models of the electronic structure of the luminescence center, i.e., of the defect or of the impurity at the lattice site. Such models as the crystal-field or ligand-field model and the various quantum mechanical (Hartree-Fock or X-α) calculations address the electronic energy levels and the transition probabilities between these electronic energy levels. Such studies are needed for understanding the transition energies and their radiative lifetimes.

We, however, are more concerned with the role that vibrations play in luminescence properties. Given that a transition occurs at some energy, what is its spectral shape, and what is the temperature dependence of this shape? Given a radiative lifetime, what is the observed transition lifetime and what is its temperature dependence? Our work begins when the electronic coordinates have been integrated out of the wavefunction, and indeed we will treat these electronic integrals as parameters.

Moreover, we treat only the simplest vibrational problem, where the total wavefunction is a product of one electronic-nuclear function and one nuclear function. Electronic degeneracies, with wavefunctions equal to sums of such products and with the thereby required Jahn-Teller distortions and more complicated vibrational energy curves, will not be treated here.

The work presented here is contained in a series of papers by the authors [1–21].

1.2 The Simplest Model: One Coordinate and Equal Force Constants

1.2.1 The Optical Band Shapes

Figure 1 shows the simplest model of a luminescence center, the single-configurational-coordinate (SCC) model, namely two parabolas representing the ground and the excited states of the center. The two parabolas are displaced along one vibrational coordinate, x, the ground state minimum being at $x = 0$ and the upper state minimum at $x = a$. Natural measures of the coordinate offset between the two parabolas are the energies $S_u \hbar \omega_u$ and $S_v \hbar \omega_v$ which are the energy differences in the parabolas between the two minima x values, namely, 0 and a, in units of the phonon energies of the parabolas. The two parabolas are also displaced in the energy coordinate. A reasonable measure of this offset is the zero-phonon energy $h\nu_{zp,vu} \equiv E_{zp,vu}$.

The curves in Fig. 1 satisfy in general the equations:

$$E_v = E_{zp,vu} - \frac{1}{2}(\hbar\omega_v - \hbar\omega_u) + S_v \hbar\omega_v \left(\frac{x}{a} - 1\right)^2$$

$$E_u = S_u \hbar\omega_u \left(\frac{x}{a}\right)^2 \tag{1.1}$$

and, for the special case of equal force constants, to which our attention will be limited in this introductory chapter:

$$E_v = E_{zp,vu} + S_0 \hbar\omega \left(\frac{x}{a} - 1\right)^2$$

$$E_u = S_0 \hbar\omega \left(\frac{x}{a}\right)^2 \tag{1.2}$$

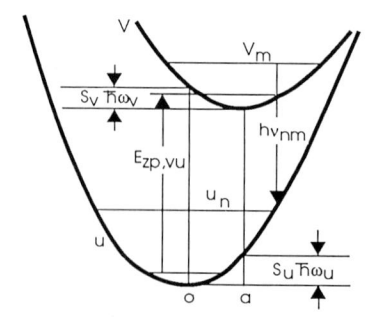

Fig. 1. The single-configurational-coordinate model

1.2 The Simplest Model: One Coordinate and Equal Force Constants

where $\hbar\omega_u = \hbar\omega_v \equiv \hbar\omega$ and $S_u = S_v \equiv S_0$. The two equal-force-constants parabolas have their crossover at:

$$\frac{x_c}{a} = \frac{1}{2}\left(\frac{E_{zp,vu} + S_0\hbar\omega}{S_0\hbar\omega}\right)$$

$$E_c = \frac{(E_{zp,vu} + S_0\hbar\omega)^2}{4S_0\hbar\omega} \quad (1.3)$$

We call the ground state vibrational wavefunctions u_n and the offset state wavefunctions v_m.

Several attributes of the performance of this luminescence center are qualitatively evident in the figure. First, neglecting small corrections due to differences in zero-point energies, there will be absorption centered at $h\nu_{abs} = h\nu_{zp,vu} + S_v\hbar\omega_v$ and emission centered at lower energy at $h\nu_{emis} = h\nu_{abs} - S_u\hbar\omega_u - S_v\hbar\omega_v$. Second, the shape of these bands will not change if the upper parabola is shifted to higher or lower energy without changing the offset. Third, the optical bands will be broad if the offset is considerable and narrow with the zero-phonon line dominant if the offset is very small. Fourth, as temperature increases and higher vibrational states are occupied, the band widths both in absorption and in emission will widen.

This book will explore the quantitative behavior of this simple center and of more general luminescence centers. For this simple equal-force-constants center, the bandshapes are given most usefully as a distribution over p, the number of phonons of energy $\hbar\omega$ created by the transition having photon energy $h\nu_{vu,p}$ in absorption, $h\nu_{uv,p}$ in emission, according to

$$p\hbar\omega \equiv (m-n)\hbar\omega = h\nu_{vu,p} - h\nu_{zp,vu}$$

$$p\hbar\omega \equiv (n-m)\hbar\omega = h\nu_{zp,vu} - h\nu_{uv,p} \quad (1.4)$$

in absorption and in emission respectively. All transitions $u_n \to v_m$ in absorption and $v_m \to u_n$ in emission with their indices following the Eq. (1.4) relationship share a common photon energy. The p index runs from $-\infty$ to ∞, with a positive value implying creating phonons. The maximum of this distribution will be near $p = S_0$.

The normalized bandshapes themselves will be given by Boltzmann-weighted Franck-Condon [22] factors, i.e., squared overlap integrals. This shape function will be

$$W_p = \sum_{n=max(0,-p)}^{\infty} (1-r)r^n |\langle v_{p+n}|u_n\rangle|^2 = \sum_{n=max(0,-p)}^{\infty} (1-r)r^n A_{p+n,n}^2$$

$$W_p = \sum_{m=max(0,-p)}^{\infty} (1-r)r^m |\langle u_{p+m}|v_m\rangle|^2 = \sum_{m=max(0,-p)}^{\infty} (1-r)r^m A_{p+m,m}^2$$

$$(1.5)$$

in absorption and in emission, respectively. Here,

$$r = \exp(-\hbar\omega/kT) \quad (1.6)$$

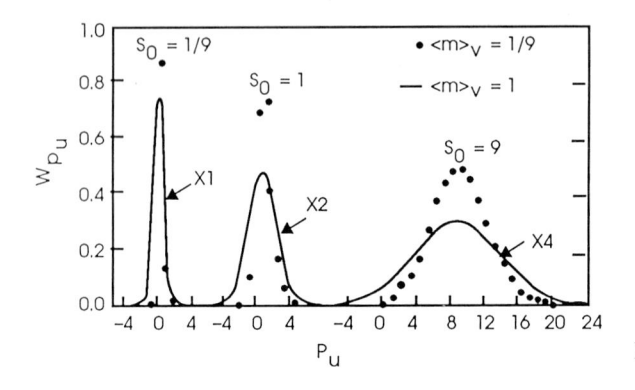

Fig. 2. W_p band shapes

is the Boltzmann factor and is related to Planck's measure of the temperature by

$$\langle m \rangle \equiv \frac{\sum_m m r^m}{\sum_m r^m} = r(1-r)\frac{d}{dr}[\sum_m r^m] = r(1-r)\frac{d}{dr}\left(\frac{1}{1-r}\right)$$

$$= \frac{r}{1-r} = \frac{1}{\exp(\hbar\omega/kT) - 1} \tag{1.7}$$

The $\langle m \rangle$ will be extensively used here as the temperature parameter.

We will use Eq. (1.5) in this introduction to calculate W_p. However, explicit expressions for W_p as functions of S_0 and $\langle m \rangle$, and that W_p is a normalized distribution, and recursion relations between W_p's with nearby indices are all derived in Chap. 4. Finally, the calculation of the A_{nm} matrix elements, using the Manneback recursion formulas which themselves are functions of S_0 and $\langle m \rangle$, is described in Chap. 3.

Figure 2 shows the general behavior of these W_p distributions for small ($S_0 = 1/9$), intermediate ($S_0 = 1$), and large ($S_0 = 9$) offsets at a low ($\langle m \rangle = 1/9$, shown in dots) and a moderate ($\langle m \rangle = 1$, shown as full curves) temperature.

At low temperature, for small offset the W_p distribution shows a strong zero-phonon line and small sidebands decreasing rapidly with phonon number; for large offset the band is broad and somewhat Gaussian; for intermediate offset, one can see distinct assymetry between positive and negative p values favoring the positive (phonon generating) transitions as expected. At high temperature, at all offsets the bands broaden and lose significantly their assymetry.

1.2.2 The Nonradiative Rate

From Fig. 1 one might also suspect that some nonradiative losses will occur after the excited state is occupied. The loss rate will be strongest when the parabolas intersect at the upper-state minimum. It might be expected that for the case pictured, the $v \to u$ nonradiative rate will increase with temperature and, as Mott [23] suggested, will be proportional to the Boltzmann factor for the availability of the crossover. The true picture is that the transition goes at some rate from

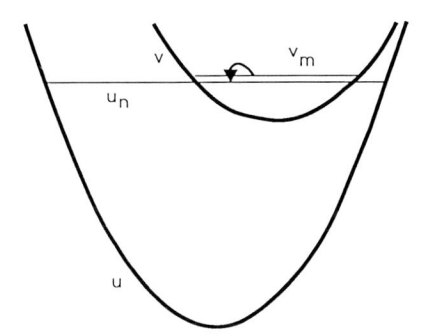

1.2 The Simplest Model: One Coordinate and Equal Force Constants

Fig. 3. Typical component of a nonradiative transition

each vibrational state m to that final state $n = p + m$ with closely the same energy, as pictured in Fig. 3.

It will be found that the same sum of thermally weighted Franck-Condon factors,

$$R_{uv,p} = A_{uv} W_p \tag{1.8}$$

with the same W_p as in Eq. (1.5) and with $A_{uv} \approx 10^{12} - 10^{14}$, will give a very reasonable picture of the nonradiative $v \to u$ transition rate, where p is determined via

$$p\hbar\omega = h\nu_{zp,vu} \tag{1.9}$$

i.e., by the analogue of Eq. (1.4) with zero photon energy. The average m through which the transition occurs, namely,

$$\langle m \rangle_p = \sum_m m(1-r)r^m \langle u_{p+m} | v_m \rangle^2 \tag{1.10}$$

is at only a small fraction of the crossover energy. We will give expressions for $\langle m \rangle_p$ in Chap. 4.

Of course, W_p and thus $R_{uv,p}$ are defined only for integer p values, i.e., only for certain relative placements of the two parabolas along the energy axis. Our practice is to interpolate logarithmically between integer p values for any intermediate placements. This practice assumes that the low frequency vibrations which must play the role of forcing energy conservation have only this interpolative effect on the nonradiative rate.

Expressions for the nonradiative rate that are more grounded theoretically are explored in this book, but the general picture will be that all these expressions, including Eq. (1.8), are indistinguishable in practical applications to experimental data.

1.2.3 Six Typical Thermal Quenching Behaviors

The nonradiative rate $A_{uv} W_p$ is found to be exceedingly sensitive to the offset S_0, to the energy offset $h\nu_{zp,vu}$, to the temperature, and to differences in the curvatures of the two parabolas. We give here six typical behaviors of thermal quenching

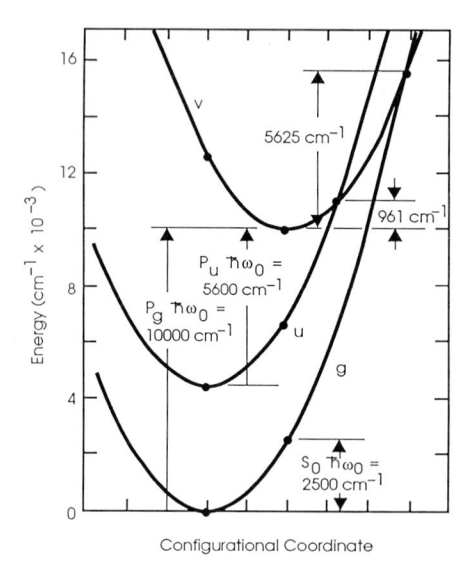

Fig. 4. Bottom and outside crossovers

for equal curvatures; our discussion of the influence of different curvatures on quenching behavior is postponed to Chap. 9. In all cases in this introductory chapter we calculate the nonradiative rate using Eq. (1.8) and show the terms in W_p according to Eq. (1.5). We will also calculate $\langle m \rangle_p$ in Eq. (1.10).

1.2.3.1 Fast Bottom Crossover

Figure 4 shows a center with a fast bottom crossover, where the crossover energy is $\approx 10^3$ cm^{-1} or less, namely, the $v \to u$ transition. There wil be no observable emission from the v state in such a center even at the lowest temperatures because the nonradiative $v \to u$ feeding overwhelms the emission rate.

Table 1 gives the description of this bottom crossover rate at four temperatures. The parameter values used in Eqs. (1.5), (1.10), and (1.8) are $\hbar\omega = 400$ cm^{-1}, $S_0 = 6.25$, $p = 14$, $A_{uv} = 10^{13}$ s^{-1}.

The transition rate for the bottom crossover is seen from the W_p values in Table 1 and Eq. (1.10) to be nearly temperature independent and fast. One can see in the four columns that the higher vibrational levels contribute more with increasing temperature. However, the crossover is at $m = 1.9$, and the transition actually takes place on the average at the $\langle m \rangle_p$ indicated, below the crossover level even at moderately high temperatures.

The bottom crossover case describes qualitatively the fast Eu^{3+} CTS $\to$ ^{5}D feeding in oxysulfides [1, 2, 11] and fast Cr^{3+} ^{4}T$_2$ $\to$ ^{2}E feeding in Al$_2$O$_3$ [24, 9].

1.2.3.2 Outside Crossover

Table 2 gives the description of a typical outside crossover, the $v \to g$ transition in Fig. 4 when the u state is eliminated. The parameter values here are $\hbar\omega = 400$ cm^{-1}, $S_0 = 6.25$, $p = 25$, $A_{uv} = 10^{13}$ s^{-1}.

1.2 The Simplest Model: One Coordinate and Equal Force Constants

Table 1. Description of a fast bottom crossover

m	$A^2_{14+m,m}$	$(1-r)r^m A^2_{14+m,m}$			
		0 K	259 K	374 K	547 K
0	0.00307	0.00307	0.00274	0.00241	0.00200
1	0.01569		0.00151	0.00265	0.00357
2	0.04002		0.00042	0.00144	0.00317
3	0.06654		0.00007	0.00052	0.00184
4	0.07842		0.00001	0.00013	0.00076
5	0.06588			0.00002	0.00022
6	0.03617				0.00005
7	0.00875				0.00001
W_{14}		0.00307	0.00475	0.00717	0.01162
$\langle m \rangle_{14}$			0.547	1.075	1.716
$\langle m \rangle$			0.122	0.273	0.536
$kT/\hbar\omega$			0.45	0.65	0.95
r			0.1084	0.2147	0.3490

Table 2. Description of an outside crossover

m	$A^2_{25+m,m}$	$10^7 \cdot (1-r)r^m A^2_{25+m,m}$			
		0 K	259 K	374 K	547 K
0	9.818-9	0.0982	0.0876	0.077	0.06
1	1.473-7		0.1423	0.248	0.34
2	1.139-6		0.1193	0.412	0.90
3	6.040-6		0.0686	0.470	1.67
4	2.468-5		0.0303	0.412	2.38
5	8.271-5		0.0111	0.296	2.79
6	2.365-4		0.0034	0.182	2.78
7	5.922-4		0.0009	0.098	2.43
8	1.324-3		0.0002	0.047	1.90
9	2.678-3			0.020	1.34
10	4.957-3			0.008	0.87
11	8.461-3			0.003	0.52
12	1.340-2			0.001	0.29
13	1.979-2				0.15
14	2.732-2				0.07
15	3.536-2				0.03
16	4.292-2				0.01
$10^7 W_{25}$		0.0982	0.4637	2.274	18.53
$\langle m \rangle_{25}$			1.707	3.548	6.10
$\langle m \rangle$			0.122	0.273	0.54
$kT/\hbar\omega$			0.45	0.65	0.95
r			0.1084	0.2147	0.3490

The outside crossover rate, as seen in Table 2, is strongly temperature dependent, rising by a factor of ~ 190 over the temperature range shown. The participation of higher vibrational levels with increasing temperature is clearly evident, the maxima in the contributions to the total rate being at $m = 0, 1, 3, 5$, at 0 K, 259 K, 374 K, and 547 K, respectively. The crossover energy is here near $m = 14$, but the measure $\langle m \rangle_{25}$ of the average m level used in the transition is only ~ 6 even at the highest temperature, where a center with a radiative rate as fast as 10^7 s^{-1} would be quenched.

The quantum efficiency of the $v \to g$ radiative transition is:

$$\eta_{gv} = \frac{R_{gv}}{R_{gv} + A_{gv} W_p} \tag{1.11}$$

and the transition lifetime is:

$$\tau_v = \frac{1}{R_{gv} + A_{gv} W_p} = \frac{1}{R_{gv}} \eta_{gv} \tag{1.12}$$

Since R_{gv} is usually nearly independent of temperature, Eq. (1.12) shows that η_{gv} and τ_v have the same temperature dependences. For a 10^4 s^{-1} radiative rate, the quenching will be almost complete even at 0 K, while for a 10^6 s^{-1} radiative rate, the quenching will be to $\eta \approx 0.70$ at 259 K, to $\eta \approx 0.30$ at 374 K, and to $\eta \approx 0.05$ at 547 K.

Radiative lifetimes range from $10^{-1} - 10^{-8}$ s. There is some correlation linking large offsets with shorter radiative lifetimes, because large offsets frequently imply significant changes in the electronic wavefunctions and therefore large parity-allowed electronic transition integrals.

This case describes qualitatively the competition between Cr^{3+} $^4T_2 \to {}^4A_2$ radiative and nonradiative transitions in Al_2O_3 [24, 25, 9].

1.2.3.3 Small-Offset Multiphonon Emission

Figure 5 shows a typical small-offset center.

Here we expect narrow line emission and absorption, with a dominant zero-phonon line and weak phonon sidebands. We anticipate from the early multiphonon-emission studies that there will be some weak nonradiative transitions. These early studies emerged independently of the treatment of nonradiative rates using the W_p function and came to an approximate form of this function.

Table 3 gives the description of the nonradiative transition for a typical small offset case, the $v \to u$ transition in Fig. 5. The parameter values here are $\hbar\omega = 400$ cm^{-1}, $S_0 = 0.09$, $p = 6$, $A_{uv} = 0.8 \times 10^{12}$ s^{-1}. The small-offset transition has the same qualitative features as the outside crossover, in that the contributions to the total rate favor the higher vibrational states with increasing temperature. The difference is quantitative. The small-offset nonradiative rate has only a modest temperature dependence, only a factor of 12 from 0 K to 547 K vs. the factor of ~ 190 for the outside crossover transition of Table 2. In the small offset case, the crossover is above $m = 100$, the transition occurs on average through levels lower

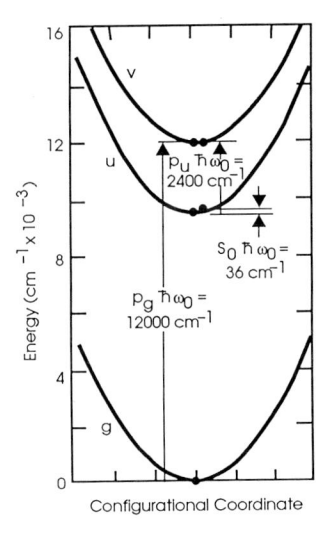

1.2 The Simplest Model: One Coordinate and Equal Force Constants

Fig. 5. Small-offset case

Table 3. Description of a small-offset transition

m	$A^2_{6+m,m}$	\multicolumn{4}{c}{$0.8 \cdot 10^9 (1-r) r^m A^2_{6+m,m}$}			
		0 K	259 K	374 K	547 K
0	6.746-10	0.540	0.481	0.424	0.35
1	4.601-9		0.356	0.621	0.84
2	1.794-8		0.150	0.520	1.14
3	5.243-8		0.048	0.326	1.16
4	1.277-7		0.013	0.171	0.99
5	2.737-7		0.003	0.079	0.74
6	5.333-7		0.001	0.033	0.50
7	9.649-7			0.013	0.32
8	1.645-6			0.005	0.19
9	2.671-6			0.002	0.11
10	4.162-6				0.06
11	6.265-6				0.03
12	9.154-6				0.02
13	1.303-5				0.01
	$0.8 \cdot 10^9 W_6$	0.540	1.052	2.194	6.46
	$\langle m \rangle_6$		0.830	1.853	3.608
	$\langle m \rangle$		0.122	0.273	0.536
	$kT/\hbar\omega$		0.45	0.65	0.95
	r		0.1084	0.2147	0.3490

than $m = 4$, and the appeal of the Mott activation energy to the crossover is totally absent.

This case describes the competitions between many 4f → 4f narrow line emissions and 4f → 4f nonradiative transitions [26–29]. Indeed, the first appeal for describing these nonradiative transitions was to the Kiel-Moos-Weber multiphonon emission rate alluded to above. We shall take up in Chap. 4 the analytic approximations which link Eq. (1.5) to this Kiel-Moos-Weber expression.

1.2.3.4 Two-Step Quenching With a Fast Second Step

Figure 6 shows a typical two-step quenching with a fast second step, namely, the $u \to v \to t$ sequence. This pattern of electronic states is frequently encountered and will be illustrated in Chap. 10, e.g., in oxysulfides and alkaline-earth flourochlorides activated with rare earths.

The center shown has narrow-line absorptions and emissions at 9 600 and 12 000 cm^{-1} and a broad band absorption near 17 700 cm^{-1}. The narrow line emission at 12 000 cm^{-1} will be precipitously quenched by the two-step process, as shown in Fig. 7. We will give now a notation for the needed rates and a derivation of the efficiency expression used to understand this temperature behavior.

The notation $(i \to f)$ will be used for the rate of the transition $i \to f$ for unit population in i, and the notation $(i \to f_1, f_2, \ldots)$ will be used for the sum of the rates of the transitions $i \to f_1, i \to f_2, \ldots$

For the equal-force constants case, it will be shown in Chap. 4 that inverse rates are related by $(u \to v)/(v \to u) = r^p$. One also knows qualitatively that the bottom crossovers $v \to u, t$ are very fast.

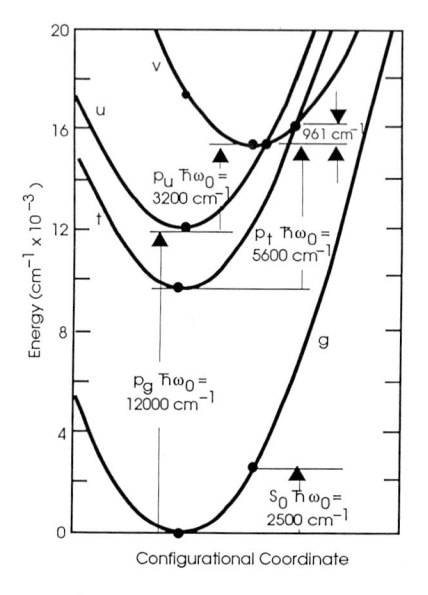

Fig. 6. Two-step quenching through a higher offset state

1.2 The Simplest Model: One Coordinate and Equal Force Constants

Since excitation leaves the u, v system only by transitions to g and t, the quantum efficiency for $u \to g$ emission is always:

$$\eta_{gu} = \frac{(u \to g)n_u}{(u \to g,t)n_u + (v \to g,t)n_v} = \left[1 + \frac{(u \to t)}{(u \to g)} + \frac{(v \to g,t)n_v}{(u \to g)\, n_u}\right]^{-1} \quad (1.13)$$

In Eq. (1.13), the populations n_u and n_v satisfy the steady-state rate equations:

$$(u \to g,t,v)n_u - (v \to u)n_v = G_u$$
$$-(u \to v)n_u + (v \to g,t,u)n_v = G_v \quad (1.14)$$

where G_u and G_v are the excitation rates into u and v, respectively.

For narrow line G_u excitation ($G_v = 0$) and broad band G_v excitation ($G_u = 0$), the population ratios are found:

$$\left.\frac{n_v}{n_u}\right|_{G_u} = \frac{(u \to v)}{(v \to g,t,u)} = \frac{(v \to u)}{(v \to g,t,u)} r^p$$

$$\left.\frac{n_v}{n_u}\right|_{G_v} = \frac{(u \to g,t,v)}{(v \to u)} = \frac{(u \to g,t)}{(v \to u)} + r^p \quad (1.15)$$

respectively. The efficiencies in Eq. (1.13) are found to be:

$$\eta_{gu}|_{G_u} = \left[1 + \frac{(u \to t)}{(u \to g)} + \frac{(v \to g,t)(u \to v)}{(u \to g)(v \to g,t,u)}\right]^{-1}$$

$$= \left[1 + \frac{(u \to t)}{(u \to g)} + \frac{(v \to g,t)(v \to u)}{(u \to g)(v \to g,t,u)} r^p\right]^{-1}$$

$$\eta_{gu}|_{G_v} = \frac{(v \to u)}{(v \to g,t,u)} \eta_{gu}|_{G_u} \quad (1.16)$$

The second of Eq. (1.16) requires after substituting the second of Eq. (1.15) into Eq. (1.13) the identity:

$$\left[\frac{(v \to u)}{(v \to g,t,u)}\right]\left[1 + \frac{(u \to t)}{(u \to g)} + \frac{(v \to g,t)}{(u \to g)} \cdot \frac{(u \to g,t,v)}{(v \to u)}\right]$$

$$= \frac{(v \to u)}{(v \to g,t,u)} + \frac{(u \to t)}{(u \to g)} \cdot \frac{(v \to u)}{(v \to g,t,u)} + \frac{(v \to g,t)}{(u \to g)} \cdot \frac{(u \to g,t,v)}{(v \to g,t,u)}$$

$$= \frac{[(v \to u)(u \to g,t) + (v \to g,t)(u \to g,t)]}{(u \to g)(v \to g,t,u)} + \frac{(v \to g,t)(u \to v)}{(u \to g)(v \to g,t,u)}$$

$$= 1 + \frac{(u \to t)}{(u \to g)} + \frac{(v \to g,t)(u \to v)}{(u \to g)(v \to g,t,u)} \quad (1.17)$$

or it may be directly derived by noting that $(v \to u)/(v \to g,t,u)$ is the fraction of G_v excitations that directly feed u and thus are equivalent to G_u excitations.

The lifetime of the u state, defined for G_u excitation, is, using the first expression in Eq. (1.13) in the second step:

$$\tau_u = \frac{n_u + n_v}{(u \to g, t)n_u + (v \to g, t)n_v}\bigg|_{G_u} = \frac{n_u + n_v}{(u \to g)n_u} \eta_{gu}|_{G_u}$$

$$= \frac{1}{(u \to g)}\left(1 + \frac{n_v}{n_u}\right)\eta_{gu}|_{G_u} \tag{1.18}$$

Since n_v/n_u is always small in the examples here:

$$\tau_u \approx \frac{1}{(u \to g)}\eta_{gu}|_{G_u} \tag{1.19}$$

and therefore τ_u and $\eta_{gu}|_{G_v}$ are both multiples of $\eta_{gu}|_{G_u}$ and it is sufficient to study $\eta_{gu}|_{G_u}$ for its temperature dependence.

In the first of Eq. (1.16) the second term in the denominator involving $(u \to t)$ is from the multiphonon emission $u \to t$ process. The third term in the denominator is from the two step $u \to v \to g,t$ processes. While one may use the W_p formula with negative p for the $(u \to v)$ rate, the expression of this term as $(v \to u)r^p$ allows a further useful approximation, namely that the factor $(v \to g, t)(v \to u)/(v \to g, t, u)$ is approximately the smaller of $(v \to g, t)$ and $(v \to u)$. For the example in Fig. 6, where the v, t crossover is higher than the v, u crossover, $(v \to g, t)$ is smaller and is further approximated as $(v \to t)$ since the bottom crossover $v \to t$ is much faster than the outside crossover $v \to g$. Then Eq. (1.16) reduces to:

$$\eta_{gu}|_{G_u} \approx \left[1 + \frac{(u \to t)}{(u \to g)} + \frac{(v \to t)}{(u \to g)}r^p\right]^{-1} \tag{1.20}$$

Figure 7 shows the precipitous quenching caused by the two step $u \to v \to g, t$ processes. We have used here the illustrative values for the $u \to g$ narrow line emission $R_{gu} = 10^3$ s^{-1}, $\hbar\omega = 400$ cm^{-1}. Those for the $v \to t$ transition are $S_0\hbar\omega = 2\,500$ cm^{-1}, $hv_{zp,vt} = p_{vt}\hbar\omega = 5\,600$ cm^{-1}, $A_{tv} = 10^{13}$ s^{-1}, $p_{vt} = 14$, $S_0 = 6.25$. Those for the $u \to t$ multiphonon emission transition are $S_0\hbar\omega = 36$ cm^{-1},

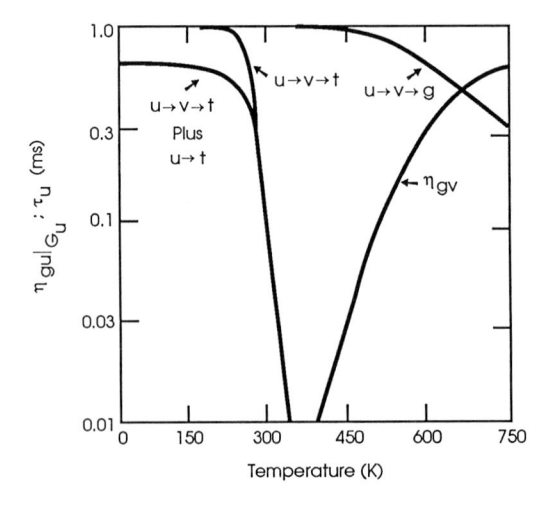

Fig. 7. Quantum efficiencies in two-step quenching processes with fast and slow second steps

1.2 The Simplest Model: One Coordinate and Equal Force Constants

$\hbar\omega = 400 \text{ cm}^{-1}$, $h\nu_{zp,ut} = p_{ut}\hbar\omega = 2400 \text{ cm}^{-1}$, $A_{tu} = 0.8 \times 10^{12} \text{ s}^{-1}$, $p_{ut} = 6$, $S_0 = 0.09$. These transitions have already been described in Tables 2 and 3. In Fig. 7, the precipitous quenching is labeled $u \to v \to t$ plus $u \to t$. The quenching induced by $u \to v \to t$ itself is also shown. The $u \to t$ process produces non-unit quantum efficiency even at the lowest temperatures.

This particular case, namely, two-step quenching through a higher offset state, involves negative p values in the $u \to v$ step and is the most favorable for the Mott-activation-energy approximate form. Indeed, one can show in general that for equal force constants the rate $A_{uv}W_p$ can be approximated by the exponential expression $A\exp(-E_x/kT)$, where:

$$E_x = [\langle m \rangle_p - \langle m \rangle]\hbar\omega$$
$$A = A_{uv} A^2_{\langle m \rangle_p} \quad (1.21)$$

Equation (1.21) is derived by forcing the values and the first derivatives with respect to $(1/kT)$ to match, and by adopting the approximation that $W_p \approx$ its largest term, which will be nearly the squared overlap integral at the m through which on average the transition takes place, namely, the $m = \langle m \rangle_p$ of Eq. (1.10).

In our example, explicit evaluation of this approximate form gives:

$$\eta_{gu}|_{G_u} \approx [1 + 0.8 \times 10^9 \, W_6 + 1.24 \times 10^8 \exp(-8.430\, \hbar\omega/kT)]^{-1} \quad (1.22)$$

In the evaluation of Eq. (1.22), the approximated activation energy is 8.430 $\hbar\omega$, of which 8 $\hbar\omega$ is the zero phonon energy, and the activation energy to the crossover would be 10.40 $\hbar\omega$. Thus the Mott activation energy is only $\sim 20\%$ too large and could be used, with an adjusted preexponential term, to give a reasonable fit to the transition rate.

This case describes qualitatively the precipitous sequential quenchings of Eu^{3+} emissions in oxysulfides [1, 2, 11], the $^5D\ Sm^{2+}$ emissions in BaClF [30, 61], $BaCl_2$, $BaBr_2$, and KCl [15, 30, 62, 63], and the $^5D_3\ Tb^{3+}$ emissions in oxysulfides [5].

1.2.3.5 Two-Step Quenching with a Slow Second Step

"Slow" here means slow compared to bottom crossover rates, i.e., $< 10^{10} \text{ s}^{-1}$. If one deletes the t curve in Fig. 6, then the $u \to g$ emissions would be competitive with such a $u \to v \to g$ transition. The particular example in this figure has as the second step the broad band $v \to g$ emission, but it is equally encountered as a $v \to g$ thermally activated outside crossover.

The analysis in Sect. 1.2.3.4 leads now to

$$\eta_{gu}|_{G_u} = \eta_{gu}|_{G_v} = [1 + \{(v \to g)/(u \to g)\}r^p]^{-1} = \frac{1}{1 + (R_{gv}/R_{gu})r^p},$$

$$\eta_{gu} + \eta_{gv} = 1. \quad (1.23)$$

This quantum efficiency is shown in Fig. 7 by the curves labeled $u \to v \to g$. Also shown is the gradual buildup with temperature of the $v \to g$ emission intensity, the curve labeled η_{gv}.

This case describes qualitatively the slow replacement of $^2E \to {}^4A_2 Cr^{3+}$ narrow line R emission in Al_2O_3 with broadband $^4T_2 \to {}^4A_2$ emission with increasing temperature [24, 25, 9]. It also explains the replacement of Sm^{2+} $^5D_0 \to {}^7F$ narrow-line emissions with broad band $4f^55d \to {}^7F$ emission [30] in $SrF_2 : Sm^{2+}$. In this latter case [31], the $4f^55d$ state is only 550 cm^{-1} above the $^5D_0$ state, and the $4f^55d \to {}^7F$ radiative rate is fast (10^7 s^{-1}) compared to the $^5D_0 \to {}^7F$ rate (10^2 s^{-1}). The replacement is then steeper than the $u \to v \to g$ curve shown in Fig. 7 and occurs at low temperature, 80 K.

1.2.3.6 Low-temperature Tunnelling Crossover

Figure 8 shows a low-temperature tunnelling transition, namely, the $u \to v$ transition which competes with $u \to g$ emission.

We describe this $u \to v$ transition in Table 4, with the parameters $\hbar\omega = 200$ cm^{-1}, $S_0 = 30.25$, $p = 2$, $A_{vu} = 10^{13}$ s^{-1}.

This $u \to v$ tunnelling transition is significantly underestimated by the Mott activation energy expression with the crossover energy at 1 307 cm^{-1}. The crossover is at $n = 6.0$ while the average n utilized is near unity when a center with radiative rate 10^3 s^{-1} is being quenched. That this is so is evident in Table 4, since the radiative rate and nonradiative rate are then equal when $10^{10} W_2 = 1$, i.e., near 57 K.

1.3 The Franck-Condon Principle for Nonradiative Rates

Condon [22] gave a clear qualitative picture of the radiative transitions in his 1928 paper. The strongest transitions with large A_{nm}^2 are between v_m, u_n states which share a region of common nuclear position and kinetic energy. We show in Fig. 9 a FC-permitted transition between two offset states and the region of common kinetic energy.

For A_{nm}^2 to be large, u_n and v_m must overlap and oscillate in phase. Since these wavefunctions satisfy the Schrödinger equation:

$$\left[\frac{\hbar^2}{2M_0}\frac{d^2}{dx^2} + \{E - V(x)\}\right]u(x) = 0 \tag{1.24}$$

the oscillation wavelength is proportional to $[E - V(x)]^{-1/2}$. Thus the need arises for a shared kinetic energy.

For such a region of common kinetic energy to exist, one can be convinced that the two v_m turning points and the two u_n turning points must be interleaved. The boundary between states with interleaved and not interleaved turning points consists of states with a shared turning point. From Eqs. (1.1) and (1.2), one gets the parametric equations

$$n + \frac{1}{2} = S_u\left(\frac{x_t}{a}\right)^2, \quad m + \frac{1}{2} = S_v\left[\left(\frac{x_t}{a}\right) - 1\right]^2 \tag{1.25}$$

1.3 The Franck-Condon Principle for Nonradiative Rates

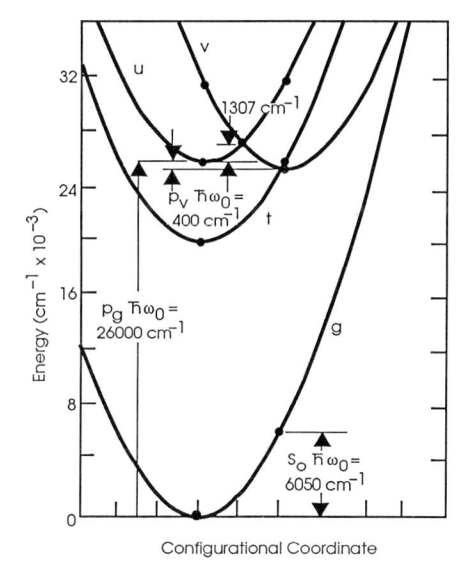

Fig. 8. Low-temperature tunnelling crossover

Table 4. Description of a tunnelling transition

		$10^{10}(1-r)r^m A_{n,2+n}^2$				
n	$A_{n,2+n}^2$	0 K	57 K	72 K	86 K	101 K
0	3.334-11	0.333	0.331	0.327	0.32	0.32
1	8.253-9		0.552	1.484	2.84	4.47
2	6.520-7		0.294	2.147	8.00	20.27
3	2.286-5		0.070	1.379	10.01	40.82
4	4.040-4		0.008	0.446	6.31	41.43
5	3.775-3		0.000	0.076	2.10	22.23
6	1.845-2			0.007	0.37	6.24
7	4.298-2			0.000	0.03	0.84
8	3.482-2				0.00	0.04
9	0.061-2					0.00
	$10^{10} W_2$	0.333	1.255	5.866	29.98	136.7
	$\langle n \rangle_2$		1.101	2.066	2.903	3.570
	$\langle n \rangle$		0.007	0.019	0.036	0.061
	$kT/\hbar\omega$		0.20	0.25	0.30	0.35
	r		0.00674	0.01832	0.03567	0.05743

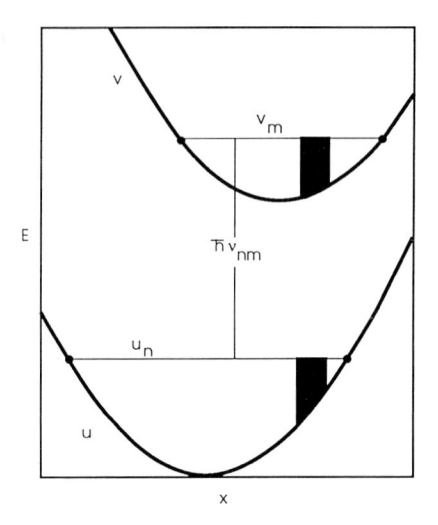

Fig. 9. A FC-allowed transition showing the shared kinetic energy

for the locus of v_m, u_n states with a shared turning point, x_t, i.e., for the boundary between large (allowed) and small (forbidden) A_{mn}^2 values. For equal force constants, the u and v subscripted quantities are equal and are given the subscript 0.

Figures 10–12 show the A_{mn}^2 matrices used for the bottom, outside, multiphonon, and tunnelling transitions of Tables 1–4. In each figure there are shown contours of constant A_{mn}^2 values labeled by their exponent. The curves containing points marked x are the turning point (TP) loci, as given by Eq. (1.25). The lines starting at the boundary, crossing the contour lines, and ending at x marks, are the set of matrix elements used for the transitions of Tables 1–4.

For example, absorptions at 0 K involve the $A_{0,m}^2$ elements and at higher temperatures get contributions from A_{nm}^2 elements with non-zero, gradually increasing n values. The emissions involve the $A_{n,0}^2$ elements at 0 K and at higher temperatures A_{nm}^2 elements with non-zero, gradually increasing m values. The largest such elements are within the TP locus curve. One can see the gradual widening of the optical bands from the shape of these loci.

The nonradiative transition rates are also understandable with these diagrams. They involve matrix elements outside these loci. The outside-crossover transitions of Table 2 involve at low temperature A_{nm}^2 elements which are generally small, and as temperature increases begin to involve larger ones nearer the TP locus curve. See for example the line marked $A_{25+m,m}^2$ in Fig. 10 corresponding to this Table 2 case. The matrix elements grow by almost six orders along this line, and it is understandable that this nonradiative transition rate grows rapidly with temperature.

The bottom crossover of Table 1 involves the matrix elements on the line labeled $A_{14+m,m}^2$ in Fig. 10. They are large even for $m = 0$, and therefore this transition will dominate over emission even at 0 K.

In general, the A_{nm}^2 values operative in a nonradiative transition are orders

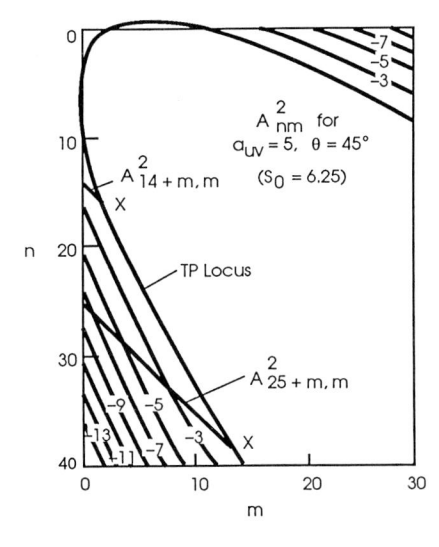

1.3 The Franck-Condon Principle for Nonradiative Rates

Fig. 10. The matrix of Franck-Condon factors for bottom and outside crossovers

of magnitude smaller than those in a radiative transition. Nevertheless, the electronic integral for the nonradiative transition is very large relative to radiative transition rates, and therefore the nonradiative transitions do compete against radiation.

For the small-offset multiphonon transition of Table 3 and Fig. 11, the contour lines are closely spaced, the region of allowed transitions is somewhat restricted, and the multiphonon transition line passes through A_{nm}^2 which do not grow so rapidly with the m index. Consequently, one can expect its temperature dependence to be somewhat more subdued than that of the outside crossover.

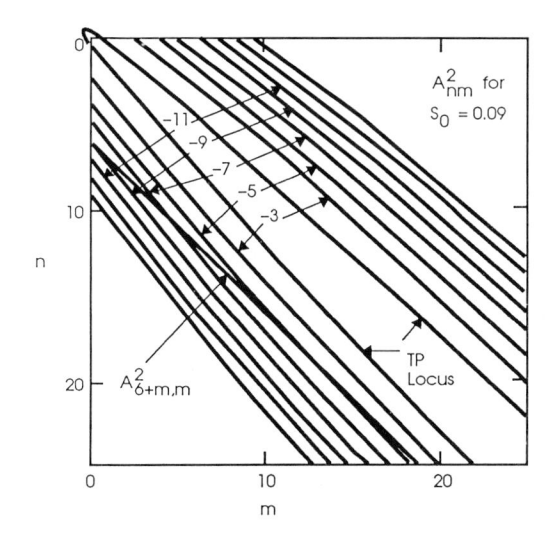

Fig. 11. The matrix of Franck-Condon factors for small-offset multiphonon transitions

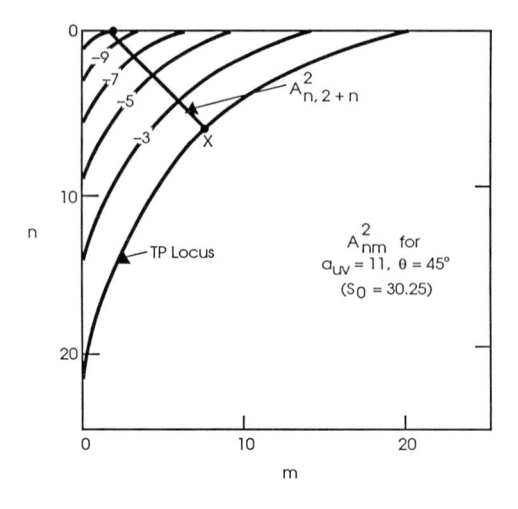

Fig. 12. The matrix of Franck-Condon factors for tunnelling transitions

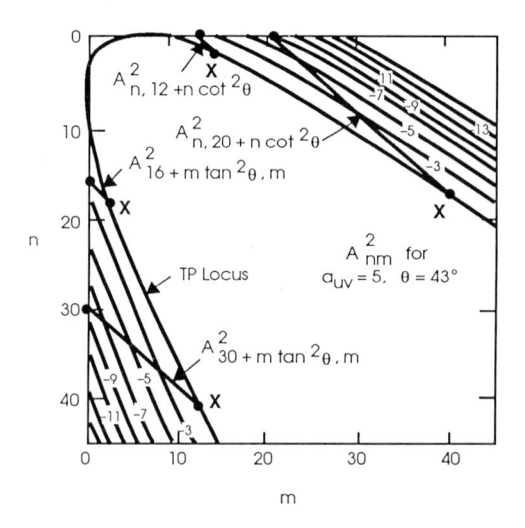

Fig. 13. The matrix of Franck-Condon factors for unequal-force-constants bottom and outside crossovers

The tunnelling transition of Table 4 involves the line labeled $A^2_{n,2+n}$ in Fig. 12. These matrix elements are somewhat small for $n = 0$ but grow enormously with increasing n. Thus the tunnelling transition will show low-temperature non-unit quantum efficiency, and will quench precipitously at relatively low temperatures.

Finally, we give in Fig. 13 the diagram corresponding to Fig. 10 for an unequal force constant case. We will discuss unequal force constants in detail in Chap. 4 and in later chapters. At the moment, we point to the similar contours and similar transition lines for both bottom and outside crossovers. The contours are no longer symmetric around the $n = m$ line.

2 Harmonic Oscillator Wavefunctions

2.1 Hermite Polynomials

The study of vibrations naturally leads to the harmonic oscillator. We will later obtain the harmonic oscillator wavefunctions in dimensionless variables. Our starting point is familiarity with Hermite polynomials, though we will derive the relation of these polynomials to the study of vibrations by demonstrating the solution of the Schrödinger equation.

We start from the generating function for these Hermite polynomials:

$$G_H(t, z_v) = e^{-t^2 + 2z_v t} = \sum_{m=0}^{\infty} \frac{t^m}{m!} H_m(z_v) \tag{2.1}$$

For familiarity one can note that:

$$H_m(z_v) = \left. \frac{\partial^m G_H}{\partial t^m} \right|_{t=0} \tag{2.2}$$

and, specifically:

$$H_0 = 1 \tag{2.3}$$

$$H_1 = 2z_v \tag{2.4}$$

We get the recursion formula for the Hermite polynomials by differentiating Eq. (2.1) with respect to t:

$$\frac{\partial G}{\partial t} = G \cdot (-2t + 2z_v) = \sum_{m=1}^{\infty} \frac{t^{m-1}}{(m-1)!} H_m(z_v)$$

$$= \sum_{m=0}^{\infty} \left[\frac{-2t^{m+1}}{m!} H_m + \frac{2z_v t^m}{m!} H_m \right] \tag{2.5}$$

The second and third expression in Eq. (2.5) can be equated using a common index:

$$0 = \sum_{k=0}^{\infty} t^k \left[\frac{-2H_{k-1}}{(k-1)!} + \frac{2z_v H_k}{k!} - \frac{H_{k+1}}{k!} \right] \tag{2.6}$$

Equation (2.6) is an infinite series in t which is identically 0 and therefore each coefficient is itself 0. Therefore:

$$-2kH_{k-1} + 2z_v H_k - H_{k+1} = 0 \tag{2.7}$$

Equation (2.7) is the desired recursion formula for the Hermite polynomials. With its use one can easily obtain:

$$H_2 = 4z_v^2 - 2, \quad H_3 = 8z_v^3 - 12z_v, \tag{2.8}$$

etc.

By differentiating Eq. (2.1) with respect to z_v, entering the sum in Eq. (2.1) for the exponential on the left-hand side, and equating the coefficients of the powers of t, one obtains the expression:

$$\frac{dH_n(z_v)}{dz_v} = 2nH_{n-1} \tag{2.9}$$

for the derivative of the Hermite functions.

2.2 Generating Function for the Harmonic Oscillator Wavefunctions

The generating function for the harmonic oscillator wavefunctions can now be written down except for a normalization factor. We will include the normalization factor immediately and show later that the value chosen for it is correct.

$$G_v(t, z_v) = e^{-t^2 + 2z_v t - z_v^2/2} = \sum_{m=0}^{\infty} \frac{t^m}{m!} H_m(z_v) e^{-z_v^2/2} = \sum_{m=0}^{\infty} \frac{t^m}{m!} (m! 2^m \pi^{1/2} x_v)^{1/2} v_m \tag{2.10}$$

In Eq. (2.10), x_v has the dimension of length and its value will be given when v_m is shown below to satisfy the Schrödinger equation. This equation defines the normalized wavefunction as:

$$v_m = H_m(z_v) e^{-z_v^2/2} (m! 2^m \pi^{1/2} x_v)^{-1/2} \tag{2.11}$$

We will first demonstrate the first three factors in the normalization constant of Eq. (2.11). We form the integral:

$$\int_{-\infty}^{\infty} e^{-t^2 + 2tz_v - (z_v^2/2)} \cdot e^{-s^2 + 2sz_v - (z_v^2/2)} dz_v$$

$$= \sum_{n,m=0}^{\infty} \frac{t^m s^n}{m! n!} \int_{-\infty}^{\infty} H_m(z_v) H_n(z_v) e^{-z^2} dz_v$$

$$= e^{2ts} \int_{-\infty}^{\infty} e^{-(z_v - t - s)^2} d(z_v - t - s) = \sqrt{\pi} e^{2ts} = \sqrt{\pi} \sum_k \frac{2^k t^k s^k}{k!} \tag{2.12}$$

2.2 Generating Function for the Harmonic Oscillator Wavefunctions 21

Equating the second and the fifth expressions in Eq. (2.12), the ones involving the summations, gives the orthogonality expressions:

$$\int_{-\infty}^{\infty} H_m(z_v) H_n(z_v) e^{-z_v^2} dz_v = 0, \qquad m \neq n$$

$$= 2^k k! \sqrt{\pi}, \qquad m = n = k \qquad (2.13)$$

One might use the expression for H_3 in Eq. (2.8) and the known definite integrals of $z_v^k e^{-z_v^2}$ to confirm this normalization constant, namely, $2^3 \cdot 3! \sqrt{\pi}$. The first three factors in parentheses in Eq. (2.10) are now derived.

The x_v factor requires the demonstration of the solution to the Schrödinger equation which will be given below. To this end, we will obtain the expressions for the z operator and the d/dz operator, from them the raising and lowering operators, from them the differential equation satisfied by v_m. Finally, we show that this equation is indeed the Schrödinger equation for the harmonic oscillator in dimensionless units.

To obtain the z operator expression, one differentiates with respect to t the first and last expression for G_v in Eq. (2.10), replaces the exponential term in the derivative with this last expression:

$$(\pi^{1/2} x_v)^{1/2} \sum_{m=0}^{\infty} \left[\frac{-2t^{m+1}}{m!} (m! 2^m)^{1/2} v_m + \frac{2t^m}{m!} (m! 2^m)^{1/2} z_v v_m \right.$$

$$\left. - \frac{m t^{m-1}}{m!} (m! 2^m)^{1/2} v_m \right] = 0 \qquad (2.14)$$

eliminates the $(\pi^{1/2} x_v)^{1/2}$ prefactor, performs the indicated divisions:

$$\sum_{m=0}^{\infty} \left[-2t^{m+1} \left(\frac{2^m}{m!} \right)^{1/2} v_m + 2t^m \left(\frac{2^m}{m!} \right)^{1/2} z_v v_m - m t^{m-1} \left(\frac{2^m}{m!} \right)^{1/2} v_m \right] = 0 \quad (2.15)$$

sets the term in t^k equal to zero:

$$-\left(\frac{2^{k+1}}{(k-1)!} \right)^{1/2} v_{k-1} + \left(\frac{2^{k+2}}{k!} \right)^{1/2} z_v v_k - (k+1) \left(\frac{2^{k+1}}{(k+1)!} \right)^{1/2} v_{k+1} = 0$$

$$(2.16)$$

and, finally, divides by $(2^{k+1}/k!)^{1/2}$, obtaining:

$$\sqrt{2} z_v v_k = \sqrt{k} v_{k-1} + \sqrt{k+1} v_{k+1} \qquad (2.17)$$

the desired expression.

A related set of algebraic steps, viz., differentiating Eq. (2.10) with respect to z_v, adding twice this expression to Eq. (2.14) and thereby eliminating the z_v term, dividing by $(\pi^{1/2} x_v)^{1/2}$, performing the indicated divisions, collecting terms in t^k and setting them equal to zero, and finally dividing by the factor $(2^{k+1}/k!)^{1/2}$, leads to the derivative expression:

$$\sqrt{2} \frac{dv_k}{dz_v} = \sqrt{k} v_{k-1} - \sqrt{k+1} v_{k+1} \qquad (2.18)$$

2 Harmonic Oscillator Wavefunctions

We now find the differential equation satisfied by the v_m. If one adds Eq. (2.17) to or subtracts it from Eq. (2.18) one gets the lowering or raising operators:

$$O_- v_k = \left(\frac{d}{dz_v} + z_v\right) v_k = (2k)^{1/2} v_{k-1} \tag{2.19}$$

$$O_+ v_k = \left(\frac{d}{dz_v} - z_v\right) v_k = -(2k+2)^{1/2} v_{k+1} \tag{2.20}$$

The desired differential equation is obtained either by the operator $O_+ O_-$ or $O_- O_+$. For example:

$$O_- O_+ v_k = \left(\frac{d}{dz_v} + z_v\right)\left(\frac{d}{dz_v} - z_v\right) v_k = \frac{d^2 v_k}{dz_v^2} - v_k - z_v^2 v_k$$

$$= \left(\frac{d}{dz_v} + z_v\right)[-(2k+2)^{1/2} v_{k+1}] = -(2k+2) v_k \tag{2.21}$$

or

$$\frac{d^2}{dz_v^2} v_k - z_v^2 v_k + (2k+1) v_k = 0 \tag{2.22}$$

This is the Schrödinger equation for the harmonic oscillator, in dimensionless units, as we will now demonstrate. This dimensioned Schrödinger equation is:

$$\left[\frac{\hbar^2}{2M}\frac{d^2}{dx^2} + E_{vm} - \tfrac{1}{2}k_v(x-a)^2\right]\Psi_m(x, k_v, a) = 0 \tag{2.23}$$

with

$$\int_{-\infty}^{\infty} \Psi_m^2(x, k_v, a)\,dx = 1 \tag{2.24}$$

In Eqs. (2.23) and (2.24), Ψ has dimensions $l^{-1/2}$; x and a have dimensions l; k_v has dimensions mt^{-2}; and $\hbar$ is Planck's constant over 2π.

Let:

$$E_{vm} = (m + \tfrac{1}{2})\hbar\omega_v \tag{2.25}$$

$$\omega_v = \left(\frac{k_v}{M}\right)^{1/2} \tag{2.26}$$

$$x_v^2 = \frac{\hbar}{\sqrt{Mk_v}} \tag{2.27}$$

$$z_v = (x - a)/x_v \tag{2.28}$$

$$\tfrac{1}{2}k_v x_v^2 = \tfrac{1}{2}\hbar\omega_v \tag{2.29}$$

Then:

$$dx = x_v\,dz_v \tag{2.30}$$

2.2 Generating Function for the Harmonic Oscillator Wavefunctions

$$\tfrac{1}{2}k_v(x-a)^2 = \tfrac{1}{2}k_v(z_v x_v)^2 = z_v^2\left(\frac{\hbar\omega_v}{2}\right) \tag{2.31}$$

$$\frac{\hbar^2}{2Mx_v^2} = \left(\frac{\hbar\omega_v}{2}\right)\left[\frac{\hbar}{Mx_v^2}\left(\frac{M}{k_v}\right)^{1/2}\right] = \left(\frac{\hbar\omega_v}{2}\right)\left[\frac{\hbar}{\sqrt{Mk_v}}\cdot\frac{1}{x_v^2}\right] = \frac{\hbar\omega_v}{2} \tag{2.32}$$

Introducing Eqs. (2.24)-(2.32) into Eq. (2.23) gives Eq. (2.22) as asserted. The dimensionality of Ψ, as needed because of the dx term in Eqs. (2.24) and (2.30), forces the introduction of the $x_v^{1/2}$ term incorporated into the normalizing factor of v_m in Eq. (2.10).

When $a = 0$ equations analogous to Eqs. (2.25)-(2.32) with $v \to u$ are posed, e.g., for Eq. (2.28),

$$z_u = x/x_u \tag{2.33}$$

Note especially that the z in Eq. (2.17) is z_v, namely, that z_v which relates Eqs. (2.22) and (2.23) correctly. We rewrite this equation both for the v and u functions:

$$\sqrt{2z_v}v_k = \sqrt{k}\,v_{k-1} + \sqrt{k+1}\,v_{k+1} \tag{2.34}$$

$$\sqrt{2z_u}u_k = \sqrt{k}\,u_{k-1} + \sqrt{k+1}\,u_{k+1} \tag{2.35}$$

This completes our derivation of the harmonic oscillator wavefunctions in the form used in this paper.

3 The Manneback Recursion Formulas

3.1 Introduction

Manneback [32] first expounded the value of recursion formulas for studying harmonic oscillator wavefunctions. We give in this section his derivation, in our notation, for the recursion formulas for the overlap integrals involved in the optical transitions and, in our model, in the nonradiative transitions also.

We will consider unequal force constants and obtain the equal-force-constants results by particularizing the unequal-force-constants results. These recursion formulas for the overlap integrals will be shown to have importance beyond their use in computations. We will in later sections use them to obtain the recursion formula for the W_p and also many functions related to W_p.

3.2 The Overlap Integral

The overlap integral in question is:

$$A_{nm} = \int_{-\infty}^{\infty} u_n(z_u) v_m(z_v)\, dx = \langle u_n | v_m \rangle \tag{3.1}$$

We recall that the dimensionality of u and v is $l^{-1/2}$ and of dx is l. We also relist the variables from Eqs. (2.28), (2.27), (2.26), and (2.32), respectively:

$$z_u = x/x_u \qquad z_v = (x - a)/x_v \tag{3.2}$$

$$x_u^2 = \hbar/(Mk_u)^{1/2} \qquad x_v^2 = \hbar/(Mk_v)^{1/2} \tag{3.3}$$

$$\omega_u = (k_u/M)^{1/2} \qquad \omega_v = (k_v/M)^{1/2} \tag{3.4}$$

$$E_u = (n + \tfrac{1}{2})\hbar\omega_u \qquad E_v = (m + \tfrac{1}{2})\hbar\omega_v \tag{3.5}$$

3.3 The Generating Function for the Overlap Integral

The Manneback recursion formulas are derived from the generating function for the overlap integrals, which in turn are derived from the generating function of the harmonic oscillator wavefunctions, applied to each electronic state:

$$G_v(z_v) = e^{-t^2+2tz_v-z_v^2/2} = \sum_{m=0}^{\infty} \frac{t^m}{m!} [m! 2^m x_v \pi^{1/2}]^{1/2} v_m(z_v) \tag{3.6}$$

$$G_u(z_u) = e^{-s^2+2sz_u-z_u^2/2} = \sum_{n=0}^{\infty} \frac{t^n}{n!} [n! 2^n x_u \pi^{1/2}]^{1/2} u_n(z_u) \tag{3.7}$$

To derive the generating function for the overlap integrals, we will define:

$$(1/x_{uv}^2) = (1/x_u^2) + (1/x_v^2) \tag{3.8}$$

$$a_{uv} = a/x_{uv} \tag{3.9}$$

$$\tan\theta = \left(\frac{k_v}{k_u}\right)^{1/4} = \left(\frac{\omega_v}{\omega_u}\right)^{1/2} = \left(\frac{1/x_v}{1/x_u}\right) = \frac{x_u}{x_v} \tag{3.10}$$

From Eqs. (3.8) and (3.10) and the trigonometric identity, $\sec^2\theta = 1 + \tan^2\theta$, one sees:

$$\frac{x_{uv}}{x_u} = \cos\theta \qquad \frac{x_{uv}}{x_v} = \sin\theta \tag{3.11}$$

We now set down the generating function for the overlap integrals. Each step of this equation will be explained immediately after the equation itself is given.

$$G_A^* = \int_{-\infty}^{\infty} G_v G_u \, dx = \sum_m \sum_n \frac{t^m s^n}{m! n!} [m! n! 2^{m+n} x_u x_v \pi]^{1/2} \int_{-\infty}^{\infty} u_n v_m \, dx$$

$$= \int_{-\infty}^{\infty} \exp\left[-t^2 + 2t\left(\frac{x-a}{x_v}\right) - \frac{1}{2}\left(\frac{x-a}{x_v}\right)^2 - s^2 + 2s\left(\frac{x}{x_u}\right)\right.$$

$$\left. - \frac{1}{2}\left(\frac{x}{x_u}\right)^2\right] dx$$

$$= \int_{-\infty}^{\infty} \exp[-t^2 + 2t(y-a_{uv})\sin\theta - \tfrac{1}{2}\sin^2\theta(y-a_{uv})^2 - s^2$$

$$+ 2sy\cos\theta - \tfrac{1}{2}\cos^2\theta y^2] x_{uv} \, dy$$

$$= \int_{-\infty}^{\infty} \exp[-\tfrac{1}{2}y^2 + y(2t\sin\theta + a_{uv}\sin^2\theta + 2s\cos\theta)$$
$$+ (-t^2 - \tfrac{1}{2}a_{uv}^2\sin^2\theta - s^2 - 2ta_{uv}\sin\theta)]x_{uv}\,dy$$
$$= \exp[-t^2 - \tfrac{1}{2}\sin^2\theta a_{uv}^2 - s^2 - 2ta_{uv}\sin\theta + \tfrac{1}{2}(2t\sin\theta$$
$$+ a_{uv}\sin^2\theta + 2s\cos\theta)^2]\cdot \int_{-\infty}^{\infty} \exp[-\tfrac{1}{2}\{y - (2t\sin\theta$$
$$+ a_{uv}\sin^2\theta + 2s\cos\theta)\}^2]x_{uv}\,dy$$
$$= \sqrt{2\pi}x_{uv}\exp[\tfrac{1}{2}a_{uv}^2(-\sin^2\theta + \sin^4\theta)]\cdot\exp[-t^2(1 - 2\sin^2\theta)$$
$$- s^2(1 - 2\cos^2\theta) + t(-2a_{uv}\sin\theta + 2a_{uv}\sin^3\theta)$$
$$+ s(2a_{uv}\sin^2\theta\cos\theta) + st(4\sin\theta\cos\theta)]$$
$$= x_{uv}\sqrt{2\pi}\exp\left(-\frac{1}{8}a_{uv}^2\sin^2 2\theta\right)e^{\delta_0}$$
$$= \sum_m \sum_n t^m s^n \left[\frac{2^{m+n}\pi}{m!n!}\right]^{1/2}\left[\frac{x_{uv}}{(\sin\theta\cos\theta)^{1/2}}\right]A_{nm}. \quad (3.12)$$

The first line of Eq. (3.12) introduces the second expressions in Eqs. (3.6) and (3.7) for G_v and G_u. The second line introduces the first expressions in these equations and uses the expressions for z_v and z_u in Eq. (3.2). The third line uses the definition of a_{uv} in Eq. (3.9), the definition:

$$y = x/x_{uv} \quad (3.13)$$

and its differential form, the θ relations of Eqs. (3.11), and the resulting relations:

$$(x - a)/x_v = (y - a_{uv})\sin\theta \quad (3.14)$$
$$x/x_u = y\cos\theta \quad (3.15)$$

The fourth line groups the terms in powers of y. The fifth line completes the square in the fourth expression and separates the terms independent of the integrating variable; the sixth line uses the integral:

$$\int_{-\infty}^{\infty} \exp[-\tfrac{1}{2}(y - b)^2]\,dy = \sqrt{2\pi} \quad (3.16)$$

and collects terms in powers of t and s. The seventh line uses the definition:

$$\delta_0 = -t^2\cos 2\theta + s^2\cos 2\theta - t(a_{uv}\sin 2\theta\cos\theta) + s(a_{uv}\sin 2\theta\sin\theta)$$
$$+ 2st\sin 2\theta \quad (3.17)$$

3.4 The Recursion Formulas for the Overlap Integrals

and the trigonometric identities:

$$\sin^4\theta - \sin^2\theta = -\tfrac{1}{4}\sin^2 2\theta \qquad (3.18)$$

$$\sin^3\theta - \sin\theta = -\tfrac{1}{2}\sin 2\theta \cos\theta \qquad (3.19)$$

$$\sin^2\theta \cos\theta = \tfrac{1}{2}\sin 2\theta \sin\theta \qquad (3.20)$$

The eighth line repeats the expression in the first line and uses the notation of Eq. (3.1), A_{nm}, for the overlap integral.

If now we equate the last two expressions of Eq. (3.12), cancel the $x_{uv}\pi^{1/2}$ term, and transfer the $(\sin\theta\cos\theta)^{1/2}$ term, we obtain:

$$G_A = \frac{(\sin\theta\cos\theta)^{1/2}}{x_{uv}\pi^{1/2}} G_A^* = \sqrt{\sin 2\theta}\exp\left[-\frac{1}{8}a_{uv}^2\sin^2 2\theta\right]e^{\delta_0}$$

$$= \sum_m\sum_n t^m s^n \left[\frac{2^{m+n}}{m!n!}\right]^{1/2} A_{nm} \qquad (3.21)$$

the desired generating function for the overlap integrals A_{nm}. In Eq. (3.21), δ_0 is given by Eq. (3.17).

3.4 The Recursion Formulas for the Overlap Integrals

We now derive two recursion formulas for the overlap integrals by taking derivatives of the generating function for them, Eq. (3.21), with respect of t and to s. After doing so, we replace the G_A with its Eq. (3.21) expression:

$$(-2t\cos 2\theta + 2s\sin 2\theta - a_{uv}\cos\theta\sin 2\theta)\sum_{m,n} t^m s^n \left[\frac{2^{m+n}}{m!n!}\right]^{1/2} A_{nm}$$

$$= \sum_{m,n} m s^n t^{m-1}\left[\frac{2^{m+n}}{m!n!}\right]^{1/2} A_{nm} \qquad (3.22)$$

$$(2s\cos 2\theta + 2t\sin 2\theta + a_{uv}\sin\theta\sin 2\theta)\sum_{m,n} t^m s^n \left[\frac{2^{m+n}}{m!n!}\right]^{1/2} A_{nm}$$

$$= \sum_{m,n} n s^{n-1} t^m \left[\frac{2^{m+n}}{m!n!}\right]^{1/2} A_{nm} \qquad (3.23)$$

We extract from each of these equations the term in $s^k t^l$, getting:

$$-2\cos 2\theta\left[\frac{2^{k+l-1}}{k!(l-1)!}\right]^{1/2} A_{k,l-1} + 2\sin 2\theta\left[\frac{2^{k+l-1}}{(k-1)!l!}\right]^{1/2} A_{k-1,l}$$

$$- a_{uv}\cos\theta\sin 2\theta\left[\frac{2^{k+l}}{k!l!}\right]^{1/2} A_{k,l} = (l+1)\left[\frac{2^{k+l+1}}{k!(l+1)!}\right]^{1/2} A_{k,l+1}$$

$$(3.24)$$

$$2\cos 2\theta \left[\frac{2^{k+l-1}}{l!(k-1)!}\right]^{1/2} A_{k-1,l} + 2\sin 2\theta \left[\frac{2^{k+l-1}}{(l-1)!k!}\right]^{1/2} A_{k,l-1}$$

$$+ a_{uv}\sin\theta\sin 2\theta \left[\frac{2^{k+l}}{k!l!}\right]^{1/2} A_{k,l} = (k+1)\left[\frac{2^{k+l+1}}{l!(k+1)!}\right]^{1/2} A_{k+1,l}$$

(3.25)

Dividing by the common factor, $(2^{l+k+1}/k!l!)^{1/2}$, we obtain the two Manneback recursion formulas:

$$-\cos 2\theta\sqrt{l}\,A_{k,l-1} + \sin 2\theta\sqrt{k}\,A_{k-1,l} - \frac{a_{uv}}{\sqrt{2}}\cos\theta\sin 2\theta\,A_{kl} = \sqrt{l+1}\,A_{k,l+1}$$

(3.26)

$$\cos 2\theta\sqrt{k}\,A_{k-1,l} + \sin 2\theta\sqrt{l}\,A_{k,l-1} + \frac{a_{uv}}{\sqrt{2}}\sin\theta\sin 2\theta\,A_{kl} = \sqrt{k+1}\,A_{k+1,l}$$

(3.27)

A better notation comes from replacing $k, l \to m, n$ and introducing:

$$k = \cos 2\theta \quad (3.28)$$

$$k_+ = \sin 2\theta \quad (3.29)$$

$$S_u^{1/2} = \frac{a_{uv}\cos\theta}{\sqrt{2}}, \quad S_v^{1/2} = \frac{a_{uv}\sin\theta}{\sqrt{2}}, \quad \frac{S_v}{S_u} = \tan^2\theta = \frac{hw_v}{\hbar\omega_u} \quad (3.30)$$

$$a_{uv} = \{2(S_u + S_v)\}^{1/2}, \quad \theta = \tan^{-1}\left[\left(\frac{S_v}{S_u}\right)^{1/2}\right] \quad (3.31)$$

Both $S_u^{1/2}$ and $S_v^{1/2}$ are signed quantities, carrying the sign of a_{uv}. The two recursion formulas then are, in a rearranged order,

$$k_+\sqrt{n}\,A_{n-1,m} = \sqrt{m+1}\,A_{n,m+1} + k_+ S_u^{1/2} A_{nm} + k\sqrt{m}\,A_{n,m-1} \quad (3.32)$$

$$k_+\sqrt{m}\,A_{n,m-1} = \sqrt{n+1}\,A_{n+1,m} - k_+ S_v^{1/2} A_{nm} - k\sqrt{n}\,A_{n-1,m} \quad (3.33)$$

The equal force constants equations are obtained from Eqs. (3.32) an (3.33) by setting $\theta = 45°$, $k = 0$, $k_+ = 1$, $S_u = S_v \equiv S = a_{uv}^2/4$

$$\sqrt{n}\,A_{n-1,m} = \sqrt{m+1}\,A_{n,m+1} + S^{1/2} A_{nm} \quad (3.34)$$

$$\sqrt{m}\,A_{n,m-1} = \sqrt{n+1}\,A_{n+1,m} - S^{1/2} A_{nm} \quad (3.35)$$

3.5 Familiarity

For familiarity, from Eq. (3.21), A_{00} for arbitrary θ is

$$A_{00} = (\sin 2\theta)^{1/2} \exp[-\tfrac{1}{8} a_{uv}^2 \sin^2 2\theta] = k_+^{1/2} \exp[-k_+(S_u S_v)^{1/2}/2] \quad (3.36)$$

3.6 The Orthonormality of the A_{nm} Matrix

and for $\theta = 45°$ is:

$$A_{00} = e^{-S/2} \tag{3.37}$$

The A_{01} and A_{10} can be obtained directly from Eq. (3.22) and (3.23), respectively, evaluated at $t = s = 0$. They are:

$$A_{01} = -(a_{uv}\cos\theta\sin 2\theta/\sqrt{2})A_{00} = -k_+ S_u^{1/2} A_{00} \tag{3.38}$$

and

$$A_{10} = (a_{uv}\sin\theta\sin 2\theta/\sqrt{2})A_{00} = k_+ S_v^{1/2} A_{00}. \tag{3.39}$$

These equations can also be obtained somewhat heuristically from Eqs. (3.26) and (3.27) or (3.32) and (3.33) by setting all A terms with negative indices to zero. This procedure is indeed always correct, and all A_{0m} and A_{n0} are readily obtained by means of three-term recursion formulas, and also A_{11} by either of two such three-term recursion formulas. Then all other rows and columns of the A matrix can be obtained in turn. This use of the recursion formulas reduces by orders of magnitude the work of getting the A matrix with respect to evaluating them from some explicit equation, such as that of Hutchisson [33].

The u_n and v_m wavefunctions are symmetric and antisymmetric when their index is even or odd, respectively. Therefore, from visualizing the two displaced wavefunctions, one can see that A_{10} and A_{01} are equal but with opposite signs when the force constants are equal, and are unequal but still have the opposite signs when the force constants are unequal. In general, these properties are true for all A_{nm} when $n - m$ is odd. When $n - m$ is even, they are identical with equal force constants and their signs are the same with unequal force constants.

These equal-force-constants relationships are valuable enough to state explicitly:

$$\langle u_\alpha | v_\beta \rangle = (-1)^{\alpha-\beta} \langle u_\beta | v_\alpha \rangle = \langle v_\beta | u_\alpha \rangle = (-1)^{\alpha-\beta} \langle v_\alpha | u_\beta \rangle \tag{3.40}$$

3.6 The Orthonormality of the A_{nm} Matrix

From Eq. (3.21) and by replacing θ with $(\pi/2) - \theta$ and a_{uv} with $-a_{uv}$, one can show that:

$$G_A(t, s, \theta, a_{uv}) = G'_A\left(s, t, \frac{\pi}{2} - \theta, -a_{uv}\right) \tag{3.41}$$

and that therefore:

$$A_{nm}(\theta, a_{uv}) = A'_{mn}\left(\frac{\pi}{2} - \theta, -a_{uv}\right) \tag{3.42}$$

The $A_{nm}(\theta, a_{uv})$ are the expansion coefficients for expanding u_n in the complete set of v_m, and the $A'_{mn}\left(\frac{\pi}{2} - \theta, a_{uv}\right)$ are the expansion coefficients for expanding

v_m in the complete set of u_n. The A matrix is thus orthonormal:

$$\sum_k A_{nk} A_{mk} = \delta(n,m)$$

$$\sum_k A_{kn} A_{km} = \delta(n,m) \qquad (3.43)$$

For example:

$$u_n = \sum_k A_{nk} v_k(\theta, a_{uv}) = \sum_k A_{nk}(\theta, a_{uv}) \sum_m A'_{km}\left(\frac{\pi}{2} - \theta, -a_{uv}\right) u_m$$

$$= \sum_{mk} A_{nk}(\theta, a_{uv}) A_{mk}(\theta, a_{uv}) u_m \qquad (3.44)$$

and therefore, since only one term in this sum survives:

$$\sum_k A_{nk} A_{mk} = \delta(n,m) \qquad (3.45)$$

3.7 Additional Equal-Force-Constants Recursion Relations

For equal force constants, one can obtain recursion formulas for other combinations of the A_{nm} from the Manneback recursion formulas, Eqs. (3.34) and (3.35). A simple diagram will elucidate the logic of this procedure.

Figure 14 diagrams Eqs. (3.34) and (3.35) on a grid of the A_{nm} matrix. The two Manneback recursion formulas relate matrix elements most picturesquely labeled as by quadrants, I and III, respectively.

The several picture equations sketched beside the grid shown combinations of these equations which share a specific matrix element in common. When this shared matrix element is eliminated, then the other orientations sketched are obtained.

To proceed according to the first picture equation, if we subtract $\sqrt{n}$ times Eq. (3.35) with $n \to n-1$ from $\sqrt{m}$ times Eq. (3.34) with $m \to m-1$, the term in

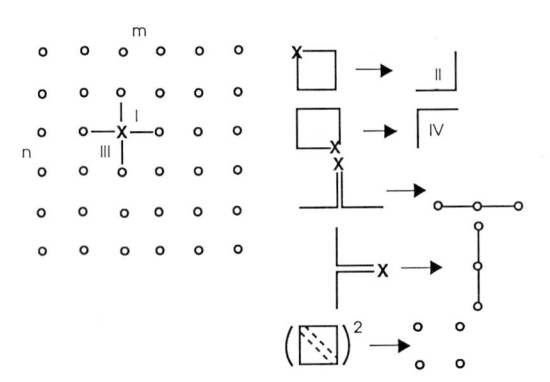

Fig. 14. Recursion relations

3.7 Additional Equal-Force-Constants Recursion Relations

$A_{n-1,m-1}$ is eliminated and we obtain:

$$(m - n)A_{n,m} + \sqrt{S}(\sqrt{m}A_{n,m-1} + \sqrt{n}A_{n-1,m}) = 0 \tag{3.46}$$

or:

$$\left(\frac{n-m}{\sqrt{S}}\right)A_{n,m} - \sqrt{m}A_{n,m-1} - \sqrt{n}A_{n-1,m} = 0 \tag{3.47}$$

Similarly, according to the second picture equation, subtracting $\sqrt{m+1}$ times Eq. (3.35) with $m \to m+1$ from $\sqrt{n+1}$ times Eq. (3.34) with $n \to n+1$ eliminates the $A_{n+1,m+1}$ term and leads to:

$$\left(\frac{n-m}{\sqrt{S}}\right)A_{n,m} = \sqrt{n+1}A_{n+1,m} + \sqrt{m+1}A_{n,m+1} \tag{3.48}$$

The third picture equation combines Eq. (3.34) and Eq. (3.47) to eliminate the term in $A_{n-1,m}$ and gives:

$$\left(\frac{n-m-S}{\sqrt{S}}\right)A_{nm} = \sqrt{m}A_{n,m-1} + \sqrt{m+1}A_{n,m+1} = \sqrt{2}\langle u_n|z_v|v_m\rangle \tag{3.49}$$

as is evident from Eq. (2.34).

The fourth picture equation combines Eq. (3.34) and Eq. (3.48) to eliminate the term in $A_{n,m+1}$ and gives:

$$\left(\frac{n-m+S}{\sqrt{S}}\right)A_{nm} = \sqrt{n+1}A_{n+1,m} + \sqrt{n}A_{n-1,m} = \sqrt{2}\langle v_m|z_u|u_n\rangle \tag{3.50}$$

We will take up the fifth picture equation when we derive the recursion formula for W_p in Sect. 4.5.

Expressions for the derivative operator analogous to Eqs. (3.49) and (3.50) can be obtained from Eq. (2.18), namely:

$$\left\langle u_n \left| \frac{d}{dz_v} \right| v_m \right\rangle = A_{n,m,d/dz_v} = \sqrt{\frac{m}{2}}A_{n,m-1} - \sqrt{\frac{m+1}{2}}A_{n,m+1} \tag{3.51}$$

$$\left\langle v_m \left| \frac{d}{dz_u} \right| u_n \right\rangle = = A'_{m,n,d/dz_u} = \sqrt{\frac{n}{2}}A_{n-1,m} - \sqrt{\frac{n+1}{2}}A_{n+1,m} \tag{3.52}$$

We can use Eqs. (3.49), (3.50), (2.34), and (2.35), for equal force constants to relate z_v matrix elements to those for z_u, thus:

$$\sqrt{2}\langle u_{p+m}|z_v|v_m\rangle = \sqrt{m}\langle u_{p+m}|v_{m-1}\rangle + \sqrt{m+1}\langle u_{p+m}|v_{m+1}\rangle$$
$$= (-1)^{p+1}[\sqrt{m}\langle u_{m-1}|v_{p+m}\rangle + \sqrt{m+1}\langle u_{m+1}|v_{p+m}\rangle]$$
$$= (-1)^{p+1}[\sqrt{m}\langle v_{p+m}|u_{m-1}\rangle + \sqrt{m+1}\langle v_{p+m}|u_{m+1}\rangle]$$
$$= \sqrt{2}(-1)^{p+1}\langle v_{p+m}|z_u|u_m\rangle \tag{3.53}$$

These relationships will be used in the discussion below of the transition rates which are only vibrationally allowed.

4 The Luminescence Center: The Single-Configurational-Coordinate Model

4.1 The Model for the Radiative Rate

We now tie these A_{nm} to a model of a luminescence center. The model of Fig. 1 need not have equal force constants. Figure 15 is drawn accurately for $\theta = 42°$, $a_{uv} = 5.886$. This set of parameters, seemingly over-particularized, is related to an equal-force-constants case with simple parameters and has roughly the largest deviation from equal force constants that we have found necessary to consider. The case involving this set of parameters will be taken up again in Sect. 9.1, where the two curves of Fig. 15 are seen again as the dotted curves of Fig. 21. In Sect. 9.1, related equal- and unequal-force constants cases are compared.

The two electronic states in the model, represented in Fig. 1 by the two parabolas labeled u for the ground state and v for the excited state, have vibrational energy levels labeled u_n and v_m, respectively. Here, the phonon energies are now different. The energies of u_0 and one u_n and of v_0 and one v_m are shown in the figure. These parabolas are shown with unequal curvatures and displaced one from the other in such a way that

$$S_u \hbar \omega_u = \tfrac{1}{2} k_u a^2 \tag{4.1}$$

is the energy of the ground state at the upper-state equilibrium coordinate, and

$$S_v \hbar \omega_v = \tfrac{1}{2} k_v a^2 \tag{4.2}$$

is the energy rise in the v parabola to the energy axis, i.e., the line $x = 0$. From Eqs. (4.1) and (4.2), the displacement a satisfies

$$a = \left(\frac{2S_v \hbar \omega_v}{k_v}\right)^{1/2} = \left(\frac{2S_u \hbar \omega_u}{k_u}\right)^{1/2} \tag{4.3}$$

The two methods of reaching the v_m vibrational level from u_0 give the energy balance equation:

$$n\hbar\omega_u + h\nu_{vm,un} = h\nu_{zp,vu} + m\hbar\omega_v \tag{4.4}$$

In Eq. (4.4), the photon energy, $h\nu_{vm,un}$, is to be read as the energy of the first indexed state minus that of the second indexed state, and similarly for the zero-photon energy, $h\nu_{zp,vu}$. Both are signed quantities. While both are pictured

4.1 The Model for the Radiative Rate

positive in Fig. 1, a much lower placement of the v state, below the u state, would change both signs. We have interest in both such placements and shall return to them in Chap. 9.

We wish now to draw attention not to the photon energy itself but to the phonon energy created during an optical transition. We then must distinguish absorption and emission, since with a given placement of the states, absorption and emission at the same photon energy will create phonons in absorption and absorb them in emission or visa-versa. We want an index which itself will give this change in signs. To this end, we rewrite Eqs. (4.4) as:

$$hv_{vm,un} - hv_{zp,vu} = p_v \hbar \omega_v = hv_{a,p_v} - hv_{zp,vu} = \left(m - n \frac{\hbar \omega_u}{\hbar \omega_v} \right) \hbar \omega_v \tag{4.5}$$

in absorption, and:

$$hv_{zp,vu} - hv_{vm,un} = p_u \hbar \omega_u = hv_{zp,vu} - hv_{e,p_u} = \left(n - m \frac{\hbar \omega_v}{\hbar \omega_u} \right) \hbar \omega_u \tag{4.6}$$

in emission. This form, Eqs. (4.5)–(4.6), stresses the phonon energy created by the transition, namely, $p_v \hbar \omega_v$ in absorption, $p_u \hbar \omega_u$ in emission. The photon energy depends not on phonon energy n and m independently but rather on the combinations which are in parentheses and which are themselves labeled p_u and p_v. For any placement of the two parabolas in Fig. 15 which maintain the same offset parameter a_{uv}, the optical bands will superpose when the zero-phonon lines are superposed. Their common distribution function will be developed below, algebraically for equal force constants, numerically for unequal.

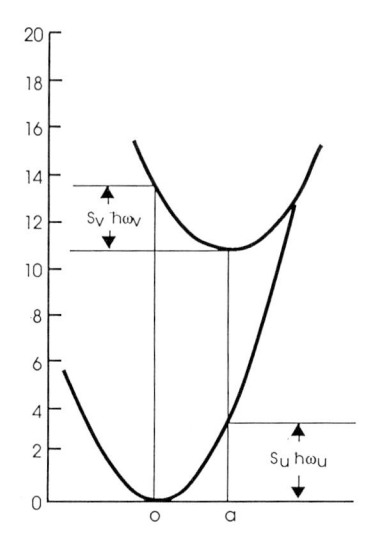

Fig. 15. The Single-Configurational-Coordinate Model with unequal force constants

4.2 The Equal-Force-Constants Radiative Rate

We now write down the equal-force constants forms of Eqs. (4.1)–(4.6):

$$S\hbar\omega = \tfrac{1}{2}ka^2 \tag{4.7}$$

$$a = \left(\frac{2S\hbar\omega}{k}\right)^{1/2} \tag{4.8}$$

$$h\nu_{vm,un} - h\nu_{zp,vu} = p_v\hbar\omega = h\nu_{p_v} - h\nu_{zp,vu} = (m-n)\hbar\omega \tag{4.9}$$

in absorption, and:

$$h\nu_{zp,vu} - h\nu_{vm,un} = p_u\hbar\omega = h\nu_{zp,vu} - h\nu_{p_u} = (n-m)\hbar\omega \tag{4.10}$$

in emission.

We give attention to that attribute of the equal-force-constants case which allows further significant analytical advances, namely, that emissions and absorptions naturally group into discrete families indexed by their optical transition energy or, alternatively, by the number of phonons generated by the transition. All absorptions from any u_n to that v_m for which $m - n = p_v$ appear at the common photon energy, $h\nu_{p_v}$, offset to higher energy from the zero-phonon line by $p_v\hbar\omega$. All emissions from any v_m to that u_n for which $n - m = p_u$ appear at the common photon energy, $h\nu_{p_u}$, offset to lower energy from the zero-phonon line by $p_u\hbar\omega$.

It is then natural to group thermally weighted squared overlap integrals, in absorption and in emission, respectively:

$$\begin{aligned} W_p &= \sum_{n=max(0,-p)} (1-r)r^n \langle v_{p+n}|u_n\rangle^2 = \sum_{n=max(0,-p)} (1-r)r^n A^2_{p+n,n} \\ W_p &= \sum_{m=max(0,-p)} (1-r)r^m \langle u_{p+m}|v_m\rangle^2 = \sum_{m=max(0,-p)} (1-r)r^m A^2_{p+m,m} \end{aligned} \tag{4.11}$$

In Eq. (4.11), r is the Boltzmann factor:

$$r = \exp(-\hbar\omega/kT) \tag{4.12}$$

with the normalization constant:

$$\sum_m r^m = \frac{1}{1-r} \tag{4.13}$$

It is clear that the expressions in absorption and in emission are equivalent, because of the properties of the A_{nm} matrix elements in Eq. (3.40).

The range of p is over all integers, including negative ones. Therefore the lower bounds of the sums in Eq. (4.11) are $n_0 \equiv max(0,-p)$ and $m_0 \equiv max(0,-p)$, as shown, since no A_{nm} index can be negative.

Equation (4.11) gives the equal-force-constants thermal-Condon factor for the radiative rate of the transition generating p phonons. It applies independently of the vertical placement of the v electronic state relative to the u state. This is the advantage for stressing the phonon energy generated rather than the optical transition photon energy. It is a normalized distribution, since:

4.2 The Equal-Force-Constants Radiative Rate

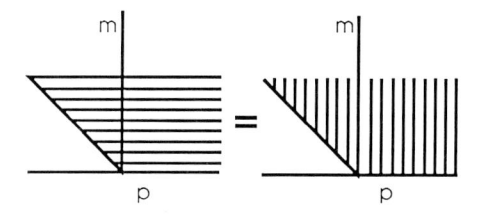

Fig. 16. Equivalent summation procedures

$$\sum_{p=-\infty}^{\infty} \sum_{m=max(0,-p)}^{\infty} (1-r_m)r^m A_{p+m,m}^2 = \sum_{m=0}^{\infty} \sum_{p=-m}^{\infty} (1-r_m)r^m A_{p+m,m}^2$$

$$= \sum_{n=p+m=0}^{\infty} \sum_{m=0}^{\infty} (1-r_m)r^m A_{n,m}^2 = \sum_{m=0}^{\infty} (1-r_m)r^m = 1 \quad (4.14)$$

The first relationship in Eq. (4.14) is inverting the order of summation, and the limits shown are clear from a diagram of the p, m plane, Fig. 16 above. Thus W_p is the normalized shape function of both the absorption and emission bands for equal force constants.

Eq. (4.11) comes from using the golden rule transition rate expression:

$$w = \frac{2\pi}{\hbar} \rho(k) |H'_{km}|^2 \quad (4.15)$$

where the wavefunctions are labeled m in the initial state and k in the final state, and $\rho(k)$ is the density of states near the final state. In our case, the wavefunctions are products of electronic-nuclear and nuclear dependent functions, and the perturbative Hamiltonian is assumed purely electronic for optical transitions. After integrating out the electronic coordinates, we assume that we are left with a simple expression for the nuclear perturbing Hamiltonian. We will often take that expression to be a constant, R_{uv}, independent of the nuclear coordinate, but we will consider also the expressions, proportional to the displacement from equilibrium, $x = x_{uv}z_u/\cos\theta = (x_{uv}z_v/\sin\theta) + a_{uv}x_{uv}$:

$$R_{vu,x}x = R_{vu,x}[x_{uv}z_u/\cos\theta] \quad (4.16)$$

in absorption;

$$R_{uv,x}x = R_{uv,x}[x_{uv}z_v/\sin\theta + a_{uv}x_{uv}] \quad (4.17)$$

in emission.

With the assumption of constancy, the transition rate from some specific initial state and generating p phonons is:

$$w_{p+m,m} = R_{uv}|\langle u_{p+m}|v_m\rangle|^2 = R_{uv}A_{p+m,m}^2$$
$$w_{p+n,n} = R_{vu}|\langle v_{p+n}|u_n\rangle|^2 = R_{vu}A_{p+n,n}^2 \quad (4.18)$$

in emission and in absorption, respectively. We now weight the contribution from each initial state by its thermal occupancy and sum the transition rates from all initial states, obtaining:

$$w_{uv} = R_{uv}W_p \qquad w_{vu} = R_{vu}W_p \quad (4.19)$$

in emission and in absorption, respectively. Here, R_{uv} is the radiative lifetime of the upper state and R_{vu} is proportional to the absorption strength or f number of the transition. The normalized shape function is for equal force constants the W_p function both for absorption and for emission.

When a transition is allowed only vibrationally, i.e., when its electronic transition integral is proportional to the factor in Eq. (4.16) in absorption or Eq. (4.17) in emission, the transition rates analogous to Eq. (4.15) are:

$$w_{p+n,n,z_u} = R_{vu,x} |\langle v_{p+n}| \frac{x_{uv} z_u}{\cos\theta} |u_n\rangle|^2$$

$$= R_{vu,x} \left(\frac{x_{uv}}{\cos\theta}\right)^2 A^2_{p+n,n,z_u} \equiv R_{vu,x} A^2_{p+n,n,z_u}$$

$$w_{p+m,m,z_v} = R_{uv,x} \langle u_{p+m}| \left[\frac{x_{uv} z_v}{\sin\theta} + x_{uv} a_{uv}\right] |v_m\rangle^2 \qquad (4.20)$$

In the first of Eq. (4.20), we have incorporated the $(x_{uv}/\cos\theta)^2$ factor into the $R_{vu,x}$ without renaming this parameter. The total absorption rate analogous to Eq. (4.19) is:

$$w_{vu,z} = R_{vu,x} V_{p,z} \to R_{vu,x} W_{p,z} \qquad (4.21)$$

where the arrow points to the equal force constants special case and, in analogy to Eq. (4.11):

$$W_{p,z} = \sum_n (1-r) r^n \langle v_{p+n}|z_u|u_n\rangle^2 = \sum_n (1-r) r^n A^2_{p+n,n,z}$$

$$= \sum_m (1-r) r^m \langle u_{p+m}|z_v|v_m\rangle^2 = \sum_m (1-r) r^m A^2_{p+m,m,z} \qquad (4.22)$$

Here we have used Eq. (3.53). Finally, using Eqs. (3.49), we can rewrite the $W_{p,z}$ term as:

$$W_{p,z} = \frac{(p-S)^2}{2S} W_p \qquad (4.23)$$

One might expect at times to encounter a transition whose electronic integral vanishes at some symmetric position $x = 0$. Then, Eqs. (4.22) and (4.23) would lead to the expectation of a cusp in the distribution function at $p = S$. However, we shall see that in multiple coordinates this cusp disappears essentially completely.

On the other hand, it is not to be expected that the electronic integral would vanish for the $v \to u$ transition at $x = a$. One might, however, be interested in an emission rate somewhat dependent upon the coordinate, as for example driven by the operator $1 + bz_v$, itself a constant times the operator in the second line of Eq. (4.20).

We shall return to the consideration of the vibrationally allowed transitions after we have the W_p recursion formula in hand, because then we shall be able to obtain its temperature-dependent moments.

4.3 The Unequal-Force-Constants Radiative Rate

When the force constants are unequal, no natural quantization of the photon energies at phonon energy intervals emerges. For example, the $u_1 \to v_m$ absorptions occur between the $u_0 \to v_m$ energy set. Nevertheless analogues for the thermal-Condon weights W_p of Eq. (4.11) are desirable, one for absorption and the other for emission. The one for absorption, called V_{p_V}, has for its energy metric the set of $u_0 \to v_m$ energies, with $\hbar\omega_v$ energy intervals. The one for emission, called U_{p_U}, uses the set of $v_0 \to u_n$ energies, with $\hbar\omega_u$ energy intervals. The indices used for the energy scale markers will be in capital letters. These distributions are obtained numerically by assigning each absorption intensity which occurs between $p_V \hbar\omega_v$ and $(p_V + 1)\hbar\omega_v$ to these two energies, weighted by $p_V + 1 - p_v$ and $p_v - p_V$ respectively, and similarly assigning each emission intensity which occurs between $p_U \hbar\omega_u$ and $(p_U + 1)\hbar\omega_u$ to these two energies, weighted by $p_U + 1 - p_u$ and $p_u - p_U$ respectively.

We have:

$$V_{p_V} = \sum_n (1 - r_u) r_u^n V_{p_V, n}^2 \tag{4.24}$$

$$U_{p_U} = \sum_m (1 - r_v) r_v^m U_{p_U, m}^2 \tag{4.25}$$

$$V_{p_V, z} = \sum_n (1 - r_u) r_u^n V_{p_V, n, z}^2 \tag{4.26}$$

$$U_{p_U, z} = \sum_m (1 - r_v) r_v^m U_{p_U, m, z}^2 \tag{4.27}$$

where

$$r_u = \exp(-\hbar\omega_u/kT) \tag{4.28}$$

$$r_v = \exp(-\hbar\omega_v/kT) \tag{4.29}$$

$$V_{p_V, n}^2 = \langle v_{i_n}|u_n\rangle^2 (1 - i_n + p_n) + \langle v_{i_n-1}|u_n\rangle^2 (i_n - p_n) \tag{4.30}$$

$$U_{p_U, m}^2 = \langle u_{i_m}|v_m\rangle^2 (1 - i_m + p_m) + \langle u_{i_m-1}|v_m\rangle^2 (i_m - p_m) \tag{4.31}$$

$$V_{p_V, n, z}^2 = \langle v_{i_n}|z|u_n\rangle^2 (1 - i_n + p_n) + \langle v_{i_n-1}|z|u_n\rangle^2 (i_n - p_n) \tag{4.32}$$

$$U_{p_U, m, z}^2 = \langle u_{i_m}|z|v_m\rangle^2 (1 - i_m + p_m) + \langle u_{i_m-1}|z|v_m\rangle^2 (i_m - p_m) \tag{4.33}$$

In Eqs. (4.30)–(4.33), p_n and p_m are noninteger numbers and i_n and i_m are integer numbers satisfying

$$p_n \hbar\omega_v - n\hbar\omega_u = p_V \hbar\omega_v,$$

$$p_m \hbar\omega_u - m\hbar\omega_v = p_U \hbar\omega_u,$$

$$i_n = \text{int}(p_n) + 1,$$

$$i_m = \text{int}(p_m) + 1,$$

$$\text{int}(-x) = -\text{int}(x), \tag{4.34}$$

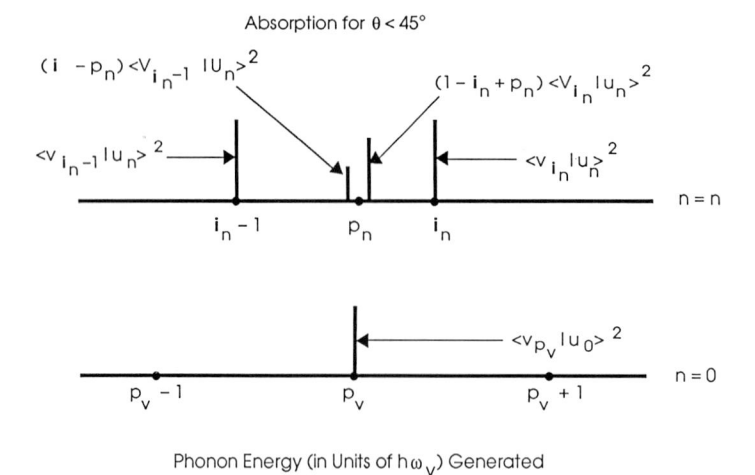

38   4 The Luminescence Center: The Single-Configurational-Coordinate Model

Fig. 17. Apportioning thermal weights

where int(x) is the truncation function giving the largest integer smaller than x when $x \geq 0$.

The simple diagram in Fig. 17 will clarify these apportionings. Only absorption is illustrated; emission would be with $n \to m$, $v \to u$ in this illustration. The weighted components at p_n are shown slightly displaced from p_n.

This figure shows the $V^2_{p_v,0}$ of (4.30) and a typical $V^2_{p_v,n}$. The $V^2_{p_v,0}$ contributions come at energies which are used to define the metric of the phonon energy generated by the transition. The $V^2_{p_v,n}$ at the energy $p_v\hbar\omega_v$ gathers its transition weight from the two nearest Franck-Condon factors, namely $\langle v_{i_n-1}|u_n\rangle^2$ and $\langle v_{i_n}|u_n\rangle^2$, and the contributions are in proportion to the energy mismatches.

These equations (4.24) to (4.34) are quite useful for the numerical evaluation of the transition rates. The unequal-force-constants A_{nm} matrix is readily obtained through the Manneback recursion formulas Eqs. (3.26) and (3.27). Then the U_{p_v}, V_{p_v}, $U_{p_v,z}$, and $V_{p_v,z}$ matrices are simple linear combinations of the A_{nm}.

4.4 The Model for the Nonradiative Rate

Figure 3 has shown a component of a nonradiative transition, from v_m to that u_n which matches closely in energy. This figure obtains both when the force constants are equal and when they are unequal. The product of nuclear and electronic derivative operators is normally invoked for driving the nonradiative transition. After integrating over the electronic coordinates and invoking the Boltzmann weighting of the initial states, we are left with the nonradiative $v \to u$ and

4.5 The W_p Recursion Formula

$u \to v$ rates, respectively:

$$R_{p,d/dz} = A_{uv} \sum_m (1 - r_m) r^m \langle u_{p+m} | d/dz | v_m \rangle^2 \equiv A_{uv} U_{p_U,d/dz}$$

$$R_{p,d/dz} = A_{vu} \sum_n (1 - r_n) r^n \langle v_{p+n} | d/dz | u_n \rangle^2 \equiv A_{vu} V_{p_V,d/dz} \tag{4.35}$$

For equal force constants, using Eqs. (3.51), (3.52), and (3.40):

$$U_{p_U,d/dz_u} = V_{p_V,d/dz_v} = W_{p,d/dz} \tag{4.36}$$

These equal-force-constants expressions will be further evaluated in Sect. 4.11.

For unequal force constants, these expressions are evaluated numerically from the $A_{nm,d/dz}$ matrix, which itself is obtained from the A_{nm} matrix using Eq. (2.18). The defining equations are analogous to Eqs. (4.24)–4.34), namely:

$$V_{p_V,d/dz} = \sum_n (1 - r_u) r_u^n V_{p_V,n,d/dz}^2$$

$$U_{p_U,d/dz} = \sum_m (1 - r_v) r_v^m U_{p_U,m,d/dz}^2$$

$$V_{p_V,n,d/dz}^2 = \langle v_{i_n} | d/dz | u_n \rangle^2 (1 - i_n + p_n) + \langle v_{i_n-1} | d/dz | u_n \rangle^2 (i_n - p_n)$$

$$U_{p_U,m,d/dz}^2 = \langle u_{i_m} | d/dz | v_m \rangle^2 (1 - i_m + p_m) + \langle u_{i_m-1} | d/dz | v_m \rangle^2 (i_m - p_m) \tag{4.37}$$

where r_u and r_v are defined in Eqs. (4.28) and (4.37) and p_U, p_V, p_i, and p_j satisfy Eq. (4.34). It is found in practice that we do not have an adequate theory of the magnitude of the electronic integral, here called A_{uv}, A_{vu}, and that an A'_{vu} and an A'_{uv} can readily be found so that, for example, $A_{uv} U_{p_U}$ and $A'_{uv} U_{p_U,d/dz}$ or $A_{vu} V_{p_V}$ and $A'_{vu} V_{p_V,d/dz}$ have practically indistinguishable temperature dependences. See Chapter 9 below. Accordingly, we often discuss nonradiative rates using Eq. (4.11) rather than Eq. (4.35), since our interest is usually elsewhere than in demonstrating which operator drives the nonradiative transition.

4.5 The W_p Recursion Formula

We return now to the fifth picture equation of Fig. 14. The open circles represent squares of the corresponding matrix element. By squaring the two recursion formulas sketched in such a way that the cross terms have common indices, one can eliminate this cross term and be left with a four-term relationship between squared A_{nm} terms. Thus, Eqs. (3.34) and (3.35) are rewritten with the $S^{1/2} A_{nm}$ term isolated to force the correct cross terms. The relationship between squares which results will lead directly to the recursion formula for the W_p. The result is:

$$(n - m)(A_{n-1,m-1}^2 - A_{nm}^2) = S(A_{n,m-1}^2 - A_{n-1,m}^2) \tag{4.38}$$

We then set $n - m = p$ and operate upon Eq. (4.38) with $\sum_m^\infty (1 - r) r^m$ with the lower m limit specified either to:

$$m_0 = max(0, -p) \qquad m_1 = max(1, -p) \qquad (4.39)$$

dependent upon whether the $m = 0$ term does not or does, respectively, lead to negative indices. The result is:

$$S \sum_{m=m_1}^{\infty} r^m(1-r)A^2_{p+m,m-1} - S \sum_{m=m_0}^{\infty} r^m(1-r)A^2_{p+m-1,m}$$

$$= p \sum_{m=m_1}^{\infty} r^m(1-r)A^2_{p+m-1,m-1} - p \sum_{m=m_0}^{\infty} r^m(1-r)A^2_{p+m,m} \qquad (4.40)$$

Now, using the definition of W_p in Eq. (4.11), namely:

$$W_p = \sum_{m=m_0}^{\infty} (1-r)r^m A^2_{p+m,m} \qquad (4.41)$$

and redefining the summation index as required, we obtain from Eq. (4.40):

$$SrW_{p+1} - SW_{p-1} = rpW_p - pW_p \qquad (4.42)$$

For example, the first sum in Eq. (4.40) uses, in turn, $p + m = p + 1 + m - 1$, $m - 1 \to m'$, and $(m = m_1) \to (m' = m_0)$. We now use, recalling Eq. (1.7):

$$\langle m \rangle = \frac{r}{1-r} \qquad r = \frac{\langle m \rangle}{\langle 1+m \rangle} \qquad 1 - r = \frac{1}{\langle 1+m \rangle} \qquad (4.43)$$

to get the desired recursion formula:

$$S\langle m \rangle W_{p+1} + pW_p - S\langle 1+m \rangle W_{p-1} = 0 \qquad (4.44)$$

The $\langle m \rangle$ is Planck's measure of the temperature.

4.6 Explicit Series Expression for the W_p Function

The explicit expression for W_p is:

$$W_p = \exp(-S\langle 2m+1 \rangle) \sum_{j=max(0,-p)}^{\infty} \frac{(S\langle m \rangle)^j (S\langle 1+m \rangle)^{p+j}}{j!(p+j)!} \qquad (4.45)$$

This can be proved by substituting Eq. (4.45) into the recursion formula Eq. (4.44) and demonstrating that indeed the result is identically zero. Thus we are to prove that:

$$S\langle m \rangle \sum_{j=max(0,-p-1)}^{\infty} \frac{(S\langle m \rangle)^j (S\langle 1+m \rangle)^{p+1+j}}{j!(p+1+j)!}$$

$$+ p \sum_{j=max(0,-p)}^{\infty} \frac{(S\langle m \rangle)^j (S\langle 1+m \rangle)^{p+j}}{j!(p+j)!}$$

$$- S\langle 1+m \rangle \sum_{j=max(0,-p+1)}^{\infty} \frac{(S\langle m \rangle)^j (S\langle 1+m \rangle)^{p-1+j}}{j!(p-1+j)!} = 0 \qquad (4.46)$$

4.7 I_p Modified Bessel Function Form for W_p

We must consider three cases separately. The first is for $p > 0$, when we extract the term from Eq. (4.46) in $(S\langle m \rangle)^k (S\langle 1 + m \rangle)^{p+k}$. Its coefficient is:

$$\left[\frac{1}{(k-1)!(p+k)!} + \frac{p}{k!(p+k)!} - \frac{1}{k!(p-1+k)!} \right] = \frac{k+p-(p+k)}{k!(p+k)!} = 0,$$

$$k > 0$$

$$\left[\frac{p}{p!} - \frac{1}{(p-1)!} \right] = 0, \quad k = 0 \tag{4.47}$$

The second case is for $p = 0$. Then the coefficient of $(S\langle m \rangle)^k (S\langle 1 + m \rangle)^k$ is:

$$\left[\frac{1}{(k-1)!(k)!} - \frac{1}{k!(k-1)!} \right] = 0 \tag{4.48}$$

The third case is for $p < 0$. Then the coefficient of $(S\langle m \rangle)^k (S\langle 1 + m \rangle)^{p+k}$ is:

$$\left[\frac{1}{(k-1)!0!} + \frac{-k}{k!0!} \right] = 0, \quad k = -p$$

$$\left[\frac{1}{(k-1)!(p+k)!} + \frac{p}{k!(p+k)!} - \frac{1}{k!(p-1+k)!} \right]$$

$$= \frac{k+p-(p+k)}{k!(p+k)!} = 0, \quad k > -p \tag{4.49}$$

Thus, in all possible cases, Eq. (4.46) is shown identically zero. It must next be proved that the expression Eq. (4.45) is the normalized solution of the recursion formula Eq. (4.44). To this end, we note that:

$$\sum_{p=-\infty}^{\infty} \sum_{j=\max(0,-p)}^{\infty} \frac{(S\langle m \rangle)^j (S\langle 1 + m \rangle)^{p+j}}{j!(p+j)!}$$

$$= \sum_{j=0}^{\infty} \sum_{p+j=k=0}^{\infty} \frac{(S\langle m \rangle)^j (S\langle 1 + m \rangle)^k}{j!(k)!}$$

$$= \exp(S\langle m \rangle + S\langle 1 + m \rangle) \tag{4.50}$$

and therefore Eq. (4.45) is the normalized solution of the recursion formula Eq. (4.44) and is a valid expression for W_p. The change in summation indices in Eq. (4.50) has been used already in Eq. (4.14) and explained using Fig. 16.

4.7 I_p Modified Bessel Function Form for W_p

The I_p modified Bessel Function can be defined as the normalized solution of its recursion formula:

$$\tfrac{1}{2} z I_{p+1} + p I_p - \tfrac{1}{2} z I_{p-1} = 0 \tag{4.51}$$

Let:

$$W_p = \left[\frac{\langle 1+m\rangle}{\langle m\rangle}\right]^{p/2} F_p(z) \tag{4.52}$$

If one substitutes Eq. (4.52) into the W_p recursion formula Eq. (4.44) one obtains:

$$S\langle m\rangle \left[\frac{\langle 1+m\rangle}{\langle m\rangle}\right]^{(p+1)/2} F_{p+1} + p\left[\frac{\langle 1+m\rangle}{\langle m\rangle}\right]^{p/2} F_p$$
$$- S\langle 1+m\rangle \left[\frac{\langle 1+m\rangle}{\langle m\rangle}\right]^{(p-1)/2} F_{p-1} = 0 \tag{4.53}$$

or:

$$S[\langle 1+m\rangle\langle m\rangle]^{1/2} F_{p+1} + pF_p - S[\langle 1+m\rangle\langle m\rangle]^{1/2} F_{p-1} = 0 \tag{4.54}$$

This is exactly Eq. (4.51) with the correspondences:

$$z \to 2S(\langle m\rangle\langle 1+m\rangle)^{1/2}$$
$$F_p \to const \cdot I_p(z)$$

The W_p expression Eq. (4.52) thus converts the W_p recursion formula into the I_p recursion formula and therefore, with the proper normalization factor:

$$W_p = \exp(-S\langle 2m+1\rangle)\left[\frac{\langle 1+m\rangle}{\langle m\rangle}\right]^{p/2} I_p(2S(\langle m\rangle\langle 1+m\rangle)^{1/2}) \tag{4.55}$$

4.8 Limiting and Approximate Forms of W_p

When $\langle m\rangle = 0$, Eq. (4.44) becomes the recursion formula for the Poisson distribution,

$$pW_p = SW_{p-1} \tag{4.56}$$

and thus:

$$W_p = e^{-S}\frac{S^p}{p!} \tag{4.57}$$

When $[S\langle m\rangle S\langle 1+m\rangle]/[\vert p\vert + 1]$ is small, then W_p is approximately the first term of Eq. (4.45):

$$W_p \approx B_p \equiv e^{-S\langle 2m+1\rangle}\frac{(S\langle 1+m\rangle)^p}{p!}, \quad p \geq 0,$$

$$\equiv e^{-S\langle 2m+1\rangle}\frac{(S\langle m\rangle)^{\vert p\vert}}{\vert p\vert!}, \quad p < 0 \tag{4.58}$$

For large S the W_p distribution approaches a Gaussian. The relation can be seen from the similarity of the W_p recursion formula for large S to the differential

4.8 Limiting and Approximate Forms of W_p

equation for a Gaussian function:

$$G(p) = \frac{1}{\sqrt{2\pi\sigma^2}} \exp[-(p-S)^2/2\sigma^2] \tag{4.59}$$

namely:

$$\sigma^2 \frac{dG}{dp} + (p-S)G = 0 \tag{4.60}$$

If we rewrite Eq. (4.44) as:

$$S\langle m \rangle [W_{p+1} - W_p] + [S\langle m \rangle + p - S\langle 1+m \rangle]W_p +$$
$$S\langle 1+m \rangle [W_p - W_{p-1}] = 0 \tag{4.61}$$

we see that the W_p recursion formula is related to the differential equation:

$$S\langle 2m+1 \rangle \frac{dW_p}{dp} + (p-S)W_p = 0 \tag{4.62}$$

i.e., to Eq. (4.60). Thus when S is large W_p appropriates a Gaussian with $\sigma^2 = S\langle 2m+1 \rangle$.

A useful relationship for the Gaussian is that between the width at half height and the σ, namely, $(\Delta p)_{G=1/2} = \sqrt{8\ln 2}\,\sigma$.

The Eq. (4.58) B_p form has the additional approximation B_p^* where the factorial is replaced by its Stirling asymptotic formula:

$$B_p^* = e^{-S\langle 2m+1 \rangle}(S\langle 1+m \rangle)^p \frac{1}{\sqrt{2\pi p}} \left(\frac{e}{p}\right)^p, \quad p > 0 \tag{4.63}$$

Finally, the I_p Bessel-function form has its asymptotic expansion:

$$I_p(x) \approx \frac{e^{y_p}}{\sqrt{2\pi y_p}} \left(\frac{x}{p + y_p}\right)^p \tag{4.64}$$

where

$$y_p = \sqrt{p^2 + x^2} \geq 1 \tag{4.65}$$

If we introduce Eq. (4.64) into Eq. (4.55), we obtain another approximate form for W_p, namely:

$$W_p \approx W_p^* \equiv e^{-S\langle 2m+1 \rangle} \left(\frac{\langle 1+m \rangle}{\langle m \rangle}\right)^{p/2} \frac{e^{y_p}}{\sqrt{2\pi y_p}} \left(\frac{x}{p+y_p}\right)^p \tag{4.66}$$

where:

$$x = 2S(\langle m \rangle \langle 1+m \rangle)^{1/2} \tag{4.67}$$

and:

$$y_p = \sqrt{p^2 + x^2} \geq 1 \tag{4.68}$$

44   4 The Luminescence Center: The Single-Configurational-Coordinate Model

4.9 The 5-W_p Formula for $W_{p,z}$

We will need another form of the $W_{p,z}$ formula, Eq. (4.23), which we will call the 5-W_p form. We get it by expanding the square in Eq. (4.23) and applying the W_p recursion formula Eq. (4.44) several times to eliminate all terms involving pW_p. The result is:

$$\begin{aligned}W_{p,z} &= \left[\frac{p^2}{2S} - p + \frac{S}{2}\right]W_p = \left(\frac{p}{2S} - 1\right)(pW_p) + \frac{1}{2}SW_p \\ &= \left(\frac{p}{2S} - 1\right)[S\langle 1+m\rangle W_{p-1} - S\langle m\rangle W_{p+1}] + \frac{1}{2}SW_p \\ &= \frac{\langle 1+m\rangle}{2}[(p-1)W_{p-1} + W_{p-1}] - S\langle 1+m\rangle W_{p-1} \\ &\quad - \frac{\langle m\rangle}{2}[(p+1)W_{p+1} - W_{p+1}] + S\langle m\rangle W_{p+1} + \frac{S}{2}W_p \\ &= \frac{\langle 1+m\rangle}{2}W_{p-1} + \frac{\langle 1+m\rangle}{2}[S\langle 1+m\rangle W_{p-2} - S\langle m\rangle W_p] \\ &\quad - S\langle 1+m\rangle W_{p-1} + \frac{\langle m\rangle}{2}W_{p+1} - \frac{\langle m\rangle}{2}[S\langle 1+m\rangle W_p \\ &\quad - S\langle m\rangle W_{p+2}] + S\langle m\rangle W_{p+1} + \frac{S}{2}W_p = L_p + S\sum_{i=-2}^{2}\alpha_{i,z}W_{p+i}\end{aligned}$$

(4.69)

where

$$L_p = \tfrac{1}{2}\{\langle 1+m\rangle W_{p-1} + \langle m\rangle W_{p+1}\}$$

$$\alpha_{-2,z} = \frac{\langle 1+m\rangle^2}{2}$$

$$\alpha_{-1,z} = -\langle 1+m\rangle$$

$$\alpha_{0,z} = \frac{1}{2} - \langle m\rangle\langle 1+m\rangle$$

$$\alpha_{1,z} = \langle m\rangle$$

$$\alpha_{2,z} = \frac{\langle m\rangle^2}{2}$$

$$\sum_i \alpha_{i,z} = 0$$

$$\sum_p W_{p,z} = \frac{1}{2}\langle 2m+1\rangle \qquad (4.70)$$

4.10 The $\langle m \rangle_p$ Formula

We define the average m through which a transition occurs:

$$\langle m \rangle_p = \frac{\sum_{m=m_0}^{\infty} m(1-r)r^m A^2_{p+m,m}}{\sum_{m=m_0}^{\infty} (1-r)r^m A^2_{p+m,m}} = \left[\frac{1}{W_p}\right] \sum_{m=m_0}^{\infty} m(1-r)r^m A^2_{p+m,m} \quad (4.71)$$

It is evident that:

$$\langle m \rangle_p W_p = r(1-r)\frac{d}{dr}\left(\frac{W_p}{1-r}\right) = r\frac{dW_p}{d\langle m \rangle}\frac{d\langle m \rangle}{dr} + rW_p\frac{1-r}{(1-r)^2} \quad (4.72)$$

However, using Eq. (4.43):

$$r\frac{d\langle m \rangle}{dr} = r\left[\frac{(1-r)+r}{(1-r)^2}\right] = \frac{r}{(1-r)^2} = \langle m \rangle \langle 1+m \rangle \quad (4.73)$$

One obtains the other needed factor most readily from the explicit formula for W_p, Eq. (4.45):

$$\frac{dW_p}{d\langle m \rangle} = \frac{d}{d\langle m \rangle}\left[e^{-S\langle 2m+1 \rangle} \sum_{j=max(0,-p)}^{\infty} \frac{(S\langle m \rangle)^j(S\langle 1+m \rangle)^{p+j}}{j!(p+j)!}\right]$$

$$= -2SW_p + e^{-S\langle 2m+1 \rangle}\left[\sum_{j=max(1,-p)}^{\infty} \frac{jS(S\langle m \rangle)^{j-1}(S\langle 1+m \rangle)^{p+j}}{j!(p+j)!}\right.$$

$$\left. + \sum_{j=max(0,-p+1)}^{\infty} \frac{S(p+j)(S\langle m \rangle)^j(S\langle 1+m \rangle)^{p+j-1}}{j!(p+j)!}\right]$$

$$= -2SW_p + Se^{-S\langle 2m+1 \rangle}$$

$$\times \left[\sum_{j-1=max(0,-p-1)}^{\infty} \frac{(S\langle m \rangle)^{j-1}(S\langle 1+m \rangle)^{p+1+j-1}}{(j-1)!(p+1+j-1)!}\right.$$

$$\left. + \sum_{j=max(0,-p+1)}^{\infty} \frac{(S\langle m \rangle)^j(S\langle 1+m \rangle)^{p-1+j}}{j!(p-1+j)!}\right]$$

$$= -2SW_p + SW_{p+1} + SW_{p-1} \quad (4.74)$$

and therefore Eq. (4.72) becomes:

$$\langle m \rangle_p W_p = S\langle m \rangle \langle 1+m \rangle [W_{p+1} - 2W_p + W_{p-1}] + \langle m \rangle W_p \quad (4.75)$$

4.11 The $W_{p,d/dz}$ Expression

We now develop the expression analogous to $W_{p,z}$ for the derivative operator, namely:

$$W_{p,d/dz} = \sum_{m=m_0}^{\infty} (1-r)r^m \langle u_{p+m}|\frac{d}{dz}|v_m \rangle^2 \quad (4.76)$$

4 The Luminescence Center: The Single-Configurational-Coordinate Model

We remember Eq. (4.22):

$$W_{p,z} = \sum_{m=m_0}^{\infty} (1-r)r^m \langle u_{p+m}|z|v_m \rangle^2 = \sum_{m=m_0}^{\infty} (1-r)r^m A_{p+m,m,z}^2$$

$$= \sum_{m=m_0}^{\infty} (1-r)r^m \left[\frac{m}{2} A_{p+m,m-1}^2 + \frac{m+1}{2} A_{p+m,m+1}^2 \right.$$

$$\left. + \sqrt{m(m+1)} A_{p+m,m-1} A_{p+m,m+1} \right] \tag{4.77}$$

Here we have used Eq. (3.49). Equations (3.49) and (3.51) can conveniently be expressed as one equation using the notation:

$$A_{p+m,m,O_\pm} = \sqrt{\frac{m}{2}} A_{p+m,m-1} \pm \sqrt{\frac{m+1}{2}} A_{p+m,m+1} \tag{4.78}$$

where:

$$O_+ = z, \qquad O_- = \frac{d}{dz}, \qquad A_{kIO} = \langle u_k|O|v_I \rangle. \tag{4.79}$$

We see directly that:

$$W_{p,z} - W_{p,d/dz} = \sum_{m=m_0}^{\infty} (1-r)r^m \left[\left\{ \sqrt{\frac{m}{2}} A_{p+m,m-1} + \sqrt{\frac{m+1}{2}} A_{p+m,m+1} \right\}^2 \right.$$

$$\left. - \left\{ \sqrt{\frac{m}{2}} A_{p+m,m-1} - \sqrt{\frac{m+1}{2}} A_{p+m,m+1} \right\}^2 \right]$$

$$= \sum_{m=m_0}^{\infty} (1-r)r^m \{2\sqrt{m(m+1)} A_{p+m,m-1} A_{p+m,m+1}\} \tag{4.80}$$

Eliminating the cross term from Eqs. (4.77) and (4.80) gives:

$$2W_{p,z} + W_{p,d/dz} - W_{p,z}$$

$$= \left[\sum_{m=m_0}^{\infty} (1-r)r^m m A_{p+m,m-1}^2 + \sum_{m=m_0}^{\infty} (1-r)r^m (1+m) A_{p+m,m+1}^2 \right] \tag{4.81}$$

Now, we use $m = m - 1 + 1$. We note that the $m = 0$ term does not contribute to the first sum of Eq. (4.81) and that therefore the lower limit for the summation index can be written $m = \max(1, -p)$ or $m - 1 = \max(0, -p - 1)$. Likewise the lower limit of the summation index of the second term can be written $m + 1 = \max(1, -p + 1)$, and since the $m + 1 = 0$ term does not contribute to this sum, we can include it in the sum and rewrite its lower limit as $m + 1 = \max(0, -p + 1)$. We get:

4.11 The $W_{p,d/dz}$ Expression

$$W_{p,z} + W_{p,d/dz} = \left\{ r \sum_{m-1=max(0,-p-1)}^{\infty} (1-r)r^{m-1}(m-1)A^2_{p+1+m-1,m-1} \right.$$

$$+ r \sum_{m-1=max(0,-p-1)}^{\infty} (1-r)r^{m-1} A^2_{p+1+m-1,m-1}$$

$$\left. + \frac{1}{r} \sum_{m+1=max(0,-p+1)}^{\infty} (1-r)r^{m+1}(m+1)A^2_{p-1+m+1,m+1} \right\}$$

$$= r\{\langle m \rangle_{p+1} W_{p+1} + W_{p+1}\} + \frac{1}{r} \langle m \rangle_{p-1} W_{p-1}$$

$$= \frac{\langle m \rangle}{\langle 1+m \rangle} \{S \langle m \rangle \langle 1+m \rangle (W_{p+2} - 2W_{p+1} + W_p)$$

$$+ \langle m \rangle W_{p+1}\} + \frac{\langle m \rangle}{\langle 1+m \rangle} \{W_{p+1}\} + \frac{\langle 1+m \rangle}{\langle m \rangle}$$

$$\times \{S \langle m \rangle \langle 1+m \rangle (W_p - 2W_{p-1} + W_{p-2}) + \langle m \rangle W_{p-1}\}$$
(4.82)

The last step introduces Eq. (4.75). Still collecting terms we get:

$$W_{p,z} + W_{p,d/dz} = \langle m \rangle W_{p+1} + \langle 1+m \rangle W_{p-1}$$
$$+ S[W_{p+2}\langle m \rangle^2 - 2W_{p+1}\langle m \rangle^2 + W_p\{\langle m \rangle^2 + \langle 1+m \rangle^2\}$$
$$- W_{p-1}\{2(\langle 1+m \rangle^2)\} + W_{p-2}\{\langle 1+m \rangle^2\}]$$
(4.83)

Using Eqs. (4.69) and (4.70) for $W_{p,z}$ and the identities:

$$\langle m \rangle^2 + \langle 1+m \rangle^2 = 2\langle m \rangle \langle 1+m \rangle + 1$$
$$\langle 1+m \rangle^2 - \langle 1+m \rangle = \langle m \rangle \langle 1+m \rangle \qquad (4.84)$$

we get finally the 5-W_p form for $W_{p,d/dz}$

$$W_{p,d/dz} = L_p + S \sum_{i=-2}^{2} \alpha_{i,d/dz} W_{p+i} \qquad (4.85)$$

with:

$$L_p = \frac{1}{2}\{\langle 1+m \rangle W_{p-1} + \langle m \rangle W_{p+1}\}$$

$$\alpha_{-2,d/dz} = \frac{\langle 1+m \rangle^2}{2}$$

$$\alpha_{-1,d/dz} = -\langle 1+m \rangle \langle 2m+1 \rangle$$

$$\alpha_{0,d/dz} = \frac{1}{2} + 3\langle m \rangle \langle 1+m \rangle$$

48   4 The Luminescence Center: The Single-Configurational-Coordinate Model

$$\alpha_{1,d/dz} = -\langle m\rangle\langle 2m+1\rangle$$

$$\alpha_{2,d/dz} = \frac{\langle m\rangle^2}{2}$$

$$\sum_i \alpha_{i,d/dz} = 0$$

$$\sum_p W_{p,d/dz} = \frac{1}{2}\langle 2m+1\rangle \tag{4.86}$$

Eqs. (4.69) and (4.85) can be combined under the notation

$$W_{p,O_\pm} = L_p + S\sum_{i=-2}^{2}\alpha_{i,\pm}W_{p+i} \tag{4.87}$$

where the $\alpha_{i,\pm}$ are defined by Eqs. (4.70) and (4.86) for $+$ and $-$, respectively.
From Eq. (4.87) and using Eqs. (4.70) and (4.86) one can derive the relation:

$$W_{p,d/dz} = W_{p,z} - 2\langle m\rangle\langle 1+m\rangle S(W_{p+1} - 2W_p + W_{p-1}) \tag{4.88}$$

4.12 The W_{-p}/W_p and Related Ratios

The ratio between inverse transition weights is obtained for all three operators under consideration. For the Condon operator:

$$W_{-p} = \exp(-p\hbar\omega/kT)W_p \tag{4.89}$$

The simplest demonstration is from Eq. (4.11) itself:

$$W_{-|p|} = \sum_{n=|p|}^{\infty}(1-r)r^n A^2_{-|p|+n,n} = r^{|p|}\sum_{n'=n-|p|=0}^{\infty}(1-r)r^{n'}A_{n',n'+|p|}$$
$$= r^{|p|}W_{|p|} = \exp(-p\hbar\omega/kT)W_{|p|} \tag{4.90}$$

The same ratio obtains for the z and d/dz operator distributions. We will give the derivation for d/dz and skip the very similar derivation for z. The derivation here uses the $5 - W_p$ forms, Eq. (4.87). These forms are correct for every p and in particular are correct for $-|p|$:

$$W_{-|p|,d/dz} = \frac{1}{2}\Big\{\langle 1+m\rangle W_{-|p|-1} + \langle m\rangle W_{-|p|+1}\Big\} + S\Big\{\frac{\langle 1+m\rangle^2}{2}W_{-|p|-2}$$
$$- \langle 1+m\rangle\langle 2m+1\rangle W_{-|p|-1} + \Big[\frac{1}{2} + 3\langle m\rangle\langle 1+m\rangle\Big]W_{-|p|}$$
$$- \langle m\rangle\langle 2m+1\rangle W_{-|p|+1} + \frac{\langle m\rangle^2}{2}W_{-|p|+2}\Big\} \tag{4.91}$$

4.13 Equal-Force-Constants Moments

If now we introduce Eq. (4.89) for each W_p, extract the factor $\exp(-|p|\hbar\omega/kT)$, and use the identity:

$$\frac{\langle 1+m \rangle}{\langle m \rangle} = \exp(\hbar\omega/kT) \tag{4.92}$$

in all but the $W_{|p|}$ term, we are led directly to the last form of Eq. (4.69) with $p = |p|$. Thus:

$$W_{-p,d/dz} = \exp(-p\hbar\omega/kT) W_{p,d/dz} \tag{4.93}$$

4.13 Equal-Force-Constants Moments

The moments of the W_p distribution are valuable tools for applying these concepts to experimental data. The first moment is readily obtained from the W_p recursion formula, merely by summing.

$$\sum_{p=-\infty}^{\infty} [S\langle m \rangle W_{p+1} + pW_p - S\langle 1+m \rangle W_{p-1}] = 0$$
$$= S\langle m \rangle + \langle p \rangle - S - S\langle m \rangle = \langle p \rangle - S \tag{4.94}$$

We get directly:

$$\langle p \rangle = S \tag{4.95}$$

The higher moments are central moments defined by:

$$\langle (p-S)^\alpha \rangle = \sum_{p=-\infty}^{\infty} (p-S)^\alpha W_p \tag{4.96}$$

These are readily evaluated using their own recursion formula:

$$\langle (p-S)^\alpha \rangle = \sum_{j=0}^{\alpha-2} \binom{\alpha-1}{j} \beta_{\alpha j} \langle (p-S)^j \rangle \tag{4.97}$$

where:

$$\begin{aligned}\beta_{\alpha j} &= S\langle 2m+1 \rangle & (\alpha - j) \text{ even} \\ &= S, & (\alpha - j) \text{ odd}\end{aligned} \tag{4.98}$$

and:

$$\binom{n}{m} = \frac{n!}{m!(n-m)!} \tag{4.99}$$

is the binomial coefficient.

Equation (4.97) is proved as follows. We form from the W_p recursion formula Eq. (4.44):

4 The Luminescence Center: The Single-Configurational-Coordinate Model

$$0 = \sum_{p=-\infty}^{\infty} (p-S)^{\alpha-1}[S\langle m\rangle W_{p+1} + (p-S)W_p + SW_p - S\langle 1+m\rangle W_{p-1}]$$

$$= \langle (p-S)^\alpha\rangle + S\langle (p-S)^{\alpha-1}\rangle + S\langle m\rangle \left[\sum_p \{(p+1-S)-1\}^{\alpha-1}W_{p+1}\right]$$

$$- S\langle 1+m\rangle\left[\sum_p\{(p-1-S)+1\}^{\alpha-1}W_{p-1}\right]$$

$$= \langle (p-S)^\alpha\rangle + S\langle (p-S)^{\alpha-1}\rangle$$

$$+ S\langle m\rangle \sum_p \sum_{j=0}^{\alpha-1}\binom{\alpha-1}{j}(p+1-S)^j(-1)^{\alpha-1-j}W_{p+1}$$

$$- S\langle 1+m\rangle\sum_p \sum_{j=0}^{\alpha-1}\binom{\alpha-1}{j}(p-1-S)^j(+1)^{\alpha-1-j}W_{p-1}$$

$$= \langle (p-S)^\alpha\rangle + S\langle (p-S)^{\alpha-1}\rangle + S\langle m\rangle \sum_{j=0}^{\alpha-1}\binom{\alpha-1}{j}(-1)^{\alpha-1-j}\langle (p-S)^j\rangle$$

$$- S\langle 1+m\rangle\sum_{j=0}^{\alpha-1}\binom{\alpha-1}{j}\langle (p-S)^j\rangle \qquad (4.100)$$

whence:

$$\sum_{j=0}^{\alpha-1}\binom{\alpha-1}{j}\langle (p-S)^j\rangle \beta_{\alpha j} = \langle (p-S)^\alpha\rangle + S\langle (p-S)^{\alpha-1}\rangle \qquad (4.101)$$

where:

$$\beta_{\alpha j} = S\langle 2m+1\rangle \qquad (\alpha - j) \text{ even}$$
$$= S \qquad (\alpha - j) \text{ odd} \qquad (4.102)$$

Note that the term on the left hand side of Eq. (4.101) with $j = \alpha - 1$ has $\beta_{\alpha j} = S$ and:

$$\binom{\alpha-1}{\alpha-1} = 1 \qquad (4.103)$$

so that this term of the sum and the second term on the right hand side cancel. Thus we obtain Eq. (4.97).

Beginning with the statement of normalization, $\langle (p-S)^0\rangle = \langle 1\rangle = 1$, and the relation from Eq. (4.95), $\langle p - S\rangle = 0$, we get in sequence from Eq. (4.97):

$$\langle (p-S)^2\rangle = \beta_{20}\binom{1}{0}\langle (p-S)^0\rangle = S\langle 2m+1\rangle$$

$$\langle (p-S)^3\rangle = \beta_{30}\binom{2}{0}\langle (p-S)^0\rangle + \beta_{31}\binom{2}{1}\langle (p-S)^1\rangle = S$$

4.14 Unequal-Force-Constants Moments

$$\langle (p-S)^4 \rangle = \beta_{40}\binom{3}{0}\langle (p-S)^0 \rangle + \beta_{41}\binom{3}{1}\langle (p-S)^1 \rangle$$
$$+ \beta_{42}\binom{3}{2}\langle (p-S)^2 \rangle$$
$$= S\langle 2m+1 \rangle + 3[S\langle 2m+1 \rangle][S\langle 2m+1 \rangle]$$
$$= S\langle 2m+1 \rangle + 3[S\langle 2m+1 \rangle]^2 \tag{4.104}$$

4.14 Unequal-Force-Constants Moments

One can also obtain rigorous moments for unequal force constants. For a $v \to u$ transition, the first moment desired of a quantity q is the average of q according to:

$$\langle q \rangle_{uv} = \sum_{m=0}^{\infty}\sum_{n=0}^{\infty} q(1-r_v) r_v^m A_{nm}^2 \tag{4.105}$$

The first moment for this $v \to u$ transition has as its q the net increase in the vibrational energy of the oscillator in units of the final-state phonon energy, p_u:

$$p_u = [(n + \tfrac{1}{2}) - \kappa_u(m + \tfrac{1}{2})] \tag{4.106}$$

where:

$$\kappa_u = \frac{\hbar\omega_v}{\hbar\omega_u} = \tan^2\theta \tag{4.107}$$

The higher moments are central moments and have for their q's the quantity:

$$p_u - \langle p_u \rangle_{uv} = (n - \langle n \rangle_{uv}) - \kappa_u(m - \langle m \rangle_{uv}) \tag{4.108}$$

raised to the appropriate power. We will state the answers first and then give their proof.

4.14.1 The Moments

The emission, $v \to u$, moments are:

$$\langle p_u \rangle_{uv} = S_u + k_\theta \langle 2m+1 \rangle_v$$

$$\langle (p_u - \langle p_u \rangle_{uv})^2 \rangle_{uv} = \left(\frac{\omega_u}{\omega_v}\right) S_u \langle 2m+1 \rangle_v + 2k_\theta^2 \langle 2m+1 \rangle_v^2$$

$$\langle (p_u - \langle p_u \rangle_{uv})^3 \rangle_{uv} = S_u + 6k_\theta \left(\frac{\omega_u}{\omega_v}\right) S_u \langle 2m+1 \rangle_v^2$$
$$+ 4k_\theta^2 \left(\frac{\omega_v}{\omega_u}\right)\langle 2m+1 \rangle_v + 8k_\theta^3 \langle 2m+1 \rangle_v^3 \tag{4.109}$$

4 The Luminescence Center: The Single-Configurational-Coordinate Model

The k'th energy moment is the k'th p moment multiplied by the final-state phonon energy, here $\hbar\omega_u$, raised to the k'th power. In Eq. (4.109), we use the notation of Eqs. (3.28) and (3.29) and the definition of k_θ, namely:

$$k = \cos 2\theta \qquad k_+ = \sin 2\theta \qquad k_\theta = \left(\frac{k}{k_+^2}\right) = \frac{\omega_u^2 - \omega_v^2}{4\omega_v\omega_u} \qquad (4.110)$$

The $u \to v$ moments are obtained from these by applying the symmetry transformation:

$$u \leftrightarrow v, \qquad m \leftrightarrow n, \qquad k_\theta \to -k_\theta \qquad (4.111)$$

which is implied in the translation of the x coordinate according to $x' \to x - a$. These $u \to v$ p moments in absorption are:

$$\langle p_v \rangle_{vu} = S_v - k_\theta \langle 2n + 1 \rangle_u$$

$$\langle (p_v - \langle p_v \rangle_{vu})^2 \rangle_{vu} = \left(\frac{\omega_v}{\omega_u}\right) S_v \langle 2n + 1 \rangle_u + 2k_\theta^2 \langle 2n + 1 \rangle_u^2$$

$$\langle (p_v - \langle p_v \rangle_{vu})^3 \rangle_{vu} = S_v - 6k_\theta \left(\frac{\omega_v}{\omega_u}\right) S_v \langle 2n + 1 \rangle_u^2$$

$$+ 4k_\theta^2 \left(\frac{\omega_u}{\omega_v}\right) \langle 2n + 1 \rangle_u - 8k_\theta^3 \langle 2n + 1 \rangle_u^3 \qquad (4.112)$$

The energy moments are these p moments multiplied by the appropriate power of $\hbar\omega_v$.

4.14.2 Preliminaries I: The (α, m, β)

For the proof of Eq. (4.109) we need the sums:

$$(\alpha, m, \beta) = \sum_{n=0}^{\infty} = n^\alpha A_{nm} A_{n, m-\beta} \qquad (4.113)$$

These sums are explicitly:

$(0, m, 0) = 1$

$(0, m, \beta) = 0, \qquad \beta \neq 0$

$(1, m, 0) = S_u + \left(\frac{1}{k_+^2}\right) \{m(1 + k^2) + k^2\}$

$(1, m, 1) = \left(\frac{1}{k_+}\right) S_u^{1/2} m^{1/2} (1 + k)$

$(1, m, 2) = \left(\frac{1}{k_+^2}\right) m^{1/2} (m - 1)^{1/2} k$

$(1, m, \beta) = 0, \qquad \beta \geq 3$

4.14 Unequal-Force-Constants Moments

$$(2, m, 0) = S_u^2 + \left(\frac{1}{k_+^2}\right) S_u \{m(4 + 4k + 4k^2) + (1 + 2k + 3k^2)\}$$

$$+ \left(\frac{1}{k_+^4}\right) \{m^2(1 + 4k^2 + k^4) + m(4k^2 + 2k^4) + (2k^2 + k^4)\}$$

$$(2, m, 1) = \left(\frac{1}{k_+}\right) S_u^{3/2} m^{1/2} (2 + 2k)$$

$$+ \left(\frac{1}{k_+^3}\right) S_u^{1/2} m^{1/2} \{m(2 + 4k + 4k^2 + 2k^3) + (-1 - k + k^2 + k^3)\}$$

$$(2, m, 2) = \left(\frac{1}{k_+^2}\right) S_u m^{1/2} (m - 1)^{1/2} (1 + 4k + k^2)$$

$$+ \left(\frac{1}{k_+^4}\right) m^{1/2} (m - 1)^{1/2} \{m(2k + 2k^3) - 2k\}$$

$$(2, m, 3) = \left(\frac{1}{k_+^3}\right) S_u^{1/2} m^{1/2} (m - 1)^{1/2} (m - 2)^{1/2} (2k + 2k^2)$$

$$(2, m, 4) = \left(\frac{1}{k_+^4}\right) m^{1/2} (m - 1)^{1/2} (m - 2)^{1/2} (m - 3)^{1/2} k^2$$

$$(2, m, \beta) = 0, \quad \beta \geq 5$$

$$(3, m, 0) = S_u^3 + \left(\frac{1}{k_+^2}\right) S_u^2 \{m(9 + 12k + 9k^2) + (3 + 6k + 6k^2)\}$$

$$+ \left(\frac{1}{k_+^4}\right) S_u \{m^2(9 + 18k + 36k^2 + 18k^3 + 9k^4)$$

$$+ m(3 + 12k + 36k^2 + 24k^3 + 15k^4) + (1 + 6k + 19k^2$$

$$+ 12k^3 + 7k^4)\}$$

$$+ \left(\frac{1}{k_+^6}\right) \{m^3(1 + 9k^2 + 9k^4 + k^6) + m^2(9k^2 + 18k^4 + 3k^6)$$

$$+ m(14k^2 + 23k^4 + 3k^6) + (4k^2 + 10k^4 + k^6)\} \tag{4.114}$$

These formulas for (α, m, β) were obtained from the equation:

$$\sum_{h=0}^{\alpha} \binom{\alpha}{h} (\alpha - h, m, \beta)$$

$$= \left(\frac{1}{k_+^2}\right) (m + 1)^{1/2} (m + 1 - \beta)^{1/2} (\alpha - 1, m + 1, \beta)$$

$$+ \left(\frac{1}{k_+}\right) S_u^{1/2} (m + 1)^{1/2} (\alpha - 1, m + 1, \beta + 1)$$

$$+ \left(\frac{k}{k_+^2}\right)(m+1)^{1/2}(m-\beta)^{1/2}(\alpha-1, m+1, \beta+2)$$

$$+ \left(\frac{1}{k_+}\right)S_u^{1/2}(m+1-\beta)^{1/2}(\alpha-1, m, \beta-1) + S_u(\alpha-1, m, \beta)$$

$$+ \left(\frac{k}{k_+}\right)S_u^{1/2}(m-\beta)^{1/2}(\alpha-1, m, \beta+1)$$

$$+ \left(\frac{k}{k_+^2}\right)m^{1/2}(m+1-\beta)^{1/2}(\alpha-1, m-1, \beta-2)$$

$$+ \left(\frac{k}{k_+}\right)S_u^{1/2}m^{1/2}(\alpha-1, m-1, \beta-1)$$

$$+ \left(\frac{k^2}{k_+^2}\right)m^{1/2}(m-\beta)^{1/2}(\alpha-1, m-1, \beta) \tag{4.115}$$

Equation (4.115) is a recursion formula giving the sum (α, m, β) in terms of $(\alpha + 9)$ similar sums with lower values of α. This recursion formula is initiated with the first two sums of Eq. (4.114), which are statements of the orthonormality of the A_{nm} matrix. The remaining sums in Eq. (4.114) then follow in sequence. It is helpful to multiply the hth term in Eq. (4.115) by the identity $1 = 1^h = [(1 - k^2)/k_+^2]^h$.

Equation (4.115) in turn is derived by operating on the first of the Manneback recursion formulas Eq. (3.32) with:

$$\sum_{n=1}^{\infty} n^{\alpha-1/2} A_{n-1, m-\beta}$$

which gives:

$$k_+ \sum_{n=1}^{\infty} n^\alpha A_{n-1, m} A_{n-1, m-\beta}$$

$$\equiv k_+ \sum_{n=0}^{\infty} (n+1)^\alpha A_{nm} A_{n, m-\beta} \equiv k_+ \sum_{h=0}^{\alpha} \binom{\alpha}{h}(\alpha - h, m, \beta)$$

$$= \sum_{n=1}^{\infty} n^{\alpha-1/2} A_{n-1, m-\beta}[(m+1)^{1/2} A_{n, m+1} + k_+ S_u^{1/2} A_{nm}$$

$$+ k(m)^{1/2} A_{n, m-1}]. \tag{4.116}$$

The last identity relation in Eq. (4.116) uses the definition Eq. (4.113). The three sums on the right-hand side of Eq. (4.116) involve a half-power in n, which can be eliminated through the use of the recursion formula Eq. (3.32). Thus we operate on Eq. (3.32), rewritten with the index m replaced by $m - \beta$, with:

$$\sum_{n=0}^{\infty} n^{\alpha-1} A_{n,\delta}$$

4.14 Unequal-Force-Constants Moments

where δ is assigned the values $m + 1$, m, $m - 1$, respectively. In turn, these give:

$$k_+ \sum_{n=1}^{\infty} n^{\alpha-1/2} A_{n-1,m-\beta} A_{n,m+1}$$

$$= (m - \beta + 1)^{1/2} \sum_{n=0}^{\infty} n^{\alpha-1} A_{n,m+1} A_{n,m-\beta+1}$$

$$+ k_+ S_u^{1/2} \sum_{n=0}^{\infty} n^{\alpha-1} A_{n,m+1} A_{n,m-\beta}$$

$$+ k(m - \beta)^{1/2} \sum_{n=0}^{\infty} n^{\alpha-1} A_{n,m+1} A_{n,m-\beta-1}$$

$$= (m - \beta + 1)^{1/2}(\alpha - 1, m + 1, \beta) + k_+ S_u^{1/2}(\alpha - 1, m + 1, \beta + 1)$$

$$+ k(m - \beta)^{1/2}(\alpha - 1, m + 1, \beta + 2)$$

$$k_+ \sum_{n=1}^{\infty} n^{\alpha-1/2} A_{n-1,m-\beta} A_{n,m} = (m - \beta + 1)^{1/2}(\alpha - 1, m, \beta - 1)$$

$$+ k_+ S_u^{1/2}(\alpha - 1, m, \beta)$$

$$+ k(m - \beta)^{1/2}(\alpha - 1, m, \beta + 1)$$

$$k_+ \sum_{n=1}^{\infty} n^{\alpha-1/2} A_{n-1,m-\beta} A_{n,m-1} = (m - \beta + 1)^{1/2}(\alpha - 1, m - 1, \beta - 2)$$

$$+ k_+ S_u^{1/2}(\alpha - 1, m - 1, \beta - 1)$$

$$+ k(m - \beta)^{1/2}(\alpha - 1, m - 1, \beta) \qquad (4.117)$$

Inserting these relations into Eq. (4.116) gives Eq. (4.115).

4.14.3 Preliminaries II: The Thermal Averages $\langle m^\alpha \rangle_v$

The thermal averages:

$$\langle m^\alpha \rangle_v = \sum_{m=0}^{\infty} (1 - r_v) r_v^m m^\alpha \qquad (4.118)$$

are evaluated by means of the recursion relation:

$$(1 - r_v)^{-1} \langle m^{\alpha+1} \rangle_v = r_v \frac{d}{dr_v} [(1 - r_v)^{-1} \langle m^\alpha \rangle_v] \qquad (4.119)$$

which is obtained by operating on Eq. (4.118) with:

$$r_v \left[\frac{d}{dr_v} (1 - r_v)^{-1} \right]$$

Equation (4.119) is initiated with:

$$\langle m^0 \rangle_v = 1 \qquad (4.120)$$

and we get, in turn:

$$\langle m \rangle_v = (1 - r_v)^{-1} r_v = \{\exp(\hbar\omega_v/kT) - 1\}^{-1}$$
$$\langle m^2 \rangle_v = 2\langle m \rangle_v^2 + \langle m \rangle_v$$
$$\langle m^3 \rangle_v = 6\langle m \rangle_v^3 + 6\langle m \rangle_v^2 + \langle m \rangle_v$$
$$\langle m^4 \rangle_v = 24\langle m \rangle_v^4 + 36\langle m \rangle_v^3 + 14\langle m \rangle_v^2 + \langle m \rangle_v \tag{4.121}$$

4.14.4 Preliminaries III: The $\langle n^\alpha m^\gamma \rangle_{uv}$

We give here certain double averages which will occur in the expressions for the desired moments:

$$\langle n^\alpha m^\gamma \rangle_{uv} = \sum_{m=0}^\infty \sum_{n=0}^\infty n^\alpha m^\gamma (1 - r_v) r_v^m A_{nm}^2$$

$$= (1 - r_v) \sum_{m=0}^\infty \left[\sum_{n=0}^\infty n^\alpha A_{nm}^2 \right] m^\gamma r_v^m$$

$$= \sum_{m=0}^\infty (\alpha, m, 0) m^\gamma (1 - r_v) r_v^m$$

$$= \langle (\alpha, m, 0) m^\gamma \rangle_v \tag{4.122}$$

which are evaluated by introducing the sums $(\alpha, m, 0)$ from Eq. (4.114) and the thermal averages $\langle m^\alpha \rangle$ from Eq. (4.121). These sums then are:

$$\langle n \rangle_{uv} = S_u + \left(\frac{1}{k_+^2}\right)[\langle m \rangle_v (1 + k^2) + k^2]$$

$$\langle m \rangle_{uv} = \langle m \rangle_v$$

$$\langle n^2 \rangle_{uv} = S_u^2 + \left(\frac{1}{k_+^2}\right) S_u [\langle m \rangle_v (4 + 4k + 4k^2) + (1 + 2k + 3k^2)]$$
$$+ \left(\frac{1}{k_+^4}\right)[\langle m \rangle_v^2 (2 + 8k^2 + 2k^4) + \langle m \rangle_v (1 + 8k^2 + 3k^4)$$
$$+ (2k^2 + k^4)]$$

$$\langle nm \rangle_{uv} = S_u \langle m \rangle_v + \left(\frac{1}{k_+^2}\right)[\langle m \rangle_v^2 (2 + 2k^2) + \langle m \rangle_v (1 + 2k^2)]$$

$$\langle m^2 \rangle_{uv} = 2\langle m \rangle_v^2 + \langle m \rangle_v$$

$$\langle n^3 \rangle_{uv} = S_u^3 + \left(\frac{1}{k_+^2}\right) S_u^2 [\langle m \rangle_v (9 + 12k + 9k^2) + (3 + 6k + 6k^2)]$$
$$+ \left(\frac{1}{k_+^4}\right) S_u [\langle m \rangle_v^2 (18 + 36k + 72k^2 + 36k^3 + 18k^4)$$
$$+ \langle m \rangle_v (12 + 30k + 72k^2 + 42k^3 + 24k^4)$$
$$+ (1 + 6k + 19k^2 + 12k^3 + 7k^4)]$$

4.14 Unequal-Force-Constants Moments

$$+ \left(\frac{1}{k_+^6}\right)[\langle m\rangle_v^3(6 + 54k^2 + 54k^4 + 6k^6)$$
$$+ \langle m\rangle_v^2(6 + 72k^2 + 90k^4 + 12k^6)$$
$$+ \langle m\rangle_v(1 + 32k^2 + 50k^4 + 7k^6) + (4k^2 + 10k^4 + k^6)]$$

$$\langle n^2 m\rangle_{uv} = S_u^2\langle m\rangle_v + \left(\frac{1}{k_+^2}\right)S_u[\langle m\rangle_v^2(8 + 8k + 8k^2)$$
$$+ \langle m\rangle_v(5 + 6k + 7k^2)] + \left(\frac{1}{k_+^4}\right)[\langle m\rangle_v^3(6 + 24k^2 + 6k^4)$$
$$+ \langle m\rangle_v^2(6 + 32k^2 + 10k^4) + \langle m\rangle_v(1 + 10k^2 + 4k^4)]$$

$$\langle nm^2\rangle_{uv} = S_u[2\langle m\rangle_v^2 + \langle m\rangle_v]$$
$$+ \left(\frac{1}{k_+^2}\right)[\langle m\rangle_v^3(6 + 6k^2) + \langle m\rangle_v^2(6 + 8k^2) + \langle m\rangle_v(1 + 2k^2)]$$

$$\langle m^3\rangle_{uv} = 6\langle m\rangle_v^3 + 6\langle m\rangle_v^2 + \langle m\rangle \qquad (4.123)$$

Several grouped terms will be needed for the moment formulas:

$$\langle n + \tfrac{1}{2}\rangle_{uv} = S_u + \left(\frac{1}{k_+^2}\right)[\langle m\rangle_v(1 + k^2) + (\tfrac{1}{2} + \tfrac{1}{2}k^2)]$$

$$\langle m + \tfrac{1}{2}\rangle_{uv} = \langle m\rangle_v + \tfrac{1}{2}$$

$$\langle n^2\rangle_{uv} - \langle n\rangle_{uv}^2 = \left(\frac{1}{k_+^2}\right)S_u[\langle m\rangle_v(2 + 4k + 2k^2) + (1 + 2k + k^2)]$$
$$+ \left(\frac{1}{k_+^4}\right)[\langle m\rangle_v^2(1 + 6k^2 + k^4)$$
$$+ \langle m\rangle_v(1 + 6k^2 + k^4) + 2k^2]$$

$$\langle nm\rangle_{uv} - \langle n\rangle_{uv}\langle m\rangle_{uv} = \left(\frac{1}{k_+^2}\right)[\langle m\rangle_v^2(1 + k^2) + \langle m\rangle_v(1 + k^2)]$$

$$\langle m^2\rangle_{uv} - \langle m\rangle_{uv}^2 = \langle m\rangle_v^2 + \langle m\rangle_v$$

$$\langle n^3\rangle_{uv} - 3\langle n^2\rangle_{uv}\langle n\rangle_{uv} + 2\langle n\rangle_{uv}^3$$
$$= \left(\frac{1}{k_+^4}\right)S_u[\langle m\rangle_v^2(6 + 24k + 36k^2 + 24k^3 + 6k^4)$$
$$+ \langle m\rangle_v(6 + 24k + 36k^2 + 24k^3 + 6k^4)$$
$$+ (1 + 6k + 10k^2 + 6k^3 + k^4)]$$
$$+ \left(\frac{1}{k_+^6}\right)[\langle m\rangle_v^3(2 + 30k^2 + 30k^4 + 2k^6)$$
$$+ \langle m\rangle_v^2(3 + 45k^2 + 45k^4 + 3k^6)$$
$$+ \langle m\rangle_v(1 + 23k^2 + 23k^4 + k^6) + (4k^2 + 4k^4)]$$

$$\langle n^2 m \rangle_{uv} - \langle n^2 \rangle_{uv} \langle m \rangle_{uv} - 2 \langle n \rangle_{uv} \langle nm \rangle_{uv} + 2 \langle n \rangle_{uv}^2 \langle m \rangle_{uv}$$
$$= \left(\frac{1}{k_+^2}\right) S_u [\langle m \rangle_v^2 (2 + 4k + 2k^2) + \langle m \rangle_v (2 + 4k + 2k^2)]$$
$$+ \left(\frac{1}{k_+^4}\right) [\langle m \rangle_v^3 (2 + 12k^2 + 2k^4) + \langle m \rangle_v^2 (3 + 18k^2 + 3k^4)$$
$$+ \langle m \rangle_v (1 + 6k^2 + k^4)]$$
$$\langle nm^2 \rangle_{uv} - 2\langle nm \rangle_{uv} \langle m \rangle_{uv} - \langle n \rangle_{uv} \langle m^2 \rangle_{uv} + 2\langle n \rangle_{uv} \langle m \rangle_{uv}^2$$
$$= \left(\frac{1}{k_+^2}\right) [\langle m \rangle_v^3 (2 + 2k^2) + \langle m \rangle_v^2 (3 + 3k^2) + \langle m \rangle_v (1 + k^2)]$$
$$\langle m^3 \rangle_{uv} - 3\langle m^2 \rangle_{uv} \langle m \rangle_{uv} + 2\langle m \rangle_{uv}^3 = [2\langle m \rangle_v^3 + 3\langle m \rangle_v^2 + \langle m \rangle_v] \quad (4.124)$$

4.14.5 The Derivation of the Moment Expressions (4.109)

The moment formulas Eq. (4.109) come from the definition of the moments:

$$\langle p_u \rangle_{uv} = \langle n + \tfrac{1}{2} \rangle_{uv} + \kappa_u \langle m + \tfrac{1}{2} \rangle_{uv}$$
$$\langle (p_u - \langle p_u \rangle_{uv})^2 \rangle_{uv} = (\langle n^2 \rangle_{uv} - \langle n \rangle_{uv}^2) - 2\kappa_u (\langle nm \rangle_{uv} - \langle n \rangle_{uv} \langle m \rangle_{uv})$$
$$+ \kappa_u^2 (\langle m^2 \rangle_{uv} - \langle m \rangle_{uv}^2)$$
$$\langle (p_u - \langle p_u \rangle_{uv})^3 \rangle_{uv}$$
$$= (\langle n^3 \rangle_{uv} - 3\langle n^2 \rangle_{uv} \langle n \rangle_{uv} + 2\langle n \rangle_{uv}^3$$
$$- 3\kappa_u \{\langle n^2 m \rangle_{uv} - \langle n^2 \rangle_{uv} \langle m \rangle_{uv} - 2\langle n \rangle_{uv} \langle nm \rangle_{uv} + 2\langle n \rangle_{uv}^2 \langle m \rangle_{uv}\}$$
$$+ 3\kappa_u^2 \{\langle nm^2 \rangle_{uv} - 2\langle nm \rangle_{uv} \langle m \rangle_{uv} - \langle n \rangle_{uv} \langle m^2 \rangle_{uv} + 2\langle n \rangle_{uv} \langle m \rangle_{uv}^2\}$$
$$- \kappa_u^3 \{\langle m^3 \rangle_{uv} - 3\langle m^2 \rangle_{uv} \langle m \rangle_{uv} + 2\langle m \rangle_{uv}^3\} \quad (4.125)$$

If the Eqs. (4.124) are introduced into Eq. (4.125), if the substitution based upon Eq. (4.107) is made, namely:

$$\kappa_u = \frac{\sin^2 \theta}{\cos^2 \theta} = \frac{1 - \cos 2\theta}{1 + \cos 2\theta} = \frac{(1 - \cos 2\theta)^2}{1 - \cos^2 2\theta} = \frac{1}{k_+^2}(1 - 2k + k^2) \quad (4.126)$$

and if the terms are combined, the moments become:

$$\langle p_u \rangle_{uv} = S_u + \left(\frac{1}{k_+^2}\right) \{2k \langle m \rangle_v + k\}$$

$$\langle (p_u - \langle p_u \rangle_{uv})^2 \rangle_{uv} = \left(\frac{1}{k_+^2}\right) S_u \{\langle m \rangle_v (2 + 4k + 2k^2) + (1 + 2k + k^2)\}$$
$$+ \left(\frac{1}{k_+^4}\right) \{\langle m \rangle_v^2 (8k^2) + \langle m \rangle_v (8k^2) + 2k^2\}$$

4.14 Unequal-Force-Constants Moments

$$\langle (p_u - \langle p_u \rangle_{uv})^3 \rangle_{uv}$$

$$= \left(\frac{1}{k_+^4}\right) S_u \{\langle m \rangle_v^2 (24k + 48k^2 + 24k^3) + \langle m \rangle_v (24k + 48k^2 + 24k^3)$$

$$+ (1 + 6k + 10k^2 + 6k^3 + k^4)\} + \left(\frac{1}{k_+^6}\right) \{\langle m \rangle_v^3 (64k^3)$$

$$+ \langle m \rangle_v^2 (96k^3) + \langle m \rangle_v (8k^2 + 32k^3 + 8k^4) + (4k^2 + 4k^4)\} \ . \quad (4.127)$$

Recognizing powers of $\langle 2m + 1 \rangle_v$ and using the substitutions:

$$\frac{1-k}{k_+} = \frac{\sin\theta}{\cos\theta} = \left(\frac{\omega_v}{\omega_u}\right)^{1/2}$$

$$\frac{1+k}{k_+} = \frac{\cos\theta}{\sin\theta} = \left(\frac{\omega_u}{\omega_v}\right)^{1/2} \quad (4.128)$$

Eqs. (4.127) can readily be simplified into Eqs. (4.109) with k_θ given by Eq. (4.110).

5 Multiple Coordinate Models of a Luminescence Center

5.1 The Einstein-Huang-Rhys-Pekar Single-Frequency Multiple-Coordinate Model

We approach the multiple coordinate models by considering the convolution (reproductive, multiplicative, combinative) properties of the W_p distribution. We assert that two W_p distributions having the same $\langle m \rangle$, i.e., the same phonon energy $\hbar\omega$, combine according to:

$$\sum_{p_1=-\infty}^{\infty} W_{p_1}(S_1, \langle m \rangle) W_{p-p_1}(S_2, \langle m \rangle) = W_p(S_1 + S_2, \langle m \rangle) \tag{5.1}$$

The proof will be that both the right hand side and the left hand side of Eq. (5.1) satisfy the same W_p recursion formula, Eq. (4.44). Thus we are to show that:

$$(S_1 + S_2)\langle m \rangle \sum_{p_1=-\infty}^{\infty} W_{p_1}(S_1, \langle m \rangle) W_{p+1-p_1}(S_2, \langle m \rangle)$$

$$+ p \sum_{p_1=-\infty}^{\infty} W_{p_1}(S_1, \langle m \rangle) W_{p-p_1}(S_2, \langle m \rangle)$$

$$- (S_1 + S_2)\langle 1 + m \rangle \sum_{p_1=-\infty}^{\infty} W_{p_1}(S_1, \langle m \rangle) W_{p-1-p_1}(S_2, \langle m \rangle) = 0 \tag{5.2}$$

for it is clear that $W_p(S_1 + S_2, \langle m \rangle)$ itself will satisfy Eq. (4.44) with $S = S_1 + S_2$. Equation (5.2) is regrouped as follows:

$$S_1 \langle m \rangle \sum_{p_1=-\infty}^{\infty} W_{p_1}(S_1, \langle m \rangle) W_{p+1-p_1}(S_2, \langle m \rangle)$$

$$+ S_2 \langle m \rangle \sum_{p_1=-\infty}^{\infty} W_{p_1}(S_1, \langle m \rangle) W_{p+1-p_1}(S_2, \langle m \rangle)$$

$$+ p_1 \sum_{p_1=-\infty}^{\infty} W_{p_1}(S_1, \langle m \rangle) W_{p-p_1}(S_2, \langle m \rangle)$$

$$+ (p - p_1) \sum_{p_1=-\infty}^{\infty} W_{p_1}(S_1, \langle m \rangle) W_{p-p_1}(S_2, \langle m \rangle)$$

$$- S_1 \langle 1 + m \rangle \sum_{p_1=-\infty}^{\infty} W_{p_1}(S_1, \langle m \rangle) W_{p-1-p_1}(S_2, \langle m \rangle)$$

$$- S_2 \langle 1 + m \rangle \sum_{p_1=-\infty}^{\infty} W_{p_1}(S_1, \langle m \rangle) W_{p-1-p_1}(S_2, \langle m \rangle) = 0 \tag{5.3}$$

The first, third, and fifth and the second, fourth, and sixth lines of Eq. (5.3) will be shown to sum independently to zero:

$$S_1 \langle m \rangle \sum_{p_1=-\infty}^{\infty} W_{p_1}(S_1, \langle m \rangle) W_{p+1-p_1}(S_2, \langle m \rangle)$$

$$+ p_1 \sum_{p_1=-\infty}^{\infty} W_{p_1}(S_1, \langle m \rangle) W_{p-p_1}(S_2, \langle m \rangle)$$

$$- S_1 \langle 1 + m \rangle \sum_{p_1=-\infty}^{\infty} W_{p_1}(S_1, \langle m \rangle) W_{p-1-p_1}(S_2, \langle m \rangle) = 0$$

$$S_2 \langle m \rangle \sum_{p_1=-\infty}^{\infty} W_{p_1}(S_1, \langle m \rangle) W_{p+1-p_1}(S_2, \langle m \rangle)$$

$$+ (p - p_1) \sum_{p_1=-\infty}^{\infty} W_{p_1}(S_1, \langle m \rangle) W_{p-p_1}(S_2, \langle m \rangle)$$

$$- S_2 \langle 1 + m \rangle \sum_{p_1=-\infty}^{\infty} W_{p_1}(S_1, \langle m \rangle) W_{p-1-p_1}(S_2, \langle m \rangle) = 0 \quad (5.4)$$

The second of the Eq. (5.4) pair sums to zero because, for every p_1, it is $W_{p_1}(S_1, \langle m \rangle)$ times an expression which is identically zero, namely, the W_p recursion formula Eq. (4.44) with $p \to p - p_1$ and with $S \to S_2$.

The first of the Eq. (5.4) pair sums to zero also. We change the summation indices in the three sums by, in order, $(p_1, p_{p+1-p_1}) \to (p + 1 - p_2, p_2)$, $(p_1, p_{p-p_1}) \to (p - p_2, p_2)$, and $(p_1, p_{p-1-p_1}) \to (p - 1 - p_2, p_2)$. We can then recognize the same pattern as above, i.e., that each term of the sum over p_2 is a factor $W_{p_2}(S_2, \langle m \rangle)$ times an expression which is identically zero, namely, Eq. (4.44) with $p \to p - p_2$ and with $S \to S_1$. Thus, the assertion Eq. (5.1) is proved.

The Huang-Rhys-Pekar model of a luminescence center is just N_{Av} (Avogadro's number) such oscillators with a common frequency, i.e., the Einstein crystal. By repetitive uses of Eq. (5.1), the HRP distribution is just the W_p distribution with its S parameter equal to the sum of the individual S_i parameters,

$$W_p(S, \langle m \rangle) = W_p(\sum S_i, \langle m \rangle) \quad (5.5)$$

5.2 The z and d/dz Multiple-Coordinate Nuclear Factors

We wish now to develop the formulas for the nuclear factor in multiple coordinates for the $z = O_+$ and $d/dz = O_-$ operators, when all coordinates have the same frequency. The result will be:

$$A^2 M_{p, O_\pm}$$

$$\equiv \sum_{\vec{p}, \vec{m}} |\langle u_{p_1+m_1} u_{p_2+m_2} \ldots u_{p_N+m_N} | A_1 O_{1_\pm} + \cdots$$

$$+ A_N O_{N_\pm} | v_{m_1} v_{m_2} \ldots v_{m_N} \rangle|^2 \cdot (1 - r)^N r^{m_1 + \cdots + m_N}$$

$$= A^2 [(1 - \gamma) W_{p, O_\pm} + \gamma L_p] \quad (5.6)$$

where:

$$A^2 = \sum_{k=1}^{N} |A_k|^2$$

$$S = \sum_{k=1}^{N} S_k$$

$$\vec{p} \to p_1, p_2, \ldots, p_N$$

$$\sum_i p_i = p$$

$$\vec{m} \to m_1, m_2, \ldots, m_N$$

$$\max(0, -p_i) \le m_i < \infty$$

$$\gamma = 1 - \frac{\left| \sum_{k=1}^{N} A_k S_k^{1/2} \right|^2}{A^2 S}$$

$$0 \le \gamma \le 1$$

$$L_p = \tfrac{1}{2}(\langle 1+m \rangle W_{p-1} + \langle m \rangle W_{p+1}) \tag{5.7}$$

where W_{p,o_+}, L_p, and the W_p functions within the L_p expression all have the arguments $(\tilde{S}, \langle m \rangle)$. The W_{p,o_+} themselves are given by Eq. (4.87).

Equation (5.6) is shown with N coordinates, and N can be any integer greater than unity, in particular, N_{Av}. The proof of Eq. (5.6) is straightforward but tedious algebra. We give first several preliminary steps.

5.2.1 Preliminaries I: the Y_p Function

We define and will prove:

$$\begin{aligned} Y_p &\equiv \sum_{m=\max(0,-p)}^{\infty} \left[\frac{m+1}{2} \right]^{1/2} \langle u_{p+m}|v_m \rangle \langle u_{p+m}|v_{m+1} \rangle (1-r) r^m \\ &= \left[\frac{S}{2} \right]^{1/2} \langle 1+m \rangle (W_{p-1} - W_p) \end{aligned} \tag{5.8}$$

The proof begins with squaring the Manneback recursion formula Eq. (3.34) with $n \to p+m$ and then using the operator:

$$\sum_{m=\max(0,-p)}^{\infty} (1-r) r^m$$

to get:

$$\begin{aligned} \sum_{m=\max(0,-p)}^{\infty} (1-r) r^m [(p+m) A_{p-1+m,m}^2 \\ = (m+1) A_{p+m,m+1}^2 + S A_{p+m,m}^2 \\ + 2\{S(m+1)\}^{1/2} A_{p+m,m} A_{p+m,m+1}] \end{aligned} \tag{5.9}$$

5.2 The z and d/dz Multiple-Coordinate Nuclear Factors

We then use Eq. (4.71), evaluated for $p \to p - 1$, the identity:

$$\sum_{m=max(0,-p)}^{\infty} (1-r)r^m(m+1)A_{p+m,m+1}^2$$

$$= \frac{1}{r}\sum_{m'=m+1=max(0,-p+1)}^{\infty}(1-r)r^{m'}(m')A_{p-1+m',m'}^2 \qquad (5.10)$$

and the definition of Y_p, Eq. (5.8), to get:

$$pW_{p-1} + \langle m \rangle_{p-1}W_{p-1} = \frac{1}{r}\langle m \rangle_{p-1}W_{p-1} + SW_p + 2^{3/2}S^{1/2}Y_p \qquad (5.11)$$

Now we introduce Eq. (4.75) and the identity:

$$1 - \frac{1}{r} = -\frac{1}{\langle m \rangle} \qquad (5.12)$$

and obtain:

$$2^{3/2}S^{1/2}Y_p = (p-1)W_{p-1} + S[-\langle 1+m \rangle W_p + 2\langle 1+m \rangle W_{p-1} - \langle 1+m \rangle W_{p-2} - W_p] \qquad (5.13)$$

If we now introduce the W_p recursion formula, Eq. (4.44) with $p \to p-1$ in order to eliminate the term in $(p-1)W_{p-1}$, and then solve for Y_p, we are led to the desired Eq. (5.8).

5.2.2 Preliminaries II: $X_{p,O_\pm}$

We define $X_{p,O_\pm}$ and give its expression in terms of W_p functions. The expression will be proved immediately afterwards.

$$X_{p,O_\pm} = \sum_{m=max(0,-p)}^{\infty} \langle u_{p+m}|v_m \rangle \langle u_{p+m}|O_\pm|v_m \rangle (1-r)r^m$$

$$= \left(\frac{S}{2}\right)^{1/2}[\langle m \rangle(W_p - W_{p+1}) \pm \langle 1+m \rangle(W_{p-1} - W_p)] \qquad (5.14)$$

The proof uses the relationship:

$$\sum_{m=max(0,-p)}^{\infty}\left(\frac{m}{2}\right)^{1/2}A_{p+m,m}A_{p+m,m-1}(1-r)r^m$$

$$= r\sum_{m'=m-1=max(0,-p-1)}^{\infty}\left(\frac{m'+1}{2}\right)^{1/2}A_{p+1+m',m'+1}A_{p+1+m',m'}(1-r)r^{m'}$$

$$= rY_{p+1} \qquad (5.15)$$

Using Eq. (4.78) for $A_{p+m,m,O_\pm}$, we obtain:

$$X_{p,O_\pm} = \sum_{m=max(0,-p)}^{\infty} \left(\frac{m}{2}\right)^{1/2} \langle u_{p+m}|v_m\rangle\langle u_{p+m}|v_{m-1}\rangle(1-r)r^m$$

$$\pm \sum_{m=max(0,-p)}^{\infty} \left(\frac{m+1}{2}\right)^{1/2} \langle u_{p+m}|v_m\rangle\langle u_{p+m}|v_{m+1}\rangle(1-r)r^m$$

$$= rY_{p+1} \pm Y_p = \frac{\langle m\rangle}{\langle 1+m\rangle}Y_{p+1} \pm Y_p \tag{5.16}$$

Introducing Y_p from Eq. (5.8) gives Eq. (5.14).

5.2.3 Preliminaries III: XX Sums

We now prove that:

$$\sum_{p_1} X_{p_1,O_{1\pm}}(S_1,\langle m\rangle)X_{p-p_1,O_{2\pm}}(S_2,\langle m\rangle)$$

$$= \sum_{p_1}(S_1 S_2)^{1/2} \sum_{j=-1}^{1} \sum_{k=-1}^{1} \frac{1}{2}\beta_{j,O_\pm}\beta_{k,O_\pm} W_{p_1+j}(S_1,\langle m\rangle)W_{p-p_1+k}(S_2,\langle m\rangle) \tag{5.17}$$

where:

$$\beta_{-1,O_\pm} = \pm\langle 1+m\rangle$$
$$\beta_{0,O_\pm} = \langle m\rangle \mp \langle 1+m\rangle$$
$$\beta_{1,O_\pm} = -\langle m\rangle \tag{5.18}$$

This relation comes from introducing Eq. (5.14) and collecting terms. The β products can be collected to give:

$$\sum_{\substack{\vec{p}\\ \sum_i p_i = p}} X_{p_1,O_{1\pm}}(S_1,\langle m\rangle)X_{p_2,O_{2\pm}}(S_2,\langle m\rangle)$$

$$= (S_1 S_2)^{1/2} \sum_{i=-2}^{2} \alpha_{i,O_\pm} W_{p+i}(S_1+S_2,\langle m\rangle)$$

$$= (S_1 S_2)^{1/2} \left\{\frac{(W_{p,O_\pm} - L_p)}{S}\right\} \tag{5.19}$$

where the $\alpha_{i,O_\pm}$ are those of Eq. (4.87), i.e., those of Eqs. (4.70) and (4.86). We have used Eq. (5.1) in the first step and Eq. (4.87) in the last step of Eq. (5.19).

The coefficient of the three $\alpha_{i,O_\pm}$ with $i < 2$ come out explicitly as:

$$\alpha_{1,O_\pm} = -\langle m\rangle^2 \pm \langle m\rangle\langle 1+m\rangle$$
$$\alpha_{0,O_\pm} = \tfrac{1}{2}\{\mp 4\langle m\rangle\langle 1+m\rangle + \langle m\rangle^2 + \langle 1+m\rangle^2\}$$
$$\alpha_{-1,O_\pm} = \pm\langle 1+m\rangle\{\langle m\rangle \mp \langle 1+m\rangle\} \tag{5.20}$$

which can be recognized as those in Eqs. (4.70) and (4.86). The $\alpha_{2,O_\pm}$ and $\alpha_{-2,O_\pm}$ are directly recognizable.

5.2 The z and d/dz Multiple-Coordinate Nuclear Factors

5.2.4 Preliminaries IV: $W_{p,O_\pm} W_p$ Sums

We prove here that:

$$\sum_{\substack{p_1,p_2 \\ p_1+p_2=p}} [|A_1|^2 W_{p_1,O_{1\pm}}(S_1,\langle m \rangle) W_{p_2}(S_2,\langle m \rangle)$$

$$+ |A_2|^2 W_{p_1}(S_1,\langle m \rangle) W_{p_2,O_{2\pm}}(S_2,\langle m \rangle)]$$

$$= [|A_1|^2 + |A_2|^2] L_p(S,\langle m \rangle)$$

$$+ [|A_1|^2 S_1 + |A_2|^2 S_2] \left[\sum_{i=-2}^{2} \alpha_{i,O_\pm} W_{p+i}(S,\langle m \rangle) \right]$$

$$= [|A_1|^2 + |A_2|^2] L_p$$

$$+ [|A_1|^2 S_1 + |A_2|^2 S_2] \left\{ \frac{(W_{p,O_\pm} - L_p)}{S} \right\} \tag{5.21}$$

where $W_{p,O_\pm}$ and L_p have the arguments $(S, \langle m \rangle)$ and where:

$$S = S_1 + S_2 \tag{5.22}$$

The proof proceeds by introducing Eq. (4.87) expressions in the $5 - W_p$ form for the two $W_{p,O_\pm}$ of Eq. (5.21), using Eq. (5.1) to combine the resulting $W_p(S_1,\langle m \rangle) W_p(S_2,\langle m \rangle)$ products, and then collecting terms.

5.2.5 Proof of Eq. (5.6) for Two Coordinates

We are now ready to prove Eq. (5.6). We do so explicitly for two coordinates, and then infer the appropriate modification for N_{Av} coordinates. We have:

$$A^2 M_{p,O_\pm} \equiv \sum_{\vec{C}} |\langle u_{p_1+m_1} u_{p_2+m_2}| A_1 O_{1\pm} + A_2 O_{2\pm} |v_{m_1} v_{m_2}\rangle|^2 (1-r)^2 r^{m_1+m_2}$$

$$= \sum_{\vec{p}} [|A_1|^2 W_{p_1,O_{1\pm}} W_{p_2} + |A_2|^2 W_{p_1} W_{p_2,O_{2\pm}}$$

$$+ (A_1 A_2^* + A_1^* A_2) X_{p_1,O_\pm} X_{p_2,O_\pm}]$$

$$= [|A_1|^2 + |A_2|^2] L_p + [|A_1|^2 S_1 + |A_2|^2 S_2] \left(\frac{W_{p,O_\pm} - L_p}{S} \right)$$

$$+ (A_1 A_2^* + A_1^* A_2)(S_1 S_2)^{1/2} \left(\frac{W_{p,O_\pm} - L_p}{S} \right)$$

$$= \left(\frac{|A_1 S_1^{1/2} + A_2 S_2^{1/2}|^2}{S} \right) W_{p,O_\pm}$$

$$+ L_p \left[A^2 - \left(\frac{|A_1 S_1^{1/2} + A_2 S_2^{1/2}|^2}{S} \right) \right]$$

$$= A^2 [(1-\gamma) W_{p,O_\pm} + \gamma L_p] \tag{5.23}$$

where:

$$A^2 = |A_1|^2 + |A_2|^2$$

$$\gamma = 1 - \left(\frac{|A_1 S_1^{1/2} + A_2 S_2^{1/2}|^2}{A^2 S} \right)$$

$$\vec{C} = \begin{pmatrix} \vec{p}, \vec{m} \\ \sum_i p_i = p \\ m_{i0} = \max(0, -p_i) \end{pmatrix} \tag{5.24}$$

The second step in Eq. (5.23) uses Eqs. (5.21) and (5.19). The extension to N_{Av} dimensions yields the same formula with:

$$A^2 = \sum_{i=1}^{N_{Av}} |A_i|^2$$

$$S = \sum_{i=1}^{N_{Av}} S_i$$

$$\gamma = 1 - \left[\frac{|\sum_{i=1}^{N_{Av}} A_i S_i^{1/2}|^2}{A^2 S} \right] \tag{5.25}$$

We will now show that:

$$0 \leq \gamma \leq 1 \tag{5.26}$$

We note that:

$$\left| \sum_i A_i S_i^{1/2} \right|^2 + \left| \sum_{i=1}^{N_{Av}-1} \sum_{k=i+1}^{N_{Av}} (A_i S_k^{1/2} - A_k S_i^{1/2}) \right|^2$$

$$= (A_1^2 + A_2^2 + A_3^2 + \cdots)(S_1 + S_2 + S_3 + \cdots) = A^2 S \tag{5.27}$$

The second term on the left hand side cancels the terms in $A_i A_k S_i^{1/2} S_k^{1/2}$ generated by the first term and adds the needed terms in $A_i^2 S_k$. Since this second term on the left hand side is non-negative, the fraction in the γ expression Eq. (5.25) is less than unity and therefore γ indeed satisfies Eq. (5.26).

5.3 Multiple-Frequency Models of a Luminescence Center

5.3.1 The Selected Model

In Sect. 3.3 the energy scale was measured in some fixed $\hbar\omega$ unit and every occurring transition was apportioned to the at most two metric energies nearest the actual transition energy. This concept is a means of combining any number of U_{p_u} distributions in emission or any number of V_{p_v} distributions in absorption. The freedom is so vast that the activity indeed can become mere curve fitting providing no physical insight.

We give here the simplest such model, namely, a set of equal-force-constants oscillators whose frequencies, $\hbar\omega_k$, are multiples of some smallest frequency, $\hbar\omega_1$. This model has $\hbar\omega_1$ as a natural energy unit and has exact distributions which can readily be calculated by recursion methods.

5.3 Multiple-Frequency Models of a Luminescence Center

It will be found necessary, in order to avoid total arbitrariness even in this model, to restrict our attention to a certain choice of the parameters of the individual equal-force-constants oscillators, namely, their S and A. We will then pose a DDESA (Discretized Debye Equal S and A) model, a choice taken to mimic a Debye frequency spectrum.

The discussion in this section will be somewhat more detailed, because this work has not appeared elsewhere.

5.3.2 Definition of the 1, z, and d/dz Operator Rates

Consider a transition between two electronic states $v \to u$ during which N_{Av} normal coordinates can change their equilibrium configurations independently. We will group the N_{Av} oscillators into K families with their frequencies:

$$\hbar\omega_k = \frac{k}{K}\hbar\omega_K \tag{5.28}$$

All the normal mode frequencies will be gathered into the nearest of this set of K possible frequencies. We will use a double index, i, j, to label the individual-mode parameters, where i labels the family and j labels the family member, running between 0 and the degeneracy of the family, g_i. The smallest frequency is that with $k = 1$.

Let $G_{p,O}^K$ be the rate of this transition driven by the operator $O = A$ (the Condon operator, for which the O subscript in G will be dropped) or $O = O_+ = \Sigma_{i,j} A_{i,j} z_{i,j}$ or $O = O_- = \Sigma_{i,j} A_{i,j} d/dz_{i,j}$:

$$G_{p,O}^K = \sum_{\vec{C}} \left\{ |\langle u_{p_{1,1}+m_{1,1}} \ldots u_{p_{K,g_K}} | O | v_{m_{1,1}} \ldots v_{m_{K,g_K}} \rangle|^2 \right. \\ \left. \times \prod_{i=1}^{K} [(1-r)^{g_i} r_i^{m_{i,1}+\cdots+m_{i,g_i}}] \right\} \tag{5.29}$$

where $\vec{C}$ denotes the set of indical constraints:

$$max(0, -p_{i,j}) \leq m_{i,j} \leq \infty$$

$$\sum_{i=1}^{K} \sum_{j=1}^{g_i} i p_{i,j} = p$$

$$-\infty \leq p_{i,j} \leq \infty$$

$$1 \leq i \leq K$$

$$1 \leq j \leq g_i \tag{5.30}$$

and where:

$$r_i = \exp(-\hbar\omega_i/kT) \tag{5.31}$$

is the Boltzmann factor for the $i'th$ frequency.

5.3.3 The Condon-Operator Distribution

For the Condon operator, we use Eq. (5.1) to obtain:

$$G_p^K(\vec{S}, \vec{\omega}_K, kT) = A^2 \sum_{p_1, p_2, \ldots, p_K} W_{p_1}(S_1, \omega_1, kT)$$
$$\times W_{p_2}(S_2, \omega_2, kT) \ldots W_{p_K}(S_K, \omega_K, kT) \quad (5.32)$$

where the summation indices are constrained by:

$$\sum_{k=1}^{K} k p_k = p \quad (5.33)$$

and:

$$S_i = \sum_{j=1}^{g_i} S_{i,j}$$

$$\omega_k = k\omega_1 \quad (5.34)$$

The W_p of Eq. (4.45) have the arguments $(S, \langle m \rangle)$, and $\langle m \rangle$ must therefore by determined for each ω_k through Eqs. (4.43) and (4.12). We now extract the A^2 factor from the G_p^K, forming the nuclear factor

$$D_p^K = A^{-2} G_p^K \quad (5.35)$$

D_p^K is obtained through $K-1$ applications of a recursion algebra, which shows in turn the accumulation of each frequency's contribution into the total distribution. The transitions with ω_k contribute to the accumulating distribution from all $k-1$ lower frequencies only at frequencies spaced by $k\omega_1$ units.

Thus let:

$$D_q^{K,l} = \sum_{p_1, p_2, \ldots, p_l} W_{p_1}(S_1, \omega_1, kT) W_{p_2}(S_2, \omega_2, kT) \ldots W_{p_l}(S_l, \omega_l, kT) \quad (5.36)$$

be the limited sum over the first l families, under the constraint:

$$\sum_{k=1}^{l} k p_k = q$$

$$1 \leq l \leq K \quad (5.37)$$

and under constraints Eq. (5.34).

If $D_m^{K,l-1}$ for all m are known, then $D_m^{K,l}$ can be generated from them by:

$$D_q^{K,l} = \sum_{p_1} D_{q-lp_1}^{K,l-1} W_{p_1}^{(l)}(S_l, \omega_l, kT)$$

$$D_q^{K,1} = W_q^{(1)}(S_1, \omega_1, kT) \equiv W_q(S_1, \langle m_1 \rangle) \quad (5.38)$$

Here $2 \leq l \leq K$ and the desired D_p^K distribution is obtained as:

$$D_{q=p}^{K,K} = D_p^K \quad (5.39)$$

All $D_q^{K,l}$ distributions sum over q to unity. There is no difficulty maintaining accuracy in this recursion procedure.

5.3 Multiple-Frequency Models of a Luminescence Center

For the special case $kT = 0$ all p_k indices in Eq. (5.36) are non-negative and furthermore each W_p in Eq. (5.38) is the Poisson distribution according to Eq. (4.57). One can then write analytic expressions for the D_p^K at $kT = 0$, namely:

$$D_0^K = \exp\left(-\sum_{k=1}^{K} S_k\right)$$

$$D_1^K = S_1 D_0^K$$

$$D_2^K = \left\{S_2 + \left(\frac{S_1^2}{2}\right)\right\} D_0^K$$

$$D_3^K = \left\{S_3 + S_2 S_1 + \left(\frac{S_1^3}{6}\right)\right\} D_0^K$$

$$D_4^K = \left\{S_4 + S_3 S_1 + \left(\frac{S_2^2}{2}\right) + \left(\frac{S_2 S_1^2}{2}\right) + \left(\frac{S_1^4}{24}\right)\right\} D_0^K \tag{5.40}$$

The degeneracies of each of the K families are arbitrary to this point.

5.3.4 The Recursion Algebra for the z and d/dz Operators

To obtain the $G_{p,O_\pm}^K$ of Eq. (5.29), one must handle the A_{ij} in the operators $O_\pm$. These A_{ij} may be complex. One must also develop a more complicated recursion algebra.

In Eq. (5.29), the only complex quantities are the A_{ij} themselves. All integrals for the 1, z, and d/dz operators of harmonic oscillator wavefunctions are real. The $G_{p,O_\pm}^K$ are sums, over all possible ways of combining K different $p_i \hbar \omega_i$ so as to yield as the net phonon energy $p \hbar \omega_1$, of squared matrix elements of an operator which is the sum of N_{Av} single-mode operators, each term of the sum being thermally weighted by its Boltzmann factor for the initial state. The $G_{p,O_\pm}^K$ are therefore double sums of terms, the first sum over these p_i, the second sum over all the inital states allowed for each of the oscillators. Each term is a product of $N_{Av} - 2$ one-dimensional overlap integrals A_{nm} of Eq. (3.1) and two one-dimensional $O_\pm$ operator integrals $A_{n,m,O_\pm}$ of Eqs. (3.51) and (3.52). There are two types of such terms, those with the two $O_\pm$ operators from the same family, i.e., having the same frequency, and those with the two $O_\pm$ operators not having the same frequency. These will be summed separately and labeled $\Psi_{p,O_\pm}^K$ and $\Phi_{p,O_\pm}^K$, respectively. We will then give for both partial sums recursion algebras which will in fact be the sequential accumulation into a running sum, called $\Psi_{p,O_\pm}^{K,l}$ and $\Phi_{p,O_\pm}^{K,l}$, respectively, of each family's contributions. Thus:

$$G_{p,O_\pm}^K = \Psi_{p,O_\pm}^K + \Phi_{p,O_\pm}^K, \tag{5.41}$$

where the right-hand side will now be specified. The bookkeeping is, regrettably, somewhat tedious.

5.3.4.1 The Ψ Functions

The $\Psi^K_{p,O_\pm}$ of Eq. (5.41) are:

$$\Psi^K_{p,O_\pm} = \sum_{\substack{p_1,\ldots,p_K \\ \Sigma k p_k = p}} \left\{ \sum_{s=1}^K \left[\sum_{\vec{m_{s,i}}} \right. \right.$$

$$\times (|\langle u_{p_{s,1}+m_{s,1}} u_{p_{s,2}+m_{s,2}} \ldots u_{p_{s,g_s}+m_{s,g_s}} | A_{s,1} O_{s,1,\pm}$$

$$+ \ldots A_{s,g_s} O_{s,g_s,\pm} | v_{m_{s,1}} v_{m_{s,2}} \ldots v_{m_{s,g_s}} \rangle)^2 (1-r_s) r_s^{m_s}$$

$$\times \left(\prod_{j=1 \neq s}^K \left\{ \sum_{\substack{\Sigma p_{j,i}=p_j \\ -\infty \leq p_{j,i} \leq \infty}} \sum_{m_{j,i}} \right. \right.$$

$$\times [|\langle u_{p_{j,1}+m_{j,1}} \ldots u_{p_{j,g_j}+m_{j,g_j}} | v_{m_{j,1}} \ldots v_{m_{j,g_j}} \rangle|]^2$$

$$\left. \left. \left. \times (1-r_j) r_j^{m_j} \right\} \right) \right] \right\}$$

$$= \sum_{\vec{C}} \left\{ \sum_{s=1}^K \left[M_p(S_s, \omega_s, kT) \prod_{j=1 \neq s}^K W_{p_j}(S_j, \omega_j, kT) \right] \right\},$$

$$\vec{C} = \begin{pmatrix} p_1, \ldots, p_K \\ \Sigma k p_k = p \end{pmatrix} \tag{5.42}$$

where

$$m_s = \sum_i m_{s,i} \qquad m_j = \sum_i m_{j,i}$$

$$\Gamma(0, -p_{s,i}) \leq m_{s,i} \leq \infty \tag{5.43}$$

where $\Gamma(0,-p)$ is the maximum of 0 and $-p$. The last step of Eq. (5.42) uses Eqs. (4.11) and (5.6).

5.3.4.2 The Ψ Recursion Algebras

The $\Psi^K_{p,O_\pm}$ are obtained by $K-1$ recursive applications of algebraic equations. Let, in analogy with Eq. (5.36), $\Psi^{K,l}_{p,O_\pm}$ be defined as in Eq. (5.42) except that the indices be restricted to only the first l families and with the $p_1, \ldots, p_l$ constrained by Eq. (5.37). We want to include terms where the operator is $O_{1,\pm}$. Therefore we define

$$\Psi^{K,1}_{q,O_\pm} = A_1^2 M_{q,O_\pm}(S_1, \omega_1, kT) \tag{5.44}$$

After this first step, each subsequent step in the accumulation will be with the $O_{j,\pm}$ operator either already included or to be included in the current step. Thus

$$\Psi^{K,l}_{q,O_\pm} = \sum_{p_1} [\Psi^{K,l-1}_{q-lp_1,O_\pm} W^{(l)}_{p_1}(S_l, \omega_l, kT)$$

$$+ D^{K,l-1}_{q-lp_1} A_l^2 M^{(l)}_{p_1,O_\pm}(S_l, \omega_l, kT)] \tag{5.45}$$

is a recursion relation allowing sequential generation of the $\Psi^{K,l}_{q,O_\pm}$ for $l = 2, \ldots,$

5.3 Multiple-Frequency Models of a Luminescence Center

K, where $\Psi_{q=p,O_\pm}^{K,K} \equiv \Psi_{p,O_\pm}^{K}$ needed in Eq. (5.41). In Eq. (5.45), the D is the D of Eqs. (5.36) and (5.38). The $\langle m_l \rangle$ are to be computed from ω_l and kT via Eqs. (4.43) and (4.12).

5.3.4.3 The Φ Functions

The second type term in Eq. (5.41) is that in which the two $O_{i,\pm}$ relate to different frequencies. Then:

$$\Phi_{p,O_\pm}^{K} = \sum_{\substack{p_1,\ldots,p_K \\ \Sigma k p_k = p}} \left\{ \sum_{s=1}^{K-1} \sum_{l=1}^{g_s} \sum_{t=s+1}^{K} \sum_{k=1}^{g_t} \left[\sum_{\vec{m}_{t,k}} \sum_{\vec{m}_{s,l}} \right. \right.$$

$$\times (|\langle u_{p_{s,l}+m_{s,l}}|A_{s,l}O_{s,l,\pm}|v_{m_{s,l}}\rangle \langle u_{p_{s,l}+m_{s,l}}|v_{m_{s,l}}\rangle$$

$$\times \langle u_{p_{t,k}+m_{t,k}}|A_{t,k}O_{t,k,\pm}|v_{m_{t,k}}\rangle \langle u_{p_{t,k}+m_{t,k}}|v_{m_{t,k}}\rangle)$$

$$\times \prod_{\substack{i=1 \neq l}}^{g_s} \langle u_{s,i}|v_{m_{s,i}}\rangle^2 \prod_{\substack{j=1 \neq k}}^{g_t} \langle u_{t,k}|v_{m_{t,k}}\rangle^2 (1-r_s)(r_s)^{m_{s,l}}$$

$$\times (1-r_t)(r_t)^{m_{t,k}} \left] \prod_{\substack{i=1 \neq s}}^{K} \prod_{j=1}^{g_i} \langle u_{i,j}|v_{m_{i,j}}\rangle^2 (1-r_i)(r_i)^{m_{i,j}} \right\}$$

$$= \sum_{\substack{p_1,\ldots,p_K \\ \Sigma k p_k = p}} \left\{ \sum_{s=1}^{K-1} \sum_{l=1}^{g_s} \sum_{t=s+1}^{K} \sum_{k=1}^{g_t} \left\{ \sum_{\vec{m}_{t,k}} \sum_{\vec{m}_{s,l}} \right. \right.$$

$$\times \left[(A_{p_{s,l}+m_{s,l},m_{s,l},O_{s,l,\pm}} A_{p_{t,k}+m_{t,k},m_{t,k}} A_{p_{t,k}+m_{t,k},m_{t,k},O_{t,k,\pm}} \right.$$

$$\times A_{p_{s,l}+m_{s,l},m_{s,l}})((1-r_s)r_s^{m_{s,l}}(1-r_t)r_t^{m_{t,k}})$$

$$\times \prod_{\substack{i=1 \neq l}}^{g_s} A_{p_{s,i}+m_{s,i},m_{s,i}}^{2} \prod_{\substack{i=1 \neq l}}^{g_t} A_{p_{t,i}+m_{t,i},m_{t,i}}^{2} (1-r_s)r_s^{m_{s,i}}$$

$$\times (1-r_t)r_t^{m_{t,i}} \left] \prod_{\substack{i=1 \neq s}}^{K} \prod_{j=1}^{g_i} A_{p_{i,j}+m_{i,j},m_{i,j}}^{2} (1-r_i)r_i^{m_{i,j}} \right\}\right\}$$

$$= \sum_{\substack{j,k=1 \\ j \neq k \\ \Sigma i p_i = p}}^{K} \left\{ \left[\prod_{\substack{i=1 \neq j,k}}^{K} W_{p_i}(S_i, \omega_i, kT) \right] \left[(\Gamma_k \Gamma_j^* + \Gamma_j \Gamma_k^*) \right. \right.$$

$$\left. \left. X_{p_j,O_\pm}(S_j, \omega_j, kT) X_{p_k,O_\pm}(S_k, \omega_k, kT) \right] \right\}. \quad (5.46)$$

In Eq. (5.46) the $p_1, \ldots, p_k$ indices are the sums of the p indices within each family, as in Eq. (5.43). These indices are constrained by Eq. (5.33). The parameters Γ_k, A_k^2, and the S_k underlying the W_{p_k} are:

$$\Gamma_k = \left(\frac{1}{\sqrt{2}}\right) \sum_{i=1}^{g_k} A_{k,i} S_{k,i}^{1/2}$$

$$A_k^2 = \sum_{i=1}^{g_k} |A_{k,i}|^2$$

$$S_k = \sum_{i=1}^{g_k} S_{k,i} \qquad (5.47)$$

The $X(S, \omega, kT)$ functions in Eq. (5.42) are related to Eqs. (5.14) and (5.16), with the $S^{1/2}$-dependent terms incorporated into the Γ's, namely,

$$X_{p,O_\pm}(S, \langle m \rangle) = \langle m \rangle (W_p - W_{p+1}) \pm \langle 1 + m \rangle (W_{p-1} - W_p). \qquad (5.48)$$

where $\langle m \rangle$ is to be determined from ω and kT via Eqs. (4.43) and (4.12).

5.3.4.4 The Φ Recursion Algebras in terms of Θ Functions

To obtain $\Phi^K_{q,O_\pm}$, one accumulates the sum of one-$O_\pm$ terms family by family. We shall call this accumulating sum $\Theta^{K,l}_{q,O_\pm}$ and use them to obtain the running sum of the $\Phi^{K,l}_{q,O_\pm}$. The desired sum is $\Phi^K_{q,O_\pm} = \Phi^{K,K}_{q,O_\pm}$. We begin with

$$\Theta^{K,l}_{q,O_\pm} = \sum_{p,\ldots,p_l} \left\{ \sum_{i=1}^{l} \left[\Gamma_i X_{p_i,O_\pm}(S_i, \omega_i, kT) \prod_{j=1 \neq i}^{l} W_{p_j}(S_j, \omega_j, kT) \right] \right\} \qquad (5.49)$$

under the constraint Eq. (5.37) on the summation indices. $\Theta^{K,l}_{q,O_\pm}$ is the running sum over the first l families of products in the large brackets of Eq. (5.49) which involve only one X function. We obtain $\Theta^{K,l}_{q,O_\pm}$ by a recursion formula which takes into account that the family involved in the X function either has or has not been included in previous steps of the accumulation. Thus

$$\Theta^{K,l}_{q,O_\pm} = \sum_{p_1} [\Theta^{K,l-1}_{q-lp_1,O_\pm} W^{(l)}_{p_1}(S_l, \omega_l, kT) + D^{K,l-1}_{q-lp_1} \Gamma_l X^{(l)}_{p_1,O_\pm}(S_l, \omega_l, kT)]$$

$$\Theta^{K,1}_{q,O_\pm} = \Gamma_1 X^{(1)}_{q,O_\pm}(S_1, \omega_1, kT) \qquad (5.50)$$

is this recursion relation allowing sequential generation of the $\Theta^{K,l}_{q,O_\pm}$ for $2 \leq l \leq K - 1$.

In turn, $\Phi^{K,l}_{q,O_\pm}$, which involves two X functions, is obtained from a recursion relation which takes into account whether the second X function has or has not entered during previous steps. Thus

$$\Phi^{K,l}_{q,O_\pm} = \sum_{p_1} [\Phi^{K,l-1}_{q-lp_1,O_\pm} W^{(l)}_{p_1}(S_l, \omega_l, kT)$$
$$+ \Theta^{K,l-1}_{q-lp_1,O_\pm} \Gamma^*_l X^{(l)}_{p_1,O_\pm}(S_l, \omega_l, kT)$$
$$+ (\Theta^{K,l-1}_{q-lp_1,O_\pm})^* \Gamma_l X^{(l)}_{p_1,O_\pm}(S_l, \omega_l, kT)]$$

$$\Phi^{K,1}_{q,O_\pm} = 0 \qquad (5.51)$$

is the recursion relation allowing sequential generation of the $\Phi^{K,l}_{q,O_\pm}$ for $2 \leq l \leq K$ and:

$$\Phi^{K,K}_{q=p,O_\pm} = \Phi^K_{p,O_\pm} \qquad (5.52)$$

5.3.5 The Discretized Debye Equal S and A Model

The Debye-model phonon spectrum has the distribution:

$$f(\omega)\,d\omega = (9N_0/\omega_K^3)\omega^2 d\omega, \qquad (5.53)$$

5.3 Multiple-Frequency Models of a Luminescence Center

where ω_K is the Debye frequency and N_0 is the number of ions per molar formula of the material. We wish now to introduce a set of allowed ω_k, according to Eq. (5.28), and then discretize all Debye modes into sets labeled by the nearest of the allowed ω_k. The degeneracies g_k and the normalized degeneracies $\bar{g}_k \equiv g_k/\Sigma_k g_k$ of these sets have the values:

$$\bar{g}_1 = \frac{27}{8K^3}, \quad \bar{g}_k = \frac{12k^2 + 1}{4K^3}, \quad \bar{g}_K = \frac{12K^2 - 6K + 1}{8K^3}, \quad \bar{g}_i = g_i/3N_0 \quad (5.54)$$

These values, Eq. (5.54), come from integrating Eq. (5.53) between $\{0, \hbar\omega_1/2\}$, $\{\hbar\omega_k - (\hbar\omega_1/2), \hbar\omega_k + (\hbar\omega_1/2)\}$, and $\{\hbar\omega_K - (\hbar\omega_1/2), \hbar\omega_K\}$, respectively, for g_1, g_k, and g_K. The normalizing factor, $3N_0$, comes from integrating Eq. (5.53) between $\{0, \hbar\omega_K\}$.

Now it is time to introduce the restriction of equal S and A. It is of course allowable within this development to use any $S(\omega_k)$ and $A(\omega_k)$ relationship as needed to fit data, but we feel at present that there is so much freedom here that the usefulness of these equations are threatened. Therefore we adopt the restrictions that all modes have the same S_0 and A_0 parameters. Because experimental spectra are approached with the concept of the Franck-Condon stabilization energy, $S\hbar\omega$, we define, using Eqs. (5.28) and (5.54),

$$S \equiv \frac{1}{\hbar\omega_K} \sum_{k=1}^{K} S_k \hbar\omega_k = \frac{S_0 \hbar\omega_0}{\hbar\omega_K} \sum_{k=1}^{K} k g_k \quad (5.55)$$

where

$$S_k = g_k S_0 \quad (5.56)$$

From Eqs. (5.55) and (5.56), we find the expression for S_k in terms of S:

$$\frac{S_k}{S} = \frac{g_k S_0}{\left(\frac{S_0}{K}\right) \sum_{k=1}^{K} k g_k} = \frac{g_k K}{\left(\sum k g_k\right)} = \frac{\bar{g}_k K}{\left(\sum k \bar{g}_k\right)} = \bar{g}_k \left(\frac{8K^4}{6K^4 + K^2 + 1}\right) \quad (5.57)$$

The last step of Eq. (5.57) uses Eq. (5.54) and the sum expressions

$$\sum_1^N k^3 = \frac{N^2(N+1)^2}{4} \qquad \sum_1^N k^2 = \frac{(N)(N+1)(2N+1)}{6} \quad (5.58)$$

For the DDESA model, in $D_{p,O_\pm}^K$, for every family k all $A_{k,i}$ are equal and all $S_{k,i}$ are equal. Therefore for each family the γ_k parameter of $M_{p_k, O_\pm}(S_k, \omega_k, kT)$ in Eq. (5.6) is zero and the $M_{p_k, O_\pm}(S_k, \omega_k, kT)$ reduce to the $W_{p_k, O_\pm}(S_k, \omega_k, kT)$ of Eq. (4.87). The Γ_k parameters needed in Eqs. (5.50) and (5.51) are those of Eq. (5.47). They are generated for this special DDESA case by setting $\gamma_k = 0$ in the seventh of Eq. (5.7) and solving, using Eq. (5.27):

$$\Gamma_k \equiv \frac{1}{\sqrt{2}} \sum_i A_{k,i} S_{k,i}^{1/2} = \frac{1}{\sqrt{2}} (A_k^2 S_k)^{1/2} \quad (5.59)$$

For the DDESA model, from the first of Eq. (5.7):

$$A_k^2 = \sum_i^{g_i} A_{ki}^2 = g_k A_0^2.\tag{5.60}$$

The assignment of the family degeneracies $\bar{g}_k$ adopted here is slightly more complicated than what might have been posed, e.g., $\bar{g}_k \propto k^2$. We have found that the $\bar{g}_k$ of Eq. (5.54) give $D_{p,o}^K$ which converge more rapidly to their $K = \infty$ limiting distribution.

The W_{p,o_+} of Eq. (4.87) are normalized to $\frac{1}{2}\langle 2m + 1 \rangle$. We obtain DDESA distributions for these operators normalized to $\frac{1}{2}\Sigma_k \bar{g}_k \langle 2m_k + 1 \rangle$ when we replace the Γ_k of Eq. (5.59) and the A_k^2 of Eq. (5.60) with

$$\hat{\Gamma}_k = \left(\frac{\bar{g}_k S_k}{2}\right)^{1/2}$$

$$\hat{A}_k^2 = \bar{g}_k \tag{5.61}$$

The DDESA distributions are invariant, as they should be on grounds of physical plausibility, under the transformation $\hat{\Gamma}_k \to -\hat{\Gamma}_k$.

6 Energy Transfer

6.1 The Model

The theories of energy transfer in the literature are concerned with the distance between the energy donor and acceptor, distributions for these distances as controlled by concentration, averaging over these distances, etc. These concerns enter the electronic factor of the transition rate. Our concern is with the nuclear factor, i.e., with the energy mismatch dependence and with the temperature dependence of these rates. We shall see in Chap. 8 that insofar as the previous authors treat specifically these two dependences they have come to approximate forms of the W_p, $W_{p,d/dz}$, or $M_{p,d/dz}$ rates. In these treatments, and in ours also, the energy donor and the energy acceptor are treated as having vibrational modes totally independent of each other.

Our model is illustrated in Fig. 18. Figure 18 shows one possible transition from donor to acceptor, namely, $d_m^+, a_l \to d_n, a_k^+$. This transition conserves energy but generates p_{da} phonons in accordance to the energy balances:

$$p_{da}\hbar\omega = [(n-m) + (k-l)]\hbar\omega = hv_{zp,d^+a} - hv_{zp,a^+a} \tag{6.1}$$

The thermal Frank-Condon weight of all $d \to a$ transitions generating p_{da} phonons is:

$$W_{p_{da}} = \sum_{m,l,n,k=0}^{\infty} (1-r_0)r_0^m(1-r_0)r_0^l \langle d_n|d_m^+\rangle^2 \langle a_k^+|a_l\rangle^2 \tag{6.2}$$

where the m, n, k, l indices are constrained to follow Eq. (6.1). Equation (6.2) is recognized through Eq. (4.41) as:

$$W_{p_{da}} = \sum_{p_d=-\infty}^{\infty} W_{p_d}(S_d, \langle m\rangle) W_{p_a}(S_a, \langle m\rangle) \tag{6.3}$$

where the indices are constrained by:

$$p_d + p_a = p_{da} \tag{6.4}$$

and is further recognized through Eq. (5.1) as itself a W_p function, namely:

$$W_{p_{da}}(S_d + S_a, \langle m\rangle) = \sum_{p_d=-\infty}^{\infty} W_{p_d}(S_d, \langle m\rangle) W_{p_a}(S_a, \langle m\rangle) \tag{6.5}$$

We will have occasion also to consider transfer between centers with different θ,

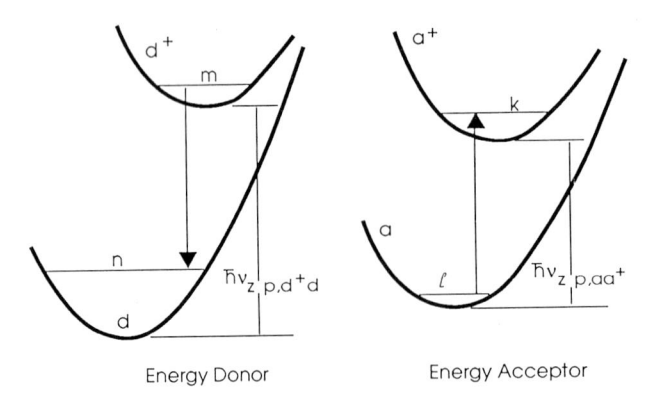

Fig. 18. The model of energy transfer from donor to acceptor

a_{uv} values. For example, we are interested in energy transfer between a broadband center whose emission is described with a U_{p_v} function of Eq. (4.37) and a narrow-line center whose absorption is described with a small-S W_p function. In general, these distributions can be combined on any phonon energy metric by interpolating each distribution at the metric energies. We choose to use the vibrational levels of the acceptor and obtain the distribution:

$$C_{p_{da}} = \sum_{p,p'} W_p U_{p'}$$

$$p\hbar\omega + p'\hbar\omega_u = p_{da}\hbar\omega \qquad (6.6)$$

to give a measure of the transfer rate.

This completes our formulas for energy transfer.

7 Compendium of Useful Equations

7.1 The Wavefunctions

The harmonic oscillator wavefunctions are defined in terms of Hermite polynomials, Eq. (2.11),

$$v_m = H_m(z_v)e^{-z^2/2}(m!2^m\pi^{1/2}x_v)^{-1/2} \tag{7.1}$$

which are given by their recursion formula, Eq. (2.7), initiated with the first two values, Eqs. (2.3) and (2.4):

$$-2kH_{k-1} + 2z_v H_k - H_{k+1} = 0$$

$$H_0 = 1$$

$$H_1 = 2z_v \tag{7.2}$$

The z and d/dz relations, Eq. (2.17) and (2.18), are:

$$\sqrt{2}z_v v_k = \sqrt{k}\,v_{k-1} + \sqrt{k+1}\,v_{k+1} \tag{7.3}$$

$$\sqrt{2}\frac{dv_k}{dz_v} = \sqrt{k}\,v_{k-1} - \sqrt{k+1}\,v_{k+1} \tag{7.4}$$

The Schrödinger equation, in dimensioned units, Eq. (2.23), and in dimensionless units, Eq. (2.22), is:

$$\left[\frac{\hbar^2}{2M}\frac{d^2}{dx^2} + E_{vm} - \tfrac{1}{2}k_v(x-a)^2\right]\Psi_m(x, k_v, a) = 0 \tag{7.5}$$

$$\frac{d^2}{dz_v^2}v_k - z_v^2 v_k + (2k+1)v_k = 0 \tag{7.6}$$

The dimensionless variables are defined through Eqs. (2.25)–(2.29), (2.31), and (2.32).

$$E_{vm} = (m + \tfrac{1}{2})\hbar\omega_v \qquad \omega_v = \left(\frac{k_v}{M}\right)^{1/2}$$

$$x_v^2 = \frac{\hbar}{\sqrt{Mk_v}} \qquad z_v = (x-a)/x_v$$

$$\tfrac{1}{2}k_v x_v^2 = \tfrac{1}{2}\hbar\omega_v \qquad \tfrac{1}{2}k_v(x-a)^2 = z_v^2\left(\frac{\hbar\omega_v}{2}\right)$$

$$\frac{\hbar^2}{2Mx_v^2} = \frac{\hbar\omega_v}{2} \tag{7.7}$$

7.2 The Manneback Recursion Formulas

The Manneback recursion formulas for unequal force constants, Eqs. (3.32) and (3.33), are:

$$k_+\sqrt{n}A_{n-1,m} = \sqrt{m+1}\,A_{n,m+1} + k_+ S_u^{1/2} A_{nm} + k\sqrt{m}\,A_{n,m-1}$$
$$k_+\sqrt{m}\,A_{n,m-1} = \sqrt{n+1}\,A_{n+1,m} - k_+ S_v^{1/2} A_{nm} - k\sqrt{n}\,A_{n-1,m} \tag{7.8}$$

where:

$$k = \cos 2\theta \qquad k_+ = \sin 2\theta$$

$$S_u^{1/2} = \frac{a_{uv}\cos\theta}{\sqrt{2}} \qquad S_v^{1/2} = \frac{a_{uv}\sin\theta}{\sqrt{2}} \tag{7.9}$$

The equal force constants equations, Eqs. (3.34) and (3.35), are:

$$\sqrt{n}\,A_{n-1,m} = \sqrt{m+1}\,A_{n,m+1} + S^{1/2} A_{nm}$$
$$\sqrt{m}\,A_{n,m-1} = \sqrt{n+1}\,A_{n+1,m} - S^{1/2} A_{nm} \tag{7.10}$$

The equal-force-constants z_v and z_u operators have the relationships, Eqs. (3.49) and (3.50):

$$\langle u_n|z_v|v_m\rangle = A_{n,m,z_v} = \sqrt{\frac{m}{2}}\,A_{n,m-1} + \sqrt{\frac{m+1}{2}}\,A_{n,m+1}$$
$$\langle v_m|z_u|u_n\rangle = A'_{m,n,z_u} = \sqrt{\frac{n}{2}}\,A_{n-1,m} + \sqrt{\frac{n+1}{2}}\,A_{n+1,m} \tag{7.11}$$

The derivative operator expressions for the $A_{nm,d/dz}$, Eqs. (3.51) and (3.52), are:

$$\left\langle u_n\left|\frac{d}{dz}\right|v_m\right\rangle = A_{n,m,d/dz_v} = \sqrt{\frac{m}{2}}\,A_{n,m-1} - \sqrt{\frac{m+1}{2}}\,A_{n,m+1}$$
$$\left\langle v_m\left|\frac{d}{dz}\right|u_n\right\rangle = A'_{m,n,d/dz_u} = \sqrt{\frac{n}{2}}\,A_{n-1,m} - \sqrt{\frac{n+1}{2}}\,A_{n+1,m} \tag{7.12}$$

7.3 The Equal-Force-Constants W_p and Related Functions in One Dimension

The W_p function is defined by Eq. (4.11):

$$W_p = \sum_m (1-r)r^m \langle u_{p+m}|v_m\rangle^2 = \sum_m (1-r)r^m A^2_{p+m,m} \tag{7.13}$$

and has the explicit expressions, Eqs. (4.45) and (4.55):

$$W_p = \exp(-S\langle 2m+1\rangle) \sum_{j=\max(0,-p)}^{\infty} \frac{(S\langle m\rangle)^j (S\langle 1+m\rangle)^{p+j}}{j!(p+j)!}$$

$$= \exp(-S\langle 2m+1\rangle)\left[\frac{\langle 1+m\rangle}{\langle m\rangle}\right]^{p/2} I_p(2S(\langle m\rangle\langle 1+m\rangle)^{1/2}) \tag{7.14}$$

The W_p satisfies the recursion relationship, Eq. (4.44):

$$S\langle m\rangle W_{p+1} + pW_p - S\langle 1+m\rangle W_{p-1} = 0 \tag{7.15}$$

At $T=0$, W_p is the Poisson distribution, Eq. (4.57):

$$W_p = e^{-S}\frac{S^p}{p!} \tag{7.16}$$

When $[S\langle m\rangle S\langle 1+m\rangle]/[1|p|+1]$ is small, then W_p has the approximate form, Eq. (4.58):

$$W_p \approx B_p \equiv e^{-S\langle 2m+1\rangle}\frac{(S\langle 1+m\rangle)^p}{p!} \quad p \geq 0$$

$$\equiv e^{-S\langle 2m+1\rangle}\frac{(S\langle m\rangle)^{|p|}}{|p|!} \quad p < 0 \tag{7.17}$$

When S is large, then W_p is approximately the Gaussian function, Eq. (4.59):

$$G(p) = \frac{1}{\sqrt{2\pi\sigma^2}}\exp[-(p-S)^2/2\sigma^2] \tag{7.18}$$

with $\sigma^2 = S\langle 2m+1\rangle$.

W_p has also the approximate form, based on the asymptotic expansion of the modified Bessel I_p function, Eqs. (4.66)–(4.68):

$$W_p \approx W_p^* \equiv e^{-S\langle 2m+1\rangle}\left(\frac{\langle 1+m\rangle}{\langle m\rangle}\right)^{p/2}\frac{e^{y_p}}{\sqrt{2\pi y_p}}\left(\frac{x}{p+y_p}\right)^p$$

$x = 2S(\langle m\rangle\langle 1+m\rangle)^{1/2}$

$y_p = \sqrt{p^2+x^2} \geq 1 \tag{7.19}$

$W_{p,z}$ can be expressed in terms of W_p via Eq. (4.23):

$$W_{p,z} = \frac{(p-S)^2}{2S} W_p \tag{7.20}$$

$W_{p,d/dz}$ is expressed in terms of $W_{p,z}$ and other W_p functions with Eq. (4.88):

$$W_{p,d/dz} = W_{p,z} - 2\langle m\rangle\langle 1+m\rangle S(W_{p+1} - 2W_p + W_{p-1}) \tag{7.21}$$

Both $W_{p,z}$ and $W_{p,d/dz}$ have 5-W_p expressions, given by Eqs. (4.69), (4.70), (4.85), and (4.86):

$$W_{p,0_\pm} = L_p + S\sum_{i=-2}^{2} \alpha_{i,\pm} W_{p+i}$$

$$\alpha_{-2,z} = \frac{\langle 1+m\rangle^2}{2} \qquad \alpha_{-2,d/dz} = \frac{\langle 1+m\rangle^2}{2}$$

$$\alpha_{-1,z} = -\langle 1+m\rangle \qquad \alpha_{-1,d/dz} = -\langle 1+m\rangle\langle 2m+1\rangle$$

$$\alpha_{0,z} = \tfrac{1}{2} - \langle m\rangle\langle 1+m\rangle \qquad \alpha_{0,d/dz} = \tfrac{1}{2} + 3\langle m\rangle\langle 1+m\rangle$$

$$\alpha_{1,z} = \langle m\rangle \qquad \alpha_{1,d/dz} = -\langle m\rangle\langle 2m+1\rangle$$

$$\alpha_{2,z} = \frac{\langle m\rangle^2}{2} \qquad \alpha_{2,d/dz} = \frac{\langle m\rangle^2}{2} \tag{7.22}$$

The average m through which a transition occurs is defined in Eq. (4.71) and an expression for it is obtained in Eq. (4.75). These are:

$$\langle m\rangle_p = \frac{\sum_{m=m_0}^{\infty} m(1-r)r^m A^2_{p+m,m}}{\sum_{m=m_0}^{\infty} (1-r)r^m A^2_{p+m,m}} = \left[\frac{1}{W_p}\right] \sum_{m=m_0}^{\infty} m(1-r)r^m A^2_{p+m,m} \tag{7.23}$$

and:

$$\langle m\rangle_p W_p = S\langle m\rangle\langle 1+m\rangle\{W_{p+1} - 2W_p + W_{p-1}\} + \langle m\rangle W_p \tag{7.24}$$

7.4 The Unequal-Force-Constants Expressions

For unequal force constants, the analogous expressions are those of Eqs. (4.24)–(4.34):

$$V_{p_V} = \sum_n (1-r_u)r_u^n V^2_{p_V,n} \qquad U_{p_U} = \sum_m (1-r_v)r_v^m U^2_{p_U,m}$$

$$V_{p_V,z} = \sum_n (1-r_u)r_u^n V^2_{p_V,n,z} \qquad U_{p_U,z} = \sum_m (1-r_v)r_v^m U^2_{p_U,m,z} \tag{7.25}$$

where:

$$r_u = \exp(-\hbar\omega_u/kT) \qquad r_v = \exp(-\hbar\omega_v/kT)$$

$$V^2_{p_V,n} = \langle v_{i_n}|u_n\rangle^2(1 - i_n + p_n) + \langle v_{i_n-1}|u_n\rangle^2(i_n - p_n)$$

$$U^2_{p_U,m} = \langle u_{i_m}|v_m\rangle^2(1 - i_m + p_m) + \langle u_{i_m-1}|v_m\rangle^2(i_m - p_m)$$

7.5 The Moments

$$V^2_{p_V,n,z} = \langle v_{i_n}|z|u_n\rangle^2(1 - i_n + p_n) + \langle v_{i_n-1}|z|u_n\rangle^2(i_n - p_n)$$
$$U^2_{p_U,m,z} = \langle u_{i_m}|z|v_m\rangle^2(1 - i_m + p_m) + \langle u_{i_m-1}|z|v_m\rangle^2(i_m - p_m) \quad (7.26)$$

where p_n and p_m are noninteger numbers and i_n and i_m are integer numbers satisfying:

$$p_n \hbar\omega_v - n\hbar\omega_u = p_V \hbar\omega_v$$
$$p_m \hbar\omega_u - m\hbar\omega_v = p_U \hbar\omega_u$$
$$i_n = \text{int}(p_n) + 1$$
$$i_m = \text{int}(p_m) + 1$$
$$\text{int}(-x) = -\text{int}(x) \quad (7.27)$$

7.5 The Moments

The equal-force-constants moments are, by Eq. (4.104):

$$\langle (p-S)^0 \rangle = \langle 1 \rangle = 1$$
$$\langle (p-S)^1 \rangle = 0$$
$$\langle (p-S)^2 \rangle = S\langle 2m+1 \rangle$$
$$\langle (p-S)^3 \rangle = S$$
$$\langle (p-S)^4 \rangle = S\langle 2m+1 \rangle + 3[S\langle 2m+1 \rangle]^2 \quad (7.28)$$

They are obtained from the moment recursion formula, Eqs. (4.97)–(4.98):

$$\langle (p-S)^\alpha \rangle = \sum_{j=0}^{\alpha-2} \binom{\alpha-1}{j} \beta_{\alpha j} \langle (p-S)^j \rangle$$
$$\beta_{\alpha j} = S\langle 2m+1 \rangle \quad (\alpha - j) \text{ even}$$
$$= S \quad (\alpha - j) \text{ odd} \quad (7.29)$$

using the first two moments to initiate the sequence of calculations.

The first three rigorous unequal-force-constants $v \to u$ emission p moments are Eqs. (4.109)–(4.110):

$$\langle p_u \rangle_{uv} = S_u + k_\theta \langle 2m+1 \rangle_v$$
$$\langle (p_u - \langle p_u \rangle_{uv})^2 \rangle_{uv} = \left(\frac{\omega_u}{\omega_v}\right) S_u \langle 2m+1 \rangle_v + 2k_\theta^2 \langle 2m+1 \rangle_v^2$$
$$\langle (p_u - \langle p_u \rangle_{uv})^3 \rangle_{uv} = S_u + 6k_\theta \left(\frac{\omega_u}{\omega_v}\right) S_u \langle 2m+1 \rangle_v^2 + 4k_\theta^2 \left(\frac{\omega_v}{\omega_u}\right) \langle 2m+1 \rangle_v$$
$$+ 8k_\theta^3 \langle 2m+1 \rangle_v^3$$

$$k = \cos 2\theta \quad k_+ = \sin 2\theta \quad k_\theta = \left(\frac{k}{k_+^2}\right) = \frac{\omega_u^2 - \omega_v^2}{4\omega_v \omega_u} \quad (7.30)$$

The $u \to v$ absorption p moments are:

$$\langle p_v \rangle_{vu} = S_v - k_\theta \langle 2n+1 \rangle_u$$

$$\langle (p_v - \langle p_v \rangle_{vu})^2 \rangle_{vu} = \left(\frac{\omega_v}{\omega_u}\right) S_v \langle 2n+1 \rangle_u + 2k_\theta^2 \langle 2n+1 \rangle_u^2$$

$$\langle (p_v - \langle p_v \rangle_{vu})^3 \rangle_{vu} = S_v - 6k_\theta \left(\frac{\omega_v}{\omega_u}\right) S_v \langle 2n+1 \rangle_u^2 + 4k_\theta^2 \left(\frac{\omega_u}{\omega_v}\right) \langle 2n+1 \rangle_u$$

$$- 8k_\theta^3 \langle 2n+1 \rangle_u^3 \tag{7.31}$$

The k'th energy moment is the k'th p moment multiplied by the final-state phonon energy, here $\hbar\omega_v$ in absorption and $\hbar\omega_u$ in emission, raised to the k'th power.

7.6 Multiple Coordinate Models of a Luminescence Center

The thermal-Condon factor for the Einstein-Huang-Rhys-Pekar model of N_{Av} oscillators having a common frequency $\hbar\omega$ and therefore a common $\langle m \rangle$ is itself a W_p function with parameters satisfying Eq. (5.5):

$$W_p(S, \langle m \rangle) = W_p(\sum S_i, \langle m \rangle). \tag{7.32}$$

The z and d/dz operator distributions are given by Eqs. (5.6) and (5.7) as:

$$A^2 M_{p,o_\pm} \equiv \sum_{\bar{p},\bar{m}} |\langle u_{p_1+m_1} u_{p_2+m_2} \cdots u_{p_N+m_N} | A_1 O_{1\pm} + \ldots A_N O_{N\pm} |$$

$$\times v_{m_1} v_{m_2} \cdots v_{m_N} \rangle|^2 \cdot (1-r)^N r^{m_1+\cdots+m_N}$$

$$= A^2[(1-\gamma)W_{p,o_\pm} + \gamma L_p] \tag{7.33}$$

where:

$$A^2 = \sum_{k=1}^{N} |A_k|^2$$

$$S = \sum_{k=1}^{N} S_k$$

$$\gamma = 1 - \frac{\left|\sum_{k=1}^{N} A_k S_k^{1/2}\right|^2}{A^2 S} \quad 0 \leq \gamma \leq 1$$

$$L_p = \tfrac{1}{2}(\langle 1+m \rangle W_{p-1} + \langle m \rangle W_{p+1}) \tag{7.34}$$

where $W_{p,o_\pm}$, L_p, and the W_p functions within the L_p expression all have the arguments $(S, \langle m \rangle)$. The $W_{p,o_\pm}$ themselves are given by Eqs. (7.20)–(7.22).

The multiple frequency model expressions for the Condon operator are defined by Eq. (5.35):

$$D_p^k = A^{-2} G_p^K \tag{7.35}$$

7.6 Multiple Coordinate Models of a Luminescence Center

where the G_p^K are defined by Eqs. (5.32) to (5.34):

$$G_p^K(\vec{S}, \vec{\omega}_K, kT) = A^2 \sum_{p_1, p_2, \ldots, p_K} W_{p_1}(S_1, \omega_1, kT)$$
$$\times W_{p_2}(S_2, \omega_2, kT) \ldots W_{p_K}(S_K, \omega_K, kT) \quad (7.36)$$

where the summation indices are constrained by:

$$\sum_{k=1}^{K} k p_k = p \quad (7.37)$$

and:

$$S_i = \sum_{j=1}^{g_j} S_{i,j}$$

$$\omega_k = k\omega_1 \quad (7.38)$$

The W_p have the normal variables $(S, \langle m \rangle)$, and $\langle m \rangle$ must be determined for each ω_k through Eqs. (4.43) and (4.12):

$$r = \exp(-\hbar\omega/kT) \qquad \langle m \rangle = \frac{r}{1-r} \quad (7.39)$$

This Condon-operator distribution is most readily obtained by the recursion algebra, Eqs. (5.38) and (5.39):

$$D_q^{K,l} = \sum_{p_1} D_{q-lp_1}^{K,l-1} W_{p_1}(S_l, \omega_l, kT)$$

$$D_q^{K,1} = W_q(S_1, \omega_1, kT) \equiv W_q(S_1, \langle m_1 \rangle)$$

$$2 \leq l \leq K$$

$$D_{q=p}^{K,K} = D_p^K \quad (7.40)$$

For the $O_\pm$ operators, the distributions are a sum of two terms, Eq. (5.41),

$$G_{p,O_\pm}^K = \Psi_{p,O_\pm}^K + \Phi_{p,O_\pm}^K \quad (7.41)$$

In Eq. (7.41), $\Psi_{p,O_\pm}^K$ is given by the recursion algebra, Eqs. (5.44) and (5.45),

$$\Psi_{q,O_\pm}^{K,1} = A_1^2 M_{q,O_\pm}(S_1, \omega_1, kT)$$

$$\Psi_{q,O_\pm}^{K,l} = \sum_{p_1} [\Psi_{q-lp_1,O_\pm}^{K,l-1} W_{p_1}^{(l)}(S_l, \omega_l, kT) + D_{q-lp_1}^{K,l-1} A_l^2 M_{p_1,O_\pm}^{(l)}(S_l, \omega_l, kT)]$$

$$\Psi_{q,O_\pm}^{K,K} \equiv \Psi_{q,O_\pm}^K \quad (7.42)$$

The $\Phi_{p,O_\pm}^K$ are obtained by the recursion algebra of Eqs. (5.50) to (5.52):

$$\Theta_{q,O_\pm}^{K,l} = \sum_{p_1} [\Theta_{q-lp_1,O_\pm}^{K,l-1} W_{p_1}^{(l)}(S_l, \omega_l, kT) + D_{q-lp_1}^{K,l-1} \Gamma_l X_{p_1,O_\pm}^{(l)}((S_l, \omega_l, kT))]$$

$$\Theta_{q,O_\pm}^{K,1} = \Gamma_1 X_{q,O_\pm}^{(1)}(S_1, \omega_1, kT)$$

$$\Phi_{q,O_\pm}^{K,l} = \sum_{p_1} [\Phi_{q-lp_1,O_\pm}^{K,l-1} W_{p_1}^{(l)}(S_l, \omega_l, kT) + \Theta_{q-lp_1,O_\pm}^{K,l-1} \Gamma_l^* X_{p_1,O_\pm}^{(l)}(S_l, \omega_l, kT)$$

$$+ (\Theta_{q-lp_1,O_\pm}^{K,l-1})^* \Gamma_l X_{p_1,O_\pm}^{(l)}(S_l, \omega_l, kT)]$$

$$\Phi_{q,O_\pm}^{K,1} = 0$$

$$\Phi_{q=p,O_\pm}^{K,K} = \Phi_{p,O_\pm}^{K} \tag{7.43}$$

In Eq. (7.43) $X_{p,O_\pm}$ is as given in Eq. (5.48). The first of Eq. (7.43) is the recursion relation which, when initiated with the second of this equation, allows sequential generation of the $\Theta_{q,O_\pm}^{K,l}$ for $2 \leq l \leq K$.

In turn, the third of this equation is the recursion relation for $\Phi_{q,O_\pm}^{K,l}$, which, when initialized as in the fourth equation, allows sequential generation of the $\Phi_{q,O_\pm}^{K,l}$ for $2 \leq l \leq K$ and leads to the desired sum in the last of this Eq. (7.43).

We adopt the Discretized Debye Equal S and A model through Eqs. (5.54) through (5.58),

$$\bar{g}_1 = \frac{27}{8K^3}, \quad \bar{g}_k = \frac{12k^2 + 1}{4K^3}, \quad \bar{g}_K = \frac{12K^2 - 6K + 1}{8K^3}, \quad \bar{g}_i = g_i/3N_0$$

$$S\hbar\omega_K = \sum_{k=1}^{K} S_k \hbar\omega_k = S_0 \hbar\omega_0 \sum_{k=1}^{K} k\bar{g}_k$$

$$S_k = \bar{g}_k S_0 = \bar{g}_k S \left\{ \frac{8K^4}{6K^4 + K^2 + 1} \right\}$$

$$\Gamma_k \equiv \frac{1}{\sqrt{2}} \sum_i A_{k,i} S_{k,i}^{1/2} = \frac{1}{\sqrt{2}} (A_k^2 S_k)^{1/2}$$

$$A_k^2 = \sum_i^{g_i} A_{ki}^2 = \bar{g}_k A_0^2$$

$$\hat{A}_k^2 = \bar{g}_k, \quad \hat{\Gamma}_k = (\bar{g}_k S_k/2)^{1/2} \tag{7.44}$$

Normalized distributions are afforded by using the $\hat{A}_k$ and $\hat{\Gamma}_k$. The normalization constant is $\Sigma_k \bar{g}_k [\frac{1}{2}\langle 2m + 1 \rangle]$.

7.7 Energy Transfer

The transition rate from an energy donor to an energy acceptor has the thermal Franck-Condon weight given by Eq. (6.2)–(6.5):

$$W_{p_{da}} = \sum_{m,l,n,k=0}^{\infty} (1 - r_0) r_0^m (1 - r_0) r_0^l \langle d_n | d_m^+ \rangle^2 \langle a_k^+ | a_l \rangle^2 \tag{7.45}$$

where the m, n, k, l indices are constrained to follow Eq. (6.1),

$$p_{da} \hbar\omega = [(n - m) + (k - l)] \hbar\omega = h\nu_{zp,d^+d} - h\nu_{zp,a^+a} \tag{7.46}$$

This nuclear factor is recognized to be

$$W_{p_{da}} = \sum_{p_d=-\infty}^{\infty} W_{p_d}(S_d, \langle m \rangle) W_{p_a}(S_a, \langle m \rangle) \tag{7.47}$$

7.7 Energy Transfer

where the indices are constrained by

$$p_d + p_a = p_{da} \tag{7.48}$$

and is further recognized as itself a W_p function, namely,

$$W_{p_{da}}(S_d + S_a, \langle m \rangle) = \sum_{p_d=-\infty}^{\infty} W_{p_d}(S_d, \langle m \rangle) W_{p_a}(S_a, \langle m \rangle) \tag{7.49}$$

8 Contact with the Theoretical Literature

8.1 Unequal-Force-Constants A_{nm}

We have already cited the explicit formulas of Hutchisson [33] for the A_{nm} for unequal force constants. These explicit formulas have not proved as useful as the Manneback [32] recursion formulas for computational ease.

8.2 Equal-Force-Constants A_{nm}

8.2.1 Explicit Formulas

For equal force constants, the explicit formulas for the A_{nm} are simpler and have appeared in the literature. Although we prefer even here the equal-force-constants Manneback equations, we give here a brief description of some explicit formulas to facilitate comparison with the literature.

The explicit formula for the A_{nm} is derived from its generating function, namely, Eq. (3.21), evaluated for $\theta = \pi/4$:

$$G_A = \exp\left(-\frac{S_0}{2}\right)\exp[(2S_0)^{1/2}(s-t) + 2st] = \sum_{n,m}\left(\frac{2^n 2^m}{n!m!}\right)^{1/2} A_{nm} s^n t^m \quad (8.1)$$

We expand the first expression for G_A in its power series:

$$G_A = \exp\left(-\frac{S_0}{2}\right)\sum_{j=0}^{\infty}\sum_{k=0}^{j}\sum_{l=0}^{j-k}\frac{s^{j-l}t^{j-k}2^{j-(k+l)/2}S_0^{(k+l)/2}(-1)^l}{k!l!(j-k-l)!} \quad (8.2)$$

and equate the coefficients of $s^n t^m$ by imposing the constraints on the summation indices of Eq. (8.2):

$$j - l = n \quad l = j - n$$
$$j - k = m \quad k = j - m \quad (8.3)$$

The summations in Eq. (8.2) then become:

$$\sum_{j=0}^{\infty}\sum_{k=0}^{j}\sum_{l=0}^{j-k} = \sum_{n=0}^{\infty}\sum_{m=0}^{\infty}\sum_{j=\max(n,m)}^{n+m} \quad (8.4)$$

8.2 Equal-Force-Constants A_{nm}

giving:

$$G_A = \exp\left(-\frac{S_0}{2}\right) \sum_{n=0}^{\infty} \sum_{m=0}^{\infty} \sum_{j=max(n,m)}^{n+m} \frac{s^n t^m 2^{(n+m)/2} S_0^{j-(n+m)/2}(-1)^{j-n}}{(j-n)!(j-m)!(n+m-j)!} \quad (8.5)$$

Equating Eq. (8.5) to the second expression for G_A in Eq. (8.1) gives the explicit expression for the A_{nm}:

$$A_{nm} e^{-(S_0/2)} \sum_{j=max(n,m)}^{n+m} \frac{(n!m!)^{1/2}(-1)^{j-n} S_0^{j-(n+m)/2}}{(j-n)!(j-m)!(n+m-j)!}$$

$$e^{-(S_0/2)} \sum_{k=0}^{min(n,m)} \frac{(n!m!)^{1/2} S_0^{((n+m)/2)-k}(-1)^{m+k}}{k!(n-k)!(m-k)!} \quad (8.6)$$

The second step in Eq. (8.6) uses:

$$k = n + m - j \quad (8.7)$$

and reverses the direction of summation via:

$$j - max(n,m) = 0 \to n + m - k - max(n,m) = 0 \to k = min(n,m)$$
$$j = n + m \to k = 0 \quad (8.8)$$

8.2.2 Laguerre Polynomial Expressions

The generating function for Laguerre polynomials is, as cited in Erdélyi et al. [34] Eq. (10.12.19):

$$G_L \equiv e^{-S_0 \tau}(1+\tau)^n = \sum_{m=0}^{\infty} L_m^{n-m}(S_0)\tau^m \quad (8.9)$$

If now we write the first expression for G_A in Eq. (8.1) as:

$$G_A = \exp\left(-\frac{S_0}{2}\right) e^{\sigma(1+\tau) - S_0 \tau} \quad (8.10)$$

where:

$$\sigma = (2S_0)^{1/2} s \qquad \tau = \left(\frac{2}{S_0}\right)^{1/2} t \quad (8.11)$$

and then expand the exponential involving $\sigma(1+\tau)$ in its power series, we obtain:

$$G_A e^{-(S_0/2)} \sum_{n=0}^{\infty} \left(\frac{\sigma^n(1+\tau)^n}{n!}\right) e^{-S_0 \tau} e^{-(S_0/2)} \sum_{n=0}^{\infty} \frac{\sigma^n}{n!} \sum_{m=0}^{\infty} L_m^{n-m}(S_0)\tau^m$$

$$e^{-(S_0/2)} \sum_n \sum_m L_m^{n-m}(S_0) \frac{(2^{n+m} S_0^{n-m})^{1/2} s^n t^m}{n!} \quad (8.12)$$

Equating the last form of Eq. (8.12) and the last form of Eq. (8.1), we find from the coefficient of $s^n t^m$:

$$A_{nm} e^{-(S_0/2)} \left(\frac{m!}{n!}\right)^{1/2} (S_0)^{(n-m)/2} L_m^{n-m}(S_0) \qquad (8.13)$$

Since:

$$m!(-S_0)^{-m} L_m^{n-m}(S_0) = n!(-S_0)^{-n} L_n^{m-n}(S_0) \qquad (8.14)$$

an equivalent formula is:

$$A_{nm} e^{-(S_0/2)} \left(\frac{n!}{m!}\right)^{1/2} (-S_0)^{(m-n)/2} L_n^{m-n}(S_0) \qquad (8.15)$$

8.2.3 Citations

The A_{nm} Laguerre formulas were first derived by Ruamps [35] and later by Wagner [36], Koide [37], and Keil [38]. Equation (8.13) is Ruamps' formula with the notation change $n \to v_2$, $m \to v_1$, $S_0^{1/2} \to 2^{-1/2} \Delta$. Equations (8.13) and (8.15) are Wagner's formulas (25) and (26) with the notation changes $n \to m$, $m \to n$, $S_0^{1/2} \to a/(2^{1/2} x_0)$. These are also Koide's formulas (14) and (15), apart from a factor $(-1)^{n-m}$ in (14) which is incorrect, with the notation changes $A_{nm} \to \langle n|U|m\rangle$, $S_0^{1/2} \to \alpha$. Ruamps, Wagner, and Koide used Laguerre polynomials restricted to a non-negative upper index. There is no necessity for such a restriction in its generating function, Eq. (8.9).

Eq. (8.15) is Keil's formula (3.7), with the notation changes $u \to a$, $v \to b$, $n \to \alpha$, $m \to \beta$, $A_{nm} \to S_{\alpha,\beta}$, $S_0^{1/2} \to a/2^{1/2}$. Because Keil's dimensionless offset a for his b parabola corresponds to the offset $-a/x_0$ for our v parabola, we believe that the proper correspondence should be not as given last above but rather $S_0^{1/2} \to -a/2^{1/2}$, and that Keil's Eq. (3.7) is incorrect by a factor of $(-1)^{\beta-\alpha}$. Fitchen [39] has summarized Keil's results, and his Eq. (5.3) is similarly incorrect by a factor of $(-1)^{n-m}$.

8.3 The W_p Formula

The W_p formula, Eq. (4.45), was first derived by Huang and Rhys [40] and Pekar [41, 42] for the Einstein model of a crystal. They both give the Eq. (4.55) Bessel function form: see Eq. (6.12) of Huang and Rhys and Eqs. (28.23) and (28.24) of Pekar's book [42]. However, both authors passed through the explicit equation (4.45): see Eqs. (4.13), (4.18), and (4.20) of Huang and Rhys and Eq. (28.24) of Pekar.

The W_p formula was first associated with the single-configurational model through the work of Lax [43, 44] and O'Rourke [45]. These authors used quantum-mechanical integral representations for the optical transition rates.

Lax began with the absorption expression, his Eqs. (2.5), (2.6):

8.3 The W_p Formula

$$I_{ba}(\nu) = h^{-1} \int \exp(-i2\pi\nu t) I_{ba}(t)\, dt$$

$$I_{ba}(t) = tr[M_{ba}^* e^{iH_b t/h} M_{ba} e^{-iH_a t/h} e^{-\beta H_a}]/tr[e^{-\beta H_a}] \tag{8.16}$$

For the Huang-Rhys model and the Condon approximation this becomes his Eqs. (7.1)–(7.2):

$$I_{ba}(t) = |M_{ba}|^2 Av[\exp(iH_b t/h)\exp(-iH_a t/h)] \tag{8.17}$$

Lax next uses the ordered operators of Dyson, Feynman, and Goldberger and Adams, for whom he gives references, which in this case to second order is, according to his Eqs. (7.7)–(7.10):

$$I_{ba}(t) = |M_{ba}|^2 \left[1 + (i/h) \int_0^t Av[\Delta E(s)]\, ds \right.$$

$$\left. - h^{-2} \int_0^t ds \int_0^s ds'\, Av[\Delta E(s')\Delta E(s)] \right] \tag{8.18}$$

He then integrates over s and s' and gets his Eq. (7.11)–(7.12):

$$I_{ba}(t) = |M_{ba}|^2 \exp[i2\pi\nu_0 t + \langle C_j^2\{i\sin\omega_j t - (2\bar{n}+1)(1-\cos\omega_j t)\}\rangle] \tag{8.19}$$

Introducing now the single-frequency Einstein-Huang-Rhys model, he is led to the integral representation of the modified Bessel function of Eq. (4.51), namely:

$$I_y(z) = \left(\frac{1}{2\pi}\right) \int_0^{2\pi} \exp(iyx + z\cos x)\, dx \tag{8.20}$$

so that Eq. (4.55) emerged.

O'Rourke started with the integral representation of the transition rate, in the Condon approximation, gained from the integral representation of the Dirac δ function, namely, his Eqs. (17), (18):

$$G_{ba}(t) = \sum_{n_j'',n_j'=0}^{\infty} \exp\left\{-\sum_{j=1}^{N}[(n_j'+\tfrac{1}{2})\beta_j - it(n_j''-n_j')\omega_j]\right\}$$

$$\times \left\{\prod_{j=1}^{N} |\langle bn_j| M_{ba}(\mathbf{R})|an_j'\rangle|^2\right\}\left\{\prod_{j=1}^{N}\sum_{n_j'=0}^{\infty} e^{-(n_j'+1/2)\beta_j}\right\}^{-1}$$

$$\beta_j \equiv \hbar\omega_j/kT \tag{8.21}$$

He used Mehler's formula for summing products of Hermite functions, namely:

$$\sum_{k=0}^{\infty} \left(\frac{e^{-(k+1/2)}}{\pi^{1/2} 2^k k!}\right) H_k(x) H_k(x') \exp\left\{-\frac{1}{2}(x^2 + x'^2)\right\}$$

$$= (2\pi \sinh \zeta)^{-1/2} \exp\left[-\frac{1}{4}\{(x + x')^2 \tanh\left(\frac{\zeta}{2}\right) + (x - x')^2 \coth\left(\frac{\zeta}{2}\right)\}\right]$$
(8.22)

and then integrated over all frequencies. He obtained his Eq. (52):

$$G_j(t) \equiv G_{ba}(t)/|M|^2$$
$$= 2^{1/2} \sinh(\beta_j/2) [\cosh \lambda_j \cosh \mu_j - 1$$
$$+ 2^{-1}(\kappa_j + \kappa_j^{-1}) \sinh \lambda_j \sinh \mu_j]^{-1/2}$$
$$\times \exp[-(\alpha_j''\alpha_j')^2 (C_j'' - C_j')^2/\Psi_j^2],$$
(8.23)

where:

$$\beta_j \equiv \hbar \omega_j'/kT \to \hbar \omega_u/kT, \qquad \lambda_j \equiv \beta_j + i\omega_j't \to \hbar \omega_u/kT + i\omega_u t,$$
$$\mu_j \equiv -i\omega_j''t \to -i\omega_v t, \qquad \kappa_j \equiv \omega_j''/\omega_j' \to \omega_v/\omega_u, \qquad \alpha_j'^2 = \omega_j'/\hbar \to \omega_u/\hbar,$$
$$\alpha_j''^2 = \omega_j''/\hbar \to \omega_v/\hbar, \qquad \alpha_j''^2 (C_j'' - C_j')^2 \to a^2/x_v^2 = 2S_v,$$
$$\Psi_j^2 = \alpha_j''^2 \coth(\lambda_j/2) + \alpha_j'^2 \coth(\mu_j/2)$$
(8.24)

The arrows show the correspondences with our notation. In checking O'Rourke's work, we find that the Ω_j^2 dividing $(C_j'' - C_j')^2$ in his Eqs. (51) and (52) should be the Ψ_j^2 given in Eq. (8.24) here. The Ω_j^2 following Λ_j^2 in his Eq. (51) is correct as it stands. O'Rourke followed Lax in recognizing the Eq. (8.20) integral representation of the I_p Bessel function and obtained our Eq. (4.55) expression for W_p involving these Bessel functions.

Curie [46], Wagner [36], Koide [37], and Keil [38] all derived the W_p formula Eq. (4.55) by using the Laguerre-polynomial forms of the overlap integrals Eqs. (8.13) and (8.15), and then using the Hille-Hardy or Myller-Lebedeff summing formula:

$$(1 - r) \sum_{m=\max(0,-p)}^{\infty} \left(\frac{m!}{(p+m)!}\right) L_m^p(x) L_m^p(y) r^m$$
$$= \exp\left[-(x+y)\left(\frac{r}{1-r}\right)\right] (xyr)^{-p/2} I_p\left\{\frac{2(xyr)^{1/2}}{1-r}\right\}$$
(8.25)

Setting in Eq. (8.25) $x = y = S$ and multiplying by $e^{-S} S^p$, one finds:

$$(1 - r) \sum_{m=\max(0,-p)}^{\infty} A_{p+m,m}^2 r^m \equiv W_p$$
$$= \exp\left[-S\left(\frac{1+r}{1-r}\right)\right] (r)^{-p/2} I_p\left(\frac{2Sr^{1/2}}{1-r}\right).$$
(8.26)

which is the r form of Eq. (4.55).

8.4 The $W_{p,d/dz}$ Formula

Mehler's equation Eq. (8.25) and the Hille-Harde Eq. (8.26) are both found in Erdélyi [34] as his Eqs. (10.12.20) and (10.13.22), respectively.

Moos and coworkers [26, 27] used as the rate of a $4f^n \to 4f^n$ nonradiative transition the expression:

$$\text{Rate} = \varepsilon^p \langle 1 + m \rangle^p \tag{8.27}$$

which is the $p \geq 0$ form of Eq. (4.58), without the explicit recognition of the $p!$ dependence of this rate.

8.4 The $W_{p,d/dz}$ Formula

To our knowledge, nobody has derived the $W_{p,d/dz}$ formula, Eq. (4.87) or (4.88), for one coordinate. Fong and his coworkers [47] have come to an approximate form of this expression, their Eq. (18), given here slightly corrected for misprints.

$$W = [(2\pi)^{1/2} \hbar^{-2} |C|^2] \sum_{v=-2}^{2} \lambda_v \left[\frac{4 \sinh(\frac{1}{2}\beta\hbar\omega)}{Lg^2\omega^2(x^2+1)^{1/2}} \right]^{1/2}$$

$$\times \exp\left[-\frac{1}{4} L g_m^2 (2n_m + 1) + (\Delta_v/\hbar\omega)\{\frac{1}{2}\beta\hbar\omega - \ln[x + (x^2+1)^{1/2}]\} \right.$$

$$\left. + \frac{Lg_m^2(x^2+1)^{1/2}}{4\sinh(\frac{1}{2}\beta\hbar\omega)} \right] \tag{8.28}$$

where:

$$\Delta_v = \varepsilon_\alpha - \varepsilon_{\alpha'} - v\hbar\omega$$

$$x = 4\Delta_v \sinh(\tfrac{1}{2}\beta\hbar\omega)/Lg_m^2\hbar\omega$$

$$\tfrac{1}{2}\lambda_{-i} = \tfrac{1}{2}\{\langle 1 + m \rangle \delta(i, -1) + \langle m \rangle \delta(i, 1)\} + (\tfrac{1}{4}Lg_m^2)\alpha_{i,d/dz} \tag{8.29}$$

where the $\alpha_{i,d/dz}$ are those of Eq. (4.86) and the δ are Kroniker δ functions. This was given as a single-frequency-multiple-coordinate-model approximate rate for the derivative operator, our Eq. (5.6). We have shown [16] that the approximations made have reduced the model to a single-coordinate derivative-operator expression, Eq. (4.87), multiplied by the electronic factor $A_{d/dz}$, with W_p expressed in its modified Bessel function form Eq. (4.55), and with these Bessel functions further approximated as in Eq. (4.64), i.e., with the W_p given as in Eq. (4.66). The correspondences needed for this identification are:

$$S \to \tfrac{1}{4} L_m g_m^2$$

$$p\hbar\omega \to \Delta_{v=0} = \varepsilon_\alpha - \varepsilon_{\alpha'}$$

$$A_{d/dz} \to \frac{4\pi}{\hbar} \frac{1}{\hbar\omega} L_m |C_{\alpha\alpha'}^m|^2 \tag{8.30}$$

and useful identities for the conversion are:

$$2\langle 1+m\rangle = \frac{\exp(\hbar\omega/2kT)}{\sinh(\hbar\omega/2kT)}$$

$$4\langle 1+m\rangle\langle m\rangle = \sinh^{-2}(\hbar\omega/2kT)$$

$$y_p = \frac{(S^2+p^2\sinh^2\beta)^{1/2}}{\sinh\beta}$$

$$x_v = \frac{(p-v)}{S}\sinh\beta$$

$$x_v^2+1 = \frac{S^2+(p-v)^2\sinh^2\beta}{S^2} = \left(\frac{y_{p-v}\sinh\beta}{S}\right)^2$$

$$\frac{Lg_m^2\sqrt{x_v^2+1}}{4\sinh\beta} = y_{p-v}$$

$$\exp(\beta\Delta_v/\hbar\omega) = [e^\beta]^{p-v}$$

$$\exp\left\{\left(\frac{\Delta_v}{\hbar\omega}\right)(-\ln[x_v+\sqrt{x_v^2+1}])\right\} = \left\{\frac{S}{(p-v+y_{p-v})\sinh\beta}\right\}^{p-v}$$

$$\left(\frac{e^\beta}{\sinh\beta}\right)^p = (2\langle 1+m\rangle)^p$$

$$x_v+\sqrt{x_v^2+1} = \left(\frac{e^\beta}{2S\langle 1+m\rangle}\right)[p-v+y_{p-v}]$$

$$\exp\left\{\frac{\Delta_v}{\hbar\omega}[\beta-\ln(x_v+\sqrt{x_v^2+1})]\right\} = \left\{\frac{2S\langle 1+m\rangle}{p-v+y_{p-v}}\right\}^{p-v}$$

$$W_p \approx \langle 1+m\rangle^p e^{-S\langle 2m+1\rangle}\frac{e^{y_p}}{\sqrt{2\pi y_p}}\left(\frac{2S}{p+y_p}\right)^p \tag{8.31}$$

The last of Eq. (8.31) is a slight revision of Eq. (4.66) using Eq. (4.67).

8.5 The Equal-Force-Constants Moments

Lax [44] has given the first four equal-force-constants Condon absorption moments Eq. (4.104) as derived by his transform method involving ordered operators.

8.6 The Unequal-Force-Constants Moments

Lax [44] and Markham [48] have derived the first two Condon absorption moments Eq. (4.112) by Lax's transform method using ordered operators [43]. Lax's two parabolas and thermal average $\langle x^2\rangle_a$ are given by his Eqs. (3.10),

8.6 The Unequal-Force-Constants Moments

(3.11), and (3.8). Comparing these with our two parabolas and thermal average $\langle 2n+1 \rangle_u$ yields the following correspondences:

$$\omega_u \to \omega_a, \quad \omega_v \to \omega_b, \quad M_0 \to M, \quad k_u \to K,$$

$$-k_v a = -M_0 \omega_v^2 a \to A, \quad k_u - k_v = M_0(\omega_u^2 - \omega_v^2) \to B$$

$$E_{vu} + k_v a^2/2 \to h\nu_{ba}, \quad \langle 2n+1 \rangle_u \to (2M\omega_a/\hbar)\langle x^2 \rangle_a$$

$$S_v \equiv \frac{a^2}{2x_v^2} = \frac{M_0 \omega_v a^2}{2\hbar} \to \frac{A^2}{2\hbar M \omega_b^3}$$

$$k_\theta \equiv \frac{\omega_u^2 - \omega_v^2}{4\omega_u \omega_v} \to \frac{B}{4\omega_a \omega_b M} \tag{8.32}$$

These substitutions give the three $u \to v$ absorption moments in Lax's notation:

$$\langle p_v \rangle_{vu} \hbar \omega_v \to \left(\frac{A^2}{2M\omega_b^2}\right) - \left(\frac{B}{2}\right)\langle x^2 \rangle_a$$

$$\langle (p_v - \langle p_v \rangle_{vu})^2 \rangle_{vu} (\hbar \omega_v)^2 \to A^2 \langle x^2 \rangle_a + \left(\frac{B^2}{2}\right)\langle x^2 \rangle_a^2$$

$$\langle (p_v - \langle p_v \rangle_{vu})^3 \rangle_{vu} (\hbar \omega_v)^3 \to \frac{A^2 \hbar^2}{2M} - 3A^2 B \langle x^2 \rangle_a^2 + \left(\frac{B^2 \hbar \omega^2}{2M}\right)\langle x^2 \rangle_a$$

$$- B^3 \langle x^2 \rangle_a^3 \tag{8.33}$$

The first two of Eq. (8.33) are indeed Lax's Eqs. (3.16) and (3.17), except that the first moment is here reduced by the parabola energy offset E_{vu} so as to give phonon moments rather than photon moments. Lax's form has arguments carrying atomic dimensions. In our form, only the $\hbar \omega_u$ has energy units, and all other parameters are dimensionless.

Markham [48] uses a notation very similar to Lax except that he uses mass-weighted coordinates. Markham's two parabolas are given by his Eq. (3.6) and (3.18), and his thermal average is $\coth(\beta_g/2) = \langle 2n+1 \rangle_g$. We choose to transform Markham → Lax via:

$$\omega_g \to \omega_a, \quad \omega_u \to \omega_b, \quad q \to M^{1/2} x, \quad \Delta \varepsilon \to h\nu_{ba}, \quad M^{1/2} \varepsilon_j \to A,$$

$$-M\varepsilon_{jj} \to B, \quad (\hbar/2M\omega_g)\coth(\beta_g/2) \to \langle x^2 \rangle_a \tag{8.34}$$

Entering Eq. (8.34) into Eq. (8.33) gives the three absorption moments in Markham's notation. The first two of these moments are indeed Markham's $g \to u$ moments (10.15) and (10.18) with the sum over j reduced to a single coordinate and with $\Delta\varepsilon$ in (10.15) given by (10.14a). Markham's (10.14a) includes an extra term $h\nu_{gu}$ which is the energy offset between the minima of the two parabolas. This term drops out for phonon moments.

Markham goes on to replace ε_j^2 in his moment formulas with a generalized Huang-Rhys-Pekar factor S defined by:

$$S \equiv \varepsilon_j^2/2\hbar\omega_{initial}\omega_{final}^2 \qquad (8.35)$$

This causes the S term in the first moment to be $S\hbar\omega_{initial}$ rather than $S_{final}\hbar\omega_{final}$ as occurs in our notation.

Keil [38] has calculated the first three Condon moments for the special case of zero offset, i.e., for $a_{uv} = S_u = S_v = 0$. Our $u \to v$ moments, Eq. (4.112), for zero offset are Keil's moment formulas, his Eq. (4.18), with the notational correspondences:

$$\langle p_v \rangle_{vu} \hbar\omega_v \to M^{(1)} + (\hbar\omega_b - \hbar\omega_a)/2, \qquad \langle (p_v - \langle p_v \rangle_{vu})^N \rangle_{vu} (\hbar\omega_v)^N \to m^{(N)},$$
$$\omega_u \to \omega_a, \qquad \omega_v \to \omega_b, \qquad k_\theta \to (\omega_a^2 - \omega_b^2)/4\omega_a\omega_b,$$
$$\langle 2n + 1 \rangle_u \to \coth(\hbar\omega_a/2kT). \qquad (8.36)$$

Keil's first moment is based on net energy changes measured from the zero-point energy levels. Ours, like Lax's and Markham's, are based on energy changes measured from the minima of the parabolas. This difference in definitions is the source of the correction term involving the zero-point-energy difference in the first of Eq. (8.36). Keil's Eq. (4.11), giving $m^{(2)}$ at $T = 0\,K$, has a misprint: the factor $(\omega_a^2 - \omega_b^2)$ should be squared.

We have also used the double-Mehler-formula transform method of O'Rourke [45] to obtain all three moments as a successful check of algebraic accuracy. O'Rourke's Eq. (52) $G_j(t)$, i.e., Eq. (8.23) above, is the generating function of the phonon moments:

$$G_j(t) = \sum_{k=0}^{\infty} \frac{(i\omega_v)^k}{k!} \langle p_v^k \rangle_{vu} t^k \qquad (8.37)$$

where p_v is given by:

$$p_v \hbar\omega_v = \hbar\omega_v(m + \tfrac{1}{2}) - \hbar\omega_u(n + \tfrac{1}{2}) \qquad (8.38)$$

We have obtained the first three moments as given in Eq. (4.112) by expanding $G_j(t)$ in Eq. (8.23) in its power series in t and equating the coefficients in t^k. When shifted to $\langle p_v \rangle_{vu}$ as center, the expressions in Eq. (4.112) do indeed result.

O'Rourke used his $G_j(t)$ and Eq. (8.37) to calculate the moments for the Huang-Rhys-Pekar model when diagonal quadratic coupling is included. Because ω_j' and ω_j'' are then infinitesimally close, the $\Omega_j^2 \to \Psi_j^2$ revision cited with respect to our Eq. (8.23) does not upset his calculation.

8.7 The Single-Frequency-Multiple-Coordinate Derivative Operator Expressions

8.7.1 Huang and Rhys

The transition rate for the derivative operator in N_{Av} coordinates has been calculated by Huang and Rhys [40]. Huang and Rhys gave as their rate:

8.7 The Single-Frequency-Multiple-Coordinate Derivative Operator Expressions

$$Rate_{p,d/dz} = \frac{h^2}{\omega_l}\{Z^2 L_p + |Y|^2[(\tfrac{1}{2} + \langle m \rangle)^2 + \tfrac{1}{2}\langle 1 + m\rangle\langle m\rangle]W_p$$
$$- |Y|^2 \langle 2m+1\rangle L_p + \tfrac{1}{4}|Y|^2[\langle m\rangle^2 W_{p+2}$$
$$+ \langle 1+m\rangle^2 W_{p-2}\} \tag{8.39}$$

This formula is Eq. (5.6) with $W_{p,d/dz}$ in Eq. (5.6) expressed in the $5 - W_p$ form Eq. (4.85), and with the correspondences:

$$A^2 = h^2 Z^2/\omega_l \qquad \gamma = 1 - |Y|^2/2Z^2 S \tag{8.40}$$

8.7.2 Perlin

Perlin [49, 50] has also calculated the derivative-operator rate, given here with a slight redefinition of his β:

$$Rate_{p,d/dz} = \frac{\pi \omega}{(E_u - E_g)^2} \exp\left(-\frac{a}{2}\coth\beta + p\beta\right)\left[\frac{1 + 2\cosh^2\beta}{4\sinh^2\beta}|b_{ug}|^2 I_p(x)\right.$$
$$+ \left\{\frac{(c_{ug} - |b_{ug}|^2 \coth\beta)}{2\sinh\beta}\right\}\{I_{p-1}(x) + I_{p+1}(x)\}$$
$$\left. + \left\{\frac{|b_{ug}|^2}{8\sinh^2\beta}\right\}\{I_{p-2}(x) + I_{p+2}(x)\}\right], \tag{8.41}$$

where:

$$\beta = \hbar\omega_0/2kT \qquad x = a/2\sinh\beta \tag{8.42}$$

This too is Eq. (5.6) with $W_{p,d/dz}$ in Eq. (5.6) expressed in the $5 - W_p$ form Eq. (4.85), and with the correspondences:

$$W_p = \exp(-\tfrac{1}{2}a\coth\beta + p\beta)I_p(a/2\sinh\beta)$$

$$S = a/2 \qquad A = 2\pi\omega c_{ug}/(E_u - E_g)^2 \qquad \gamma = 1 - \frac{|b_{ug}|^2}{c_{ug}a} \tag{8.43}$$

Perlin's form is not exactly Eq. (8.41) but rather this equation with $p \to -p$. This difference is due, we believe, to Perlin's $p\hbar\omega$ being defined as the energy absorbed by a transition while we define it as the energy emitted by the transition. At any rate, both Perlin and we recognize the relationship $M_{-p,d/dz} = \exp(-p\hbar\omega/kT)M_{p,d/dz}$, and it is clear that the transition which absorbs phonons is slower than the transition which emits phonons. This requires Perlin's p to be a negative number. We have therefore translated Perlin's expression to our more natural definition of p.

8.7.3 Miyakawa and Dexter

Miyakawa and Dexter [51] have also calculated the d/dz operator rate in multiple coordinates. They used a model with a spectrum of phonon energies but, in the end, particularized to a single phonon energy. Thus, their final nuclear factor should have been the single-frequency multiple coordinate rate Eq. (5.6). It was not.

We are uncertain how to read the temperature dependence of their rate expression because Miyakawa and Dexter's coupling parameter g is referred to in text as a constant and yet seems to be defined in their Eq. (3.16) as increasing in temperature as our quantity $S\langle 2m + 1 \rangle$. At 0 K their g is certainly our S, and their formula for $M_{p,d/dz}$, offered for small S, is, according to their Eq. (5.6):

$$M_{p,d/dz}(0\ K) = \left(1 - \frac{p}{S}\right)^2 \frac{S^p}{p!} \tag{8.44}$$

At 0 K, W_p reduces to the Poisson distribution Eq. (4.57), and the exact $M_{p,d/dz}$ rate obtained here and by Huang and Rhys and Perlin is:

$$M_{p,d/dz}(0\ K) = \left(\frac{1}{2S}\right)\{(1 - \gamma)(p - S)^2 + \gamma p\} e^{-S} \frac{S^p}{p!} \tag{8.45}$$

Miyakawa and Dexter's expression is a multiple of the 0 K one dimensional weight $W_{p,d/dz}$, obtained by taking $\gamma = 0$ in Eq. (8.45). We conclude that they were led by their approximations to miss the γ parameter which characterizes the multiple coordinate expression.

8.8 Multiple-Frequency Rates

8.8.1 Perlin's Condon-Operator Distribution

8.8.1.1 The Distribution

Perlin [49] also gave an approximate expression, his Eq. (3.28), for the Condon-operator multiple frequency model when every frequency is a multiple of some lowest frequency called $\bar{\omega}$, namely:

$$I_{ug}(\Omega) \approx |M_{ug}|^2 \left(\frac{2\pi}{\varphi_p''(w_0)}\right)^{1/2} \exp[\varphi_p(w_0)] \tag{8.46}$$

where, his Eq. (3.12):

$$\varphi_p(w) = -pw - \frac{1}{2}\sum_k \Delta_k^2 \coth\left(\frac{\beta_k}{2}\right) + \frac{1}{2}\sum_k \Delta_k^2 \frac{\cosh\left(w\omega_k + \frac{\beta_k}{2}\right)}{\sinh\frac{\beta_k}{2}} \tag{8.47}$$

w_0 is the w at which $\varphi_p' = 0$, namely, the root of his Eq. (3.25):

$$\frac{1}{2}\sum_k \Delta_k^2 \omega_k \left\{\frac{\sinh[w\omega_k + (\beta_k/2)]}{\sinh(\beta_k/2)}\right\} - p = 0 \tag{8.48}$$

8.8 Multiple-Frequency Rates

and where:

$$\beta_k = \hbar\omega_k/kT \tag{8.49}$$

This expression Eq. (8.48) is an approximate form of Eq. (5.29). It is derived by using the generating function for the G_p functions, Perlin's Eq. (3.4), somewhat rewritten here:

$$P(z) = \sum_p \left[\frac{G_p}{p!}\right] z^p \tag{8.50}$$

We give here our understanding of the use of the integral representation of the generating function and of the application of the saddle-point approximation to its evaluation. Perlin missed a factor of 2π and he introduced some confusion over the sign of p. We will elucidate both these issues in our development of the saddle-point approximation to Eq. (5.35). We will also derive our expansion of v_m for small offset in terms of the u_n set, because these are used by Perlin to approximate the generating function.

8.8.1.2 Cauchy's Integral Theorem and its Consequences

The saddle-point approximation, as we will use it here, uses Cauchy's integral theorem, namely:

$$\oint f(z)\, dz = 0 \tag{8.51}$$

where $f(z)$ is analytic at all points within the closed integration path denoted by the circle. The test that $f(z) = u(x, y) + iv(x, y)$ is analytic is the Cauchy-Riemann conditions:

$$\frac{\partial u}{\partial x} = \frac{\partial v}{\partial y} \qquad \frac{\partial u}{\partial y} = -\frac{\partial v}{\partial x} \tag{8.52}$$

For familarity, it may be helpful to work out that $e^z = e^x \cos y + ie^x \sin y$ and z^m are analytic functions, that if $f(z)$ is analytic so also is $1/f(z)$ unless $f(z) = 0$, and that if $f(z)$ and $g(z)$ are analytic so also are $f(z)g(z)$ and $f(z) + g(z)$.

Some consequences of Eq. (8.52) are, for a function and its derivatives:

$$f(z_0) = \frac{1}{2\pi i} \oint \frac{f(z)}{(z - z_0)}\, dz$$

$$f^{(1)}(z_0) = \frac{1}{2\pi i} \oint \frac{f(z)}{(z - z_0)^2}\, dz$$

$$f^{(n)}(z_0) = \frac{n!}{2\pi i} \oint \frac{f(z)}{(z - z_0)^{n+1}}\, dz \tag{8.53}$$

The first of Eq. (8.53) is derived by recognizing that the integrand on the right hand side is non-analytic at $z = z_0$, constructing a path as in Fig. 19 which excludes this point, recognizing that the integration from point A to B cancels

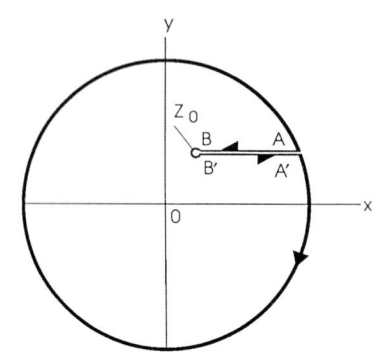

Fig. 19. The integration path for the first of Eq. (8.53)

that from point B' to A', obtaining the integral around the infinitesmal circle surrounding $z = z_0$ by transforming to polar coordinates, and then invoking the Cauchy integral theorem, Eq. (8.51).

We also need the basic equation in the theory of residues:

$$\oint \left[\frac{A_m}{(z-z_0)^m} + \frac{A_{m-1}}{(z-z_0)^{m-1}} + \cdots + \frac{A_1}{(z-z_0)} + g(z) \right] dz = 2\pi i A_1 \quad (8.54)$$

where $g(z)$ is analytic throughout the interior of the closed curve. Equation (8.54) is proved by determining the integrals:

$$\oint \frac{dz}{z^m} = (2\pi i)\delta_{m,-1} \quad (8.55)$$

8.8.1.3 The Saddle Point Approximation

Of all statements of the saddle-point approximation known to us, that of Fuchs and Levin [52] is the clearest. We will quote Theorem 12 of this reference.

"Let $a(p)$ and $b(p)$ be regular in a certain domain, which contains the point p_0, at which $b'(p_0) = 0$, and let L be a path, which lies in this domain and passes through the point p_0 in such a way that on a certain segment of this path, which contains the point p_0, Im $b(p) = \beta_1 = const$. Also, let Re $b(p) <$ Re $b(p_0) - \delta$, $(\delta > 0)$, everywhere on L outside this segment. Then as $\tau \to \infty$:

$$e^{-\tau b(p_0)} \int_L a(p) e^{\tau b(p)} dp \sim \left(\frac{2\pi}{|b''(p_0)|} \right)^{1/2} a(p_0) e^{i\theta} \tau^{-1/2} \quad (8.56)$$

if the integral converges for all sufficiently large τ. In other words:

$$\lim_{\tau \to \infty} \left[\tau^{1/2} \left\{ e^{-\tau b(p_0)} \int_L a(p) e^{\tau b(p)} dp - \left(\frac{2\pi}{|b''(p_0)|} \right)^{1/2} a(p_0) e^{i\theta} \tau^{-1/2} \right\} \right] = 0 \quad (8.57)$$

or:

8.8 Multiple-Frequency Rates

$$\lim_{\tau \to \infty} \left[\tau^{1/2} e^{-\tau b(p_0)} \int_L a(p) e^{\tau b(p)} dp \right] = \left(\frac{2\pi}{|b''(p_0)|} \right)^{1/2} a(p_0) e^{i\theta} \quad (8.58)$$

where $e^{i\theta} = (dp/d\lambda)_{p=p_0}$, and λ is the length of the arc of the path L, measured from the point p_0 in the direction of integration."

The value of θ gives the sign of the integral and allows its sign to depend upon the direction of integration. Perhaps it will be helpful to point out that if a circle is drawn through the saddle point, $e^{i\theta} = i$ if the integration proceeds through the saddle point in a counter-clockwise direction along this circle, and $e^{i\theta} = -i$ if in a clockwise direction.

Equation (8.58) is usually used with $a(p) = const$ and with $\tau = 1$. Fuchs and Levin give in their exposition of the saddle-point method a derivation also of the further terms of this approximation. These added terms are seldom used, because they involve higher derivatives of $b(p)$ and of $a(p)$ which are available often only with severe algebraic or computational difficulty.

8.8.1.4 The Use of the Saddle Point Approximation Here

Let us consider then evaluating $\int \{P(z)/z^{p+1}\} dz$ where $P(z)$ is the generating function, Eq. (8.50). Each G_p in this generating function is a positive number. Because G is normalized, $P(1) < e$ and in general $P(n) < e^n$. Also the derivatives $P^{(n)}(z) = \Sigma_p\{G_{p+n}z^p/n!\}$ along the positive real axis are all positive. Consequently $g(z) = P(z)/z^{n+1}$ has a minimum at some positive $x = x_s = (n+1)P(x_s)/P^{(1)}(x_s)$. It is a consequence of having such a minimum that simultaneously this function has a maximum value for excursions in the direction perpendicular to the real axis, namely, along the imaginary (y) axis, i.e., that x_s is a saddle point.

Let us then use Cauchy's integral expression, Eq. (8.51), along the closed integration path shown in Fig. 20, namely, from point $A = (-R, -\varepsilon)$ counter-clockwise around a circle to point $B = (-R, \varepsilon)$; then at constant $y = \varepsilon$ to the point $C = (x_\mu, \varepsilon)$ at which the saddle-point integration will begin; then, along a path which holds $\text{Re}[g(z)] = const = g_s - \mu$, to point $D = (x_s, y_s)$; then, along the path which holds $\text{Im}[g(z)] = const = 0$, through the saddle point $E = z_s =$

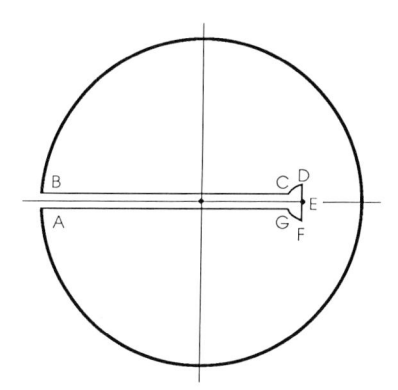

Fig. 20. The integration path for the saddle point approximation

100   8 Contact with the Theoretical Literature

$(x_s, 0)$, at which $g(z_s) = g_s$ and $g''(z_s) = g_s''$ to point $F = (x_s, -y_s)$ where again $\text{Re}[g(z)] = g_s - \mu$; then to point $G = (x_\mu, -\varepsilon)$ along a path which again holds $\text{Re}[g(z)] = const = g_s - \mu$; then to the original point $A = (-R, -\varepsilon)$.

Within this enclosed region, $g(z)$ is analytic and Cauchy's theorem, Eq. (8.51), applies, i.e., the line integral around the total path is zero. The integral along CDEFG is to be approximated by the saddle point method. We are not prepared to find this path for the general $g(z)$, but we will do so for the case that $g(z)$ can be approximated as a quadratic in z:

$$g(z) = \gamma_0 + \gamma_1 z + \frac{\gamma_2}{2} z^2 = u(z) + iv(z)$$

$$u(z) = g_s + \frac{g_s'' x_s^2}{2} - g_s'' x_s x + \frac{g_s''}{2}(x^2 - y^2)$$

$$v(z) = -g_s'' x_s y + g_s'' xy$$

$$z = x + iy \tag{8.59}$$

Equation (8.59) is an analytic function, obeying the Cauchy-Riemann conditions. It has $dg(z)/dz = 0$ at $z_s = (x_s, 0)$ as desired. It also gives $g(z)$ as a real function along the x axis. This quadratic form seems appropriate for our case because we are to obtain a formula involving only g and its curvature. In this case of a quadratic, we have for the various points in the integration path:

$$g(z_s) = g_s$$

$$g''(z_s) = g_s''$$

$$u(x_s, y_s) = u(x_s, -y_s) = u(x_v, \varepsilon) = u(x_v, -\varepsilon) = g_s - \mu$$

$$v(x_s, y) = 0$$

$$y_s = \left(\frac{2\mu}{g_s''}\right)^{1/2}$$

$$x_\mu = x_s - \left(\frac{\varepsilon^2 - 2\mu}{g_s''}\right)^{1/2} \tag{8.60}$$

The integration path portrayed in Fig. 20 is grossly distorted, in that all displacements from the real axis, the $x_s - x_\mu$ displacement, the decrease μ in $\text{Re}[g(x_s, y)]$ along the $x = x_s$ line integral are all infinitesimals, whose values are fixed by the values of the δ and ε, themselves infinitesimals, adopted.

As p decreases towards zero z_s tends towards zero. The generating function Eq. (8.50) does not allow negative p.

We get then, denoting by Γ the counterclockwise path around the origin from A to B and by Λ the path CDEFG:

$$\int_\Gamma \Sigma_p \left[\frac{G_p}{p!}\right] \frac{z^p}{z^{p+1}} dz + \int_\Lambda e^{g(z)} dz = 2\pi i G_p + \int_\Lambda e^{g(z)} dz = 0. \tag{8.61}$$

8.8 Multiple-Frequency Rates

If now we integrate around the large circle, using Eq. (8.54), note that the integral from B to C cancels that from G to A, use the saddle-point method for the integral for Λ, note that the integral through the saddle point is in the clockwise direction and therefore $e^{i\theta} = -i$, and set $\tau = 1$, we get:

$$2\pi i G_p + e^{g(x_s)}(-i)\left(\frac{2\pi}{g''(x_s)}\right)^{1/2} = 0 \qquad (8.62)$$

or:

$$G_p = \frac{e^{g(x_s)}}{\sqrt{2\pi\{g''(x_s)\}}} \qquad (8.63)$$

Equation (8.63) is the saddle-point approximation for G_p. This expression differs from Perlin's expression by a factor of $1/(2\pi)$. There is no confusion that p is a positive number. Perlin must have missed the factor 2π in the first term of Eq. (8.62). The question is left open whether $\tau = 1$ is sufficiently large to obtain close to the asymptotic limiting behavior of the saddle-point integral. We shall find that for the Condon operator, this value is somewhat satisfactory, whereas for the derivative operator it is less so.

8.8.1.5 The integral of dz/z

One might question how paths other than the one given in Fig. 20 will afford the same answer. For example, if we ended the horizontal lines at some point along the negative real axis, say at the points $(-x, \pm\varepsilon)$, then the line integral around this new closed path would indeed equal $2\pi i G_p$, but we would have no method of determining its value. If we ended these lines at some point along the positive real axis, say at $(+x, \pm\varepsilon)$, then the integral around the full path would be zero because we have excluded the pole; the integral around the circle from A to B would equal $2\pi i G_p$; but the integral from (x, ε) to $(x, -\varepsilon)$ would not be infinitesimally close to zero but rather infinitesimally close to $-2\pi i G_p$. Again, we would not be able to evaluate the G_p. This value for the integral across the positive x axis is a consequence of the multiple-valued nature of $\int dz/z$ defining $\ln z$, such that its principal value is found between $0 \le \theta \le 2\pi$; the value $-\theta$ is not within the range of the principal value, and $\int dz/z$ from θ to $-\theta$ is $-2\pi i$.

The path CDEFGC is usable and gives Eq. (8.63). The full line integral is zero; the saddle-point path integral as in Eq. (8.62), and the integral from G to C gives $-2\pi i G_p$. We can then determine G_p as in Eq. (8.63).

Another possible path would follow CDEFGC but with the G to C integral following a counterclockwise circle. However, we are led to the impractical approximation that $G_p \equiv 2G_p$, for, because we now are encircling a pole, we must use Eq. (8.54) directly with two nonzero contributions to the full path integral coming from integrating through the saddle-point and integration around the circle.

A usable path would be to integrate along the saddle point direction out to where $\text{Re}[g(z)] < 0$, then to follow a path holding this constant through the imaginary axis to an intersection at the real negative axis. Then the full saddle point integration path would itself encircle the pole at the origin of the Argand diagram and Eq. (8.54) would apply to give Eq. (8.62).

8.8.1.6 Expansion of v_m for Small Offset

We use Eq. (2.11) for v_m with z_v given by Eq. (2.28), and we use Eq. (2.18) for dv_m/dz_v.

If we expand Eq. (2.11) in a power series in a, we get

$$v_m = v_m|_{a=0} + \left.\frac{dv}{da}\right|_{a=0} \cdot a + \frac{1}{2}\left.\frac{d^2v_m}{da^2}\right|_{a=0} \cdot a^2 \tag{8.64}$$

where the terms are in general the ones involving v below, and for equal force constants are the terms involving u pointed to by arrows:

$$v_m|_{a=0} = \left.\frac{\exp(-(x-a)^2/2x_v)H_m\{(x-a)/x_v\}}{\sqrt{m!2^m\pi^{1/2}x_v}}\right|_{a=0}$$

$$= \frac{\exp(-x^2/2x_v)H_m\{x/x_v\}}{\sqrt{m!2^m\pi^{1/2}x_v}} \to u_m$$

$$\left.\frac{dv_m}{da}\right|_{a=0} = -\left.\frac{dv}{dz_v}\right|_{a=0} = -\left(\frac{m}{2}\right)^{1/2} v_{m-1} + \left(\frac{m+1}{2}\right)^{1/2} v_{k+1}$$

$$\to -\left(\frac{m}{2}\right)^{1/2} u_{m-1} + \left(\frac{m+1}{2}\right)^{1/2} u_{m+1}$$

$$\frac{d^2v_m}{da^2} = \frac{d^2v_m}{dz^2} = \frac{\sqrt{m(m-1)}}{2}v_{m-2} - \left(m+\frac{1}{2}\right)v_m + \frac{\sqrt{(m+1)(m+2)}}{2}v_{m+2}$$

$$\to \frac{\sqrt{m(m-1)}}{2}u_{m-2} - \left(m+\frac{1}{2}\right)u_m + \frac{\sqrt{(m+1)(m+2)}}{2}u_{m+2} \tag{8.65}$$

For Perlin's expression for the derivative operator, which we will address below, we will need:

$$\left.\frac{dv_m}{dz_v}\right|_{a=0} = \left(\frac{m}{2}\right)^{1/2} v_{m-1} - \left(\frac{m+1}{2}\right)^{1/2} v_{m+1} \to \left(\frac{m}{2}\right)^{1/2} u_{m-1}$$

$$- \left(\frac{m+1}{2}\right)^{1/2} u_{m+1}$$

$$\left.\frac{d}{da}\left(\frac{dv_m}{dz}\right)\right|_{a=0} = -\left.\frac{d^2v_m}{dz_v^2}\right|_{a=0} = \frac{-\{m(m-1)\}^{1/2}}{2}v_{m-2} + \left(m+\frac{1}{2}\right)v_m$$

$$- \frac{\{(m+1)(m+2)\}^{1/2}}{2}v_{m+2}$$

$$\to \frac{-\{m(m-1)\}^{1/2}}{2}u_{m-2} + \left(m+\frac{1}{2}\right)u_m$$

$$- \frac{\{(m+1)(m+2)\}^{1/2}}{2}u_{m+2},$$

8.8 Multiple-Frequency Rates

$$\frac{d^2}{da^2}\left(\frac{dv_m}{dz}\right) = \frac{d^3 v_m}{dz^3}$$

$$= \left(\frac{m(m+1)(m-2)}{2^3}\right)^{1/2} v_{m-3} + \left(\frac{m(m+1)(m-1)}{2^3}\right)^{1/2} v_{m-1}$$

$$- \left(m+\frac{1}{2}\right)\left(\frac{m}{2}\right)^{1/2} v_{m-1} + \left(m+\frac{1}{2}\right)\left(\frac{m+1}{2}\right)^{1/2} v_{m+1}$$

$$+ \left(\frac{(m+1)(m+2)(m-2)}{2^3}\right)^{1/2} v_{m+1}$$

$$- \left(\frac{(m+1)(m+2)(m+3)}{2^3}\right)^{1/2} v_{m+3}$$

$$\rightarrow \left(\frac{m(m+1)(m-2)}{2^3}\right)^{1/2} u_{m-3} + \left(\frac{m(m+1)(m-1)}{2^3}\right)^{1/2} u_{m-1}$$

$$- \left(m+\frac{1}{2}\right)\left(\frac{m}{2}\right)^{1/2} u_{m-1} + \left(m+\frac{1}{2}\right)\left(\frac{m+1}{2}\right)^{1/2} u_{m+1}$$

$$+ \left(\frac{(m+1)(m+2)(m-2)}{2^3}\right)^{1/2} u_{m+1}$$

$$- \left(\frac{(m+1)(m+2)(m+3)}{2^3}\right)^{1/2} u_{m+3} \quad (8.66)$$

From Eqs. (8.65) and (8.66) we find:

$$\langle u_n | v_n \rangle \approx 1 - \frac{a^2}{2}(n+\tfrac{1}{2})$$

$$\langle u_{n+1} | v_n \rangle \approx a\left(\frac{n+1}{2}\right)^{1/2}$$

$$\langle u_{n-1} | v_n \rangle \approx -a\left(\frac{n}{2}\right)^{1/2}$$

$$\langle u_n | \frac{d}{dz_v} | v_n \rangle \approx a(n+\tfrac{1}{2})$$

$$\langle u_{n+1} | \frac{d}{dz_v} | v_n \rangle \approx -\left(\frac{n+1}{2}\right)^{1/2}$$

$$\langle u_{n-1} | \frac{d}{dz_v} | v_n \rangle \approx \left(\frac{n}{2}\right)^{1/2} \quad (8.67)$$

Equations (8.67) are Perlin's Equations (3.8) and (3.44) with the correspondence $\Delta_k \rightarrow -a$, except for the sign for the fifth of these, which we believe to be a misprint, since we cannot trace its consequences further into Perlin's exposition.

8.8.1.7 Perlin's Derivation

Perlin started from the expression for the probability per second of an optical absorption, averaged over the initial distribution and summed over the final states, his Eq. (2.8), namely:

$$l_{u,g}(\Omega) = |M_{ug}|^2 Av(n) \sum_{(n')} \prod_k |\langle n'_k | n_k \rangle|^2 , \qquad (8.68)$$

which is his notation for $|M_{ug}|^2 G_p^k$ as given in Eq. (5.29). He then approximated all the $\langle n'_k | n_k \rangle$ in the generating function, Eq. (8.50), via his Eq. (3.8), our Eq. (8.64), used $1 - x \to e^{-x}$ with x all the non-unity terms accumulating in the coefficients of this sum, used:

$$\frac{\cosh\left(\omega j_k + \frac{\beta_k}{2}\right)}{\sinh\left(\frac{\beta_k}{2}\right)} = \frac{e^{\omega j_k + \beta_k} + e^{-\omega j_k}}{e^{\beta_k} - 1} \qquad (8.69)$$

and obtained the exponential expression for $P(z)$, his Eq. (3.9):

$$P(z) = \exp\left\{-\frac{1}{2} \sum_k \Delta_k^2 (n + \tfrac{1}{2}) + \frac{1}{2} \sum_k \Delta_k^2 [(n_k + 1) z^{j_k} + n_k z^{-j_k}]\right\} \qquad (8.70)$$

The derivatives of $P(z)$ evaluated at $z = 0$ give the desired nuclear factors, and these derivatives are used in the form Eq. (8.54) with $z_0 = 0$. Perlin then averages thermally over the initial states, uses:

$$z = e^w \qquad dz = e^w dw \qquad (8.71)$$

and then uses the Cauchy theorem to get an integral expression for $l_{ug}(\Omega)$, his Eqs. (3.6) and (3.11):

$$l_{ug}(\Omega) = |M_{ug}|^2 Av(n) \frac{1}{2\pi i} \oint \frac{P(z) dz}{z^{p+1}} = \frac{|M_{ug}|^2}{2\pi i} \oint dw \exp[\varphi_p(w)] \qquad (8.72)$$

In Eq. (8.72) φ is the φ of Eq. (8.47) and the line integral in the second expression in Eq. (8.72) is the second integral of Eq. (8.61) for this case. The thermal averages merely replace $n, m \to \bar{n}, \bar{m} \equiv \langle n \rangle, \langle m \rangle$. He finally applies the saddle point approximation, his Eq. (3.28) and our (8.63), to Eq. (8.72) to obtain his Eq. (3.28), our Eq. (8.46).

8.8.2 The Correspondence between Perlin's and Our Multiple Frequency Expression

As a vehicle for comparison between Perlin and our expression for the multiple frequency Condon operator rate, namely, D_p^k of Eqs. (5.32)–(5.35), we have used as $\bar{\omega}$ the ω_1 of the K-equally-spaced-frequencies model of Sect. 5.3. This model is algebraically allowed for these Perlin expressions.

There are some preliminary minor details to be taken care of. The $\sqrt{2\pi}$ factor

8.8 Multiple-Frequency Rates

in Perlin's Eq. (3.28) shown in the numerator should be, as previously stated, in the denominator. Moreover, in the form given by Perlin, his underlying expressions (3.27), (3.25), and (3.12) are dimensionally uninterpretable. The problem is that, in his Eq. (3.4), the $P(z)$ generating function is given as a series in z^{ω_x} where ω_x has the dimensions $\sec^{-1}$. If, instead, one uses z^{j_x} where $j_x = \omega_x/\bar{\omega}$, one obtains the usable expression Eq. (8.70). The correspondence here is between Perlin's $\frac{1}{2}\Lambda_x$ and our $S_{i,j}$. To evaluate D_p^{K*}, as for D_p^k itself, it is sufficient to specify S, $\hbar\omega_K$, and kT, using Eqs. (5.54) and (5.57). Introducing the DDESA model parameters into these expressions, we obtain for the Condon operator:

$$D_p^{K*} \equiv \frac{l_{ug}(\Omega)}{|M_{ug}|^2} = \frac{1}{\sqrt{2\pi\varphi_p''(w_0)}} \exp(\varphi_p(w_0)) \qquad (8.73)$$

where w_0 is the root of his Eq. (3.51) for this case, namely:

$$\sum_{k=1}^{K} kS_k \left\{ \frac{\sinh[kw + (\beta_k/2)]}{\sinh(\beta_k/2)} \right\} - p = 0 \qquad (8.74)$$

and where:

$$\varphi_p = -pw + \sum_k S_k \left\{ \cosh[kw + (\beta_k/2)] - \frac{\cosh(\beta_k/2)}{\sinh(\beta_k/2)} \right\} \qquad (8.75)$$

where:

$$\beta_k = \hbar\omega_k/kT \qquad \Omega = p\hbar\omega \qquad (8.76)$$

We have programmed the evaluation of Eq. (8.73) by numerically solving Eq. (8.74), evaluating Eq. (8.75) at the root value found for w_0, and then evaluating the second derivative of Eq. (8.75) at this w value.

Table 5 gives an illustration of the D_p^{K*}, and the exact D_p^K distributions for $K = 3$. For p large enough, the two distributions differ only by a percent. For p near zero, larger deviations occur. At small kT, Eq. (8.73) becomes unusable. In this table, values are given for $K = 3$, $S = 6.25$, $\hbar\omega_K = 400 \text{ cm}^{-1}$, which we shall use later for illustrating nonradiative rates in luminescence centers. The values listed as $a - b$ are to be read as $a \cdot 10^{-b}$.

Note that our values satisfy the expected relations:

$$D_{-p}^K = \exp(-p\hbar\omega/kT)D_p^K \qquad (8.77)$$

8.8.3 Perlin's Multiple-Coordinate Derivative Operator Expression

Perlin's expression, his Eq. (3.55), for the nuclear factor for the multiple-frequency, d/dz operator model is:

$$P_{ug} = \frac{1}{\sqrt{2\pi\varphi_p''(w_0)}} \frac{F(w_0)}{h^2\bar{\omega}} \exp[\varphi_p(w_0)] \qquad (8.78)$$

where w_0 is the root of his Eq. (3.25), our Eq. (8.74) above, with $\Omega = 0$, that is, $p = p_0$, and:

Table 5. Comparison of Perlin's and our Condon expressions

p	D_p^3	D_p^{3*}	D_p^{-3}/D_p^3	$\exp(-p\hbar\omega_K/3kT)$
$kT = 200$ cm^{-1}				
−20	7.567-8			
−10	3.772-5			
0	3.006-3	2.954-3		
10	2.964-2	2.935-2	1.273-3	1.273-3
20	4.672-2	4.665-2	1.620-6	1.620-6
30	1.808-2	1.811-2		
40	2.378-3	2.384-3		
50	1.322-4	1.326-4		
60	3.620-6	3.630-6		
70	5.437-8	5.451-8		
$kT = 50$ cm^{-1}				
−10	7.725-14			
−5	1.038-8			
0	2.817-4	1.64-4		
5	6.373-3	6.366-3	1.620-6	1.620-6
10	2.945-2	2.996-2	2.623-12	2.623-12
15	5.480-2	5.559-2		
20	5.469-2	5.529-2		
25	3.41-2	3.443-2		
30	1.464-2	1.475-2		
35	4.595-3	4.624-3		
$kT = 10$ cm^{-1}				
0	2.8434-4			
1	2.9022-4			
2	1.2015-3			
3	2.1037-3			
4	3.5111-3			

$$F(w) = \left| \sum_k L_k \Delta_k \left\{ \frac{\cosh\left(w\omega_k + \frac{\beta_k}{2}\right) - \cosh\frac{\beta_k}{2}}{2\sinh\frac{\beta_k}{2}} \right\} \right|^2$$

$$+ \frac{1}{2}\sum_k |L_k|^2 \left\{ \frac{\cosh\left(w\omega_k + \frac{\beta_k}{2}\right)}{\sinh\frac{\beta_k}{2}} \right\} \quad (8.79)$$

This uses the $a(p) = F(w)$ form of the saddle-point approximation Eq. (8.58) with $\tau = 1$. The correction explained above from ω_k to $\omega_k/\bar{\omega}$ is needed. In addition,

8.8 Multiple-Frequency Rates

the confusion concerning the sign of p referred to with respect to Eq. (8.63) was resolved here by replacing $-p \to p$. As we have also stated with respect to Eq. (8.67), we do not agree with the sign in the second of Perlin's Eq. (3.44). From the fifth of Eq. (8.67), this sign should be negative. Nevertheless, we cannot trace the consequences of this error and believe it to be a misprint. Finally, we will show that $\tau = 1$ is on the verge of inadequacy for this case.

8.8.4 The Correspondence between Perlin's and Our Multiple Frequency Derivative Expression

For the DDESA model, following Perlin's instructions, we now need the additional equations:

$$D^{K^*}_{p,d/dz} = \frac{1}{\sqrt{2\pi \varphi_p''(w_0)}} \frac{F(w_0)}{h^2 \bar{\omega}} \exp[\varphi_p(w_0)] \tag{8.80}$$

where:

$$\frac{F(w)}{h^2 \bar{\omega}} = \frac{1}{2\pi} \left\{ \sum_{k=1}^{K} \sqrt{k} \Lambda_k \left[\frac{\cosh[kw + (\beta_k/2)] - \cosh(\beta_k/2)}{\sinh(\beta_k/2)} \right] \right\}^2$$
$$+ \frac{1}{2} \left\{ \sum_{k=1}^{K} k A_k^2 \left[\frac{\cosh[kw + (\beta_k/2)]}{\sinh(\beta_k/2)} \right] \right\} \tag{8.81}$$

where:

$$\Lambda_k = \sum_j A_{k,j} S_{k,j}^{1/2} \qquad \beta_k = \hbar \omega_k / kT \tag{8.82}$$

For DDESA-model assumptions, Λ_k is determined by the requirement that γ_k in Eq. (5.24) be zero, that is:

$$\Lambda_k = \sqrt{A_k^2 S_k} \tag{8.83}$$

and S_k and A_k^2 are determined through Eq. (5.54). We have computed Eq. (8.80) for the same cases used in the Condon-operator demonstration of correspondences.

The results are shown in Table 6. They are not as close as are Perlin's Condon-operator expression to its exact counterpart. It is possible that going to larger τ values or including higher-order terms in the asymptotic expansion might improve this agreement, at the cost of severe algebraic and computational complexity. It is also possible that minimizing the function $\varphi_{p_0}(w) + \ln F(w)$ would lead to a different w_0 than Perlin used and might give a closer approximation. Whatever the case, it is not easy to judge the adequacy of the saddle-point method approximate results.

8.8.5 Mostoller, Ganguly, and Wood

Mostoller, Ganguly, and Wood [53] have given the 0 K Condon operator nuclear factors. Except for a factor of 2π their Eq. (29) and our Eq. (5.40) are equivalent.

Table 6. Comparison of Perlin's and our derivative expressions

p	$D^3_{p,d/dz}$	$D^{3*}_{p,d/dz}$	$D^{-3}_{p,d/dz}/D^3_{p,d/dz}$	$\exp(-p\hbar\omega_K/3kT)$
$kT = 200$ cm^{-1}				
−20	3.1009-8			
−10	3.660-5			
0	4.263-3	2.49-3		
10	2.876-2	3.37-2	1.273-3	1.273-3
20	2.445-2	8.93-2	1.620-6	1.620-6
30	2.007-2	5.29-2		
40	7.536-3	9.63-3		
50	9.076-4	6.93-4		
60	4.434-5	2.35-5		
70	1.052-6	4.18-7		
$kT = 50$ cm^{-1}				
−10	6.158-14			
−5	2.045-8			
0	1.111-3	−1.79-4		
5	1.263-2	−3.94-3	1.620-6	1.620-6
10	2.348-2	6.97-4	2.623-12	2.623-12
15	1.004-2	3.74-2		
20	3.966-2	7.33-2		
25	1.512-2	6.80-2		
30	1.873-2	3.87-2		
35	1.182-2	1.51-2		

8.9 Energy Transfer

8.9.1 Förster and Dexter

Förster [54] and Dexter [55] have described the energy transfer rate as proportional to the overlap of the donor emission and acceptor absorption bands. This is precisely Eq. (6.3). It is necessary to show that for:

$$p_d + p_a = p_{da} \tag{8.84}$$

the emission frequency and absorption frequency are indeed identical. From Eqs. (4.5) and (4.6) this is seen to be true, since:

$$p_d \hbar\omega = h\nu_{zp,d^+d} - h\nu_{d,p_d}$$
$$p_a \hbar\omega = h\nu_{a,p_a} - h\nu_{3p,a^+a} \tag{8.85}$$

Adding the two Eq. (8.85), transferring the zero-point energies to the left-hand side, and imposing the Eq. (6.1) constraint gives the two frequencies identical.

8.9.2 Miyakawa and Dexter

Miyakawa and Dexter [51] give as the energy transfer rate their Eq. (6.15). At T = 0 K this rate becomes the Poisson distribution:

$$W_{p,MD} = \exp(-S_{da})S_{da}^p/p! \tag{8.86}$$

This is the $T = 0 \, K$ form of Eq. (6.5), invoking Eq. (4.57). As discussed with respect to Miyakawa and Dexter's handling of single-center distributions, e.g., Eq. (8.45) above, we are unable to understand their temperature dependences and constrain ourselves to discussing their rates for T = 0 K.

9 Representative Luminescence Centers

9.1 Equal- and Unequal-Force-Constants Bandshapes and Nonradiative Transitions

9.1.1 Bandshapes

As one searches for the most perceptive fit of these model functions to all experimental data, one often wishes to describe the same absorption and emission spectra both with equal and with unequal force constants. Such descriptions might keep the sum of the first moments and the product of the second moments constant, i.e., using the first two of Eqs. (4.109) and (4.112) truncated to only their first terms and evaluated at 0 K, would have

$$S_v \hbar \omega_v + S_u \hbar \omega_u = 2 S \hbar \omega$$

$$\left[\frac{\omega_v}{\omega_u} S_v (\hbar \omega_v)^2 \right] \left[\frac{\omega_u}{\omega_v} S_u (\hbar \omega_u)^2 \right] = [S(\hbar \omega)^2]^2 \tag{9.1}$$

We use the $(\theta, a_{uv}) \leftrightarrow (S_u, S_v, \omega_u, \omega_v)$ relationships in Eqs. (3.10), (3.30), and (3.31) to get:

$$\hbar \omega_u = \hbar \omega \left[\frac{1 + \tan^4 \theta}{2 \tan^3 \theta} \right]$$

$$\hbar \omega_v = \hbar \omega_u \tan^2 \theta$$

$$S_u = S \left[\frac{4 \tan^3 \theta}{(1 + \tan^4 \theta)^2} \right]$$

$$S_v = S \left[\frac{4 \tan^5 \theta}{(1 + \tan^4 \theta)^2} \right]$$

$$a_{uv}(\theta) = a_{uv}(\theta = 45°) \left[\frac{2 \tan^3 \theta (1 + \tan^2 \theta)}{(1 + \tan^4 \theta)^2} \right]^{1/2}$$

$$h\nu_{zp, vu}(\theta) - h\nu_{zp, vu}(\theta = 45°) = S_u \hbar \omega_u - S \hbar \omega + \tfrac{1}{2}(\hbar \omega_v - \hbar \omega_u) \tag{9.2}$$

Figure 21 shows a typical example of these similar centers. Both equal-force constant curves and $\theta = 42°$ curves are drawn to follow Eqs. (9.1), (9.2). The offset parameters are $a_{uv} = 6$ for $\theta = 45°$ and $a_{uv} = 5.886$ for $\theta = 42°$. The phonon

9.1 Equal- and Unequal-Force-Constants Bandshapes

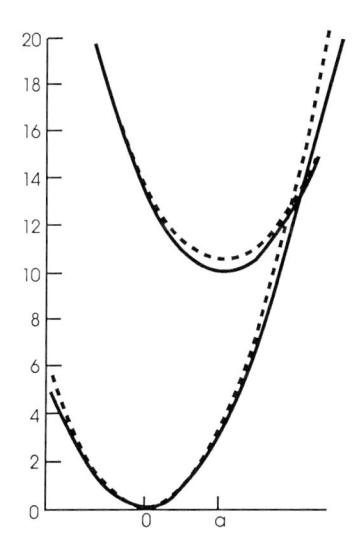

Fig. 21. Single-configurational models with similar optical bands

energies are $\hbar\omega = 400$ cm^{-1} for $\theta = 45°$ and $\hbar\omega_u = 1.135\,\hbar\omega = 454.1$ cm^{-1}, $\hbar\omega_v = 0.920\,\hbar\omega = 368.1$ cm^{-1} for $\theta = 42°$. Thus $S = 9$ for $\theta = 45°$ and $S_u = 9.568$, $S_v = 7.757$ for $\theta = 42°$. Through these choices of the parameters, the sum of the first moments is thus maintained at $18\,\hbar\omega$ and the product of the second moments is maintained at $81\,(\hbar\omega)^4$. The last of Eq. (9.2) keeps the energy difference between the two parabolas constant at the upper-state minimum.

The zero-phonon line position is shown in the figure at 10 000 for $\theta = 45°$ and at $h\nu_{zp,vu}(\theta = 42°) - h\nu_{zp,vu}(\theta = 45°) = 702$ cm^{-1} for $\theta = 42°$, or $h\nu_{zp,vu}(\theta = 42°) = 10\,702$ cm$^{-1} = 23.57\,\hbar\omega_u$.

Figure 22 gives the U_{p_u} and V_{p_v} band shapes at 0 K for $\theta = 45°$ (points) and for $\theta = 42°$ (curves). When, as here, the sum of the first moments and the product of the second moments are maintained equal, then for $\theta = 42°$ the absorption is narrower and the emission is broader, and the zero-phonon energy is shifted slightly to higher energy. The small differences between these lineshapes are often

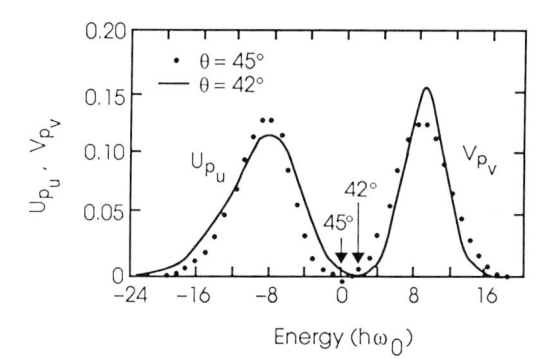

Fig. 22. Band shapes for equal and unequal force constants cases

of secondary importance compared to the differences predicted for other properties such as the thermal quenching behavior or the feeding fractions from an offset state into small-offset states.

We will take up the feeding fraction behavior with respect to oxysulfides: Eu systems in Chap. 10.

9.1.2 Nonradiative Rates

The $v \to u$ nonradiative transitions for these two cases, as described using the Condon operator, are shown in Table 7 for $\theta = 42°$ and in Table 8 for $\theta = 45°$.

Table 7 gives in its second and third columns the $U^2_{23+m,m}$ and $U^2_{24+m,m}$ matrix elements for the $\theta = 42°$, $a_{uv} = 5.886$ case. These in turn were calculated from the A^2_{nm} matrix using Eq. (4.31). The U matrix elements are then shown thermally weighted by the initial-state Boltzmann factor, normalized, according to Eq. (4.25). We use $\hbar\omega_v = 368.1$ cm^{-1} to connect T with $kT/\hbar\omega_v$. These weightings are shown in pairs in the remaining columns at the three temperatures specified. The sums, Eq. (4.25), are given below for both p_u indices, and the logarithmic interpolations are shown for $p_u = 23.627$, as needed according to the discussion on page 111. Table 8 gives the corresponding equal-force-constants case, for $\theta = 45°$, $a_{uv} = 6$. Only the $A^2_{25+m,m}$ elements enter into Eq. (4.11), the terms of which are given in the table. The frequency is $\hbar\omega = 400$ cm^{-1}. The sensitivity of the nonradiative rate to the θ can be demonstrated by considering a 1 ms. radiative lifetime. Although, as seen in Fig. 22, these two centers have very similar optical band shapes and positions, for $\theta = 45°$ the emission would be with

Table 7. Outside crossover for unequal force constants case

m	$U^2_{23+m,m}$	$U^2_{24+m,m}$	\multicolumn{2}{c}{$10^3(1-r_v)r_v^m U^2_{23+m,m}$, $10^3(1-r_v)r_v^m U^2_{24+m,m}$}				
			0 K		238.4 K		503.3 K	
0	0.0004003	0.0002006	0.4003	0.2006	0.3569	0.1789	0.2606	0.1305
1	0.003321	0.001847			0.3209	0.1784	0.7545	0.4196
2	0.01261	0.007848			0.1320	0.0822	1.0000	0.6223
3	0.02917	0.02059			0.0331	0.0234	0.8073	0.5699
4	0.04514	0.03703			0.0056	0.0046	0.4360	0.3577
5	0.04706	0.04699			0.0006	0.0006	0.1587	0.1584
6	0.02929	0.03973					0.0345	0.0467
7	0.007418	0.01907					0.0030	0.0078
8	0.001916	0.002804					0.0003	0.0004
9	0.01449	0.004427					0.0007	0.0002
10	0.02657	0.1866					0.0005	0.0003
	$10^3 U_{23}$, $10^3 U_{24}$		0.4003	0.2006	0.8492	0.4681	3.4565	2.3142
	$10^3 U_{23.57}$		0.302		0.696		2.372	
	$\langle m \rangle_v$		0		0.1215		0.5361	
	$kT/\hbar\omega_v$		0		.45		.95	

9.2 One- and N_{Av}-Dimensional Bandshapes

Table 8. Outside crossover for equal force constants case

m	$A^2_{25+m,m}$	$10^3(1-r_0)r_0^m A^2_{25+m,m}$		
		0 K	259 K	546.9 K
0	0.000005711	0.0057	0.0051	0.0037
1	0.00006349		0.0061	0.0144
2	0.0003589		0.0038	0.0285
3	0.001370		0.0015	0.0379
4	0.003960		0.0005	0.0383
5	0.009197		0.0001	0.0310
6	0.01778			0.0209
7	0.02924			0.0120
8	0.04137			0.0059
9	0.05058			0.0025
10	0.05315			0.0009
11	0.04719			0.0003
12	0.03398			0.0001
	$10^3 W_{25}$	0.0057	0.0172	0.1964
	$\langle m \rangle_v$	0	.1215	.5361
	$kT/\hbar\omega_v$	0	.45	.95

$\eta \approx 0.84$ at 546.9 K, whereas for $\theta = 42°$ the emission would be at only $\eta \approx 0.80$ even at 0 K and would fall to $\eta \approx 0.28$ at 503.3 K. This sensitivity to θ is typical of outside crossovers and tunnelling crossovers.

9.2 One- and N_{Av}-Dimensional Bandshapes

Because W_p describes both the one-dimensional and the N_{Av}-dimensional single-frequency (Einstein model) bandshapes, no appeal to experiment can be expected to differentiate between these two models.

The multiple-frequency model does lead to different, but very similar, bandshapes. We will give examples of small, intermediate, and large offsets in Tables 9–11. In Table 9 $W_p(S, \langle m \rangle)$ are given for the small offset case, namely, $S = 0.05$, $\hbar\omega = 300$ cm^{-1}, at three temperatures as noted in the table. We also show for the DDESA-model $D_p^K(S, \omega_K, kT)$ of Eq. (5.35) with $K = 6$, $S = 0.05$, $\hbar\omega_K = 300$ cm^{-1}, and with the S apportioned according to Eq. (5.55) among the 6 frequencies. We also show $D_p^K(S, \omega_K, kT)$ for $K = 3$, $S = 0.05$, $\hbar\omega_K = 300$ cm^{-1}, and with the S apportioned according to Eq. (5.55) among the 3 frequencies. These distributions are labeled in the table.

In order to relate these three distributions, we show also in this table groupings of the D_p^K values so that the grouped D_p^6 intensities mimic the D_p^3 distribution,

Table 9. Comparison of one- and multi-dimensional bandshapes for small offset

p	D_{p-3}^6	D_{p-2}^6	D_{p-1}^6	D_p^6	D_{p+1}^6	D_{p+2}^6	p	$\Sigma_{p-1}^{6,3}$	$\Sigma_p^{6,3}$	$\Sigma_{p+1}^{6,3}$	D_{p-1}^3	D_p^3	D_{p+1}^3	p	$\Sigma_p^{3,1}$	W_p
\multicolumn{17}{l}{$kT = 75$ cm^{-1}, $kT/\hbar\omega = 0.25$}
−6	0	0	0	3	8	10	−3	0	7	20	0	5	21	−1	26	9
0	12	13	10	9247	20	48	0	24	9262	103	27	9263	104	0	9395	9495
6	90	148	223	145	3	4	3	304	258	8	296	263	8	1	567	484
12	5	5	4	1	0	0	6	10	3	0	8	4	0	2	12	12
\multicolumn{17}{l}{$kT = 225$ cm^{-1}, $kT/\hbar\omega = 0.75$}
−12	0	0	0	0	1	1	−6	0	1	2	0	5	2	−2	3	2
−6	1	2	2	48	98	90	−3	3	98	178	3	87	180	−1	270	165
0	78	62	42	8529	53	96	0	122	8577	198	133	8574	208	0	8915	9189
6	152	218	297	181	8	10	3	443	333	20	438	330	19	1	787	624
12	11	10	6	2	0	0	6	18	5	0	17	7	1	2	25	21
18	0	0	0	0	0	0	9	0	0	0	8	0	0	3	0	1
\multicolumn{17}{l}{$kT = 450$ cm^{-1}, $kT/\hbar\omega = 1.50$}
−12	0	0	0	1	4	7	−6	0	3	14	0	4	12	−2	16	12
−6	9	10	9	126	236	204	−3	19	248	406	19	231	410	−1	660	453
0	168	130	92	7412	103	163	0	260	7510	332	281	7503	351	0	8135	8606
6	235	318	412	246	21	24	3	641	463	46	639	450	45	1	1134	882
12	24	21	14	5	1	1	6	40	13	2	37	14	2	2	53	45
18	1	1	0	0	0	0	9	2	0	0	1	0	0	3	1	2

9.2 One- and N_{Av}-Dimensional Bandshapes

and the grouped D_p^3 intensities mimic the W_p distribution. These groupings are defined and labeled in the table as:

$$\Sigma_p^{6,3} = D_{2p}^6 + \tfrac{1}{2}(D_{2p-1}^6 + D_{2p+1}^6)$$
$$\Sigma_p^{3,1} = D_{3p-1}^3 + D_{3p}^3 + D_{3p+1}^3 \tag{9.3}$$

to mimic D_p^3 and W_p, respectively. Such groupings preserve the normalization of all distributions. Values are given in units of 10^{-4} in Table 9. One sees first that $\Sigma_p^{3,1}$ and W_p are very similar but not identical at all three temperatures. See the last two columns of the table. The $\Sigma_p^{3,1}$ distribution is slightly broader. Likewise the $\Sigma_p^{6,3}$ and D_p^3 distributions are very similar: see the two triplets of columns 9–11 and 12–14. Finally, the D_p^6 distribution itself, in the 2nd through 7th columns, shows oscillatory fine structure, especially on the high p side of D_0^6. Table 10 shows the same distributions at two temperatures for the intermediate offset case, $S = 3$, $\hbar\omega_K = 400$ cm^{-1}. The values of D_p^6, D_p^3 and $\Sigma_p^{6,3}$ are in units of 10^{-5}, and the $\Sigma_p^{3,1}$ and W_p are in units of 10^{-4}. The same behavior as in the small-offset case is visible here, with a trend toward even closer tracking of $\Sigma_p^{3,1}$ and W_p, and of $\Sigma_p^{6,3}$ and D_p^3. The fine structure in D_p^6 is confined more to lower temperatures and to lower p values.

Table 11 shows some of these distributions for the large offset case $S = 28$, $\hbar\omega_K = 170$ cm^{-1}, and $kT = 85$ cm^{-1}. The values are in units of 10^{-4}. This case is shown only for one low temperature. Here the differentiation of $\Sigma_p^{3,1}$ and W_p is virtually impossible and all structure has disappeared from the largest two orders of the D_p^3 distribution.

Figure 23 shows D_p^{12}, $\Sigma_p^{12,1}$, and W_p distributions plotted logarithmically against p for the small-offset case, $S = 0.25$, $\hbar\omega = 400$ cm^{-1}, $kT = 10$ cm^{-1}. The $\Sigma_p^{12,1}$ is defined as:

$$\Sigma_p^{12,1} = \sum_{j=-5}^{5} D_{12p+j}^{12} + \frac{1}{2}(D_{12p+6}^{12} + D_{12p-6}^{12}) \tag{9.4}$$

The line curve in this figure is for an alternative apportioning of the S among the S_k, namely, $S_k = \bar{g}_k KS/k$. It is included to show that the fine structure predicted is somewhat insensitive to the detailed model assumptions.

Figure 24 shows the same distributions plotted logarithmically for the large offset case, $S = 8$, $\hbar\omega = 400$ cm^{-1}, $kT = 10$ cm^{-1}.

Once again these D distributions are all very similar but not identical to the W_p distributions. One can see in the figures and in the tables that the $\Sigma_p^{n,1}$ distributions are somewhat narrower in the wings, somewhat broader at the peak, than their corresponding W_p distributions, but the differences are not striking.

The fine structure is apparent at low p values for the small-offset case and even for the large-offset case, although for large offset it is submerged by three orders from the peak amplitude.

Table 10. Comparison of one- and multi-dimensional bandshapes for moderate offset

p	D^6_{p-3}	D^6_{p-2}	D^6_{p-1}	D^6_p	D^6_{p+1}	D^6_{p+2}	p	$\Sigma^{6,3}_{p-1}$	$\Sigma^{6,3}_p$	$\Sigma^{6,3}_{p+1}$	D^3_{p-1}	D^3_p	D^3_{p+1}	p	$\Sigma^{3,1}_p$	W_p
\multicolumn{17}{l}{$kT = 10$ cm^{-1}, $kT/\hbar\omega = 0.025$}
0	0	0	0	1866	116	425	0	0	1924	965	0	1985	973	0	296	498
6	964	1767	2908	2499	1592	2614	3	3703	4749	5218	3769	5047	5173	1	1399	1494
12	3615	4256	4071	3473	3707	4463	6	8099	7362	8684	7830	7873	8191	2	2389	2240
18	4734	4490	4044	3892	4051	4070	9	8879	7940	7987	8850	8051	7599	3	2450	2240
24	3783	3395	3115	2973	2825	2571	12	6844	5943	5119	6927	5859	5012	4	1780	1680
30	2270	2013	1820	1650	1463	1269	15	4058	3292	2547	4115	3257	2566	5	994	1008
36	1095	952	831	718	610	515	18	1915	1438	1037	1954	1457	1072	6	448	504
42	435	368	310	258	213	176	21	741	520	355	769	543	378	7	168	216
48	145	120	98	79	64	52	24	241	160	105	258	173	115	8	55	81
54	42	33	27	21	17	13	27	68	43	26	75	48	31	9	15	27
60	10	8	6	5	4	3	30	16	10	6	19	12	7	10	4	8
66	2	2	1	1	1	1	33	3	2	1	4	3	2	11	1	2
72	0						36				1	1	0	12	0	1

9.2 One- and N_{Av}-Dimensional Bandshapes

Table 10. Continued
$kT = 200 \text{ cm}^{-1}, kT/\hbar\omega = 0.50$

	0	0	1	1	1	2	−12	1	2	4	1	2	4	−4	1	1
−24	0	0	1	1	1	2	−12	1	2	4	1	2	4	−4	1	1
−18	3	4	6	9	12	17	−9	9	18	35	8	17	34	−3	6	5
−12	24	33	44	60	81	109	−6	67	122	222	654	120	216	−2	40	36
−6	142	182	231	303	390	475	−3	368	613	958	366	608	961	−1	194	189
0	576	696	839	1089	1171	1355	0	1403	2094	2723	1416	2096	2759	0	627	667
6	1566	1803	2065	2237	2381	2620	3	3618	4460	5235	3645	4496	5273	1	1341	1397
12	2849	3042	3169	3253	3368	3485	6	6051	6522	6941	6048	6563	6913	2	1952	1952
18	3544	3546	3513	3479	3438	3362	9	7074	6954	6705	7057	6937	6649	3	2064	2005
24	3247	3111	2970	2827	2671	2500	12	6220	5648	4997	6190	5605	4961	4	1675	1614
30	2324	2151	1983	1818	1656	1498	15	4305	3637	3001	4283	3617	2992	5	1089	1065
36	1349	1211	1081	960	848	746	18	2426	1924	1497	2424	1927	1505	6	586	593
42	653	570	494	427	368	315	21	1143	858	634	1154	870	646	7	267	286
48	269	229	194	164	138	151	24	460	330	268	472	340	241	8	105	121
54	96	80	66	55	45	37	27	161	111	74	169	116	79	9	36	46
60	30	25	20	16	13	11	30	50	33	22	53	35	23	10	12	16
66	9	7	5	4	3	3	33	14	8	5	15	16	6	11	3	5
72	2	2	1	1	1	1	36	4	2	1	4	2	1	12	1	1
78	0						39				1	1	0	13	0	0

Table 11. Comparison of one- and multi-dimensional bandshapes for large offset

p	D^3_{p-1}	D^3_p	D^3_{p+1}	p	$\Sigma^{3,1}_p$	W_p
27	1	1	1	9	3	3
30	1	1	2	10	4	5
33	2	3	3	11	8	9
36	4	4	5	12	13	16
39	6	7	9	13	22	26
42	10	12	14	14	36	40
45	16	18	21	15	55	61
48	23	27	30	16	80	88
51	34	38	43	17	115	125
54	48	53	59	18	160	170
57	65	71	78	19	214	224
60	85	92	99	20	276	285
63	107	115	123	21	345	352
66	131	139	147	22	417	421
69	155	163	171	23	489	488
72	178	185	192	24	555	549
75	198	204	209	25	611	599
78	213	217	221	26	651	636
81	223	225	226	27	674	656
84	227	227	226	28	680	658
87	224	221	218	29	663	642
90	215	210	205	30	630	611
93	200	194	188	31	582	566
96	181	174	167	32	522	511
99	160	153	145	33	458	450
102	137	130	122	34	389	387
105	115	108	101	35	324	325
108	94	87	81	36	262	266
111	75	69	63	37	207	213
114	58	53	48	38	159	167
117	44	40	36	39	120	128
120	32	29	26	40	87	96
123	23	21	18	41	62	70
126	16	15	13	42	44	50
129	11	10	9	43	30	35
132	8	7	6	44	21	24
135	5	4	4	45	13	17
138	3	3	2	46	8	11
141	2	2	1	47	5	7
144	1	1	1	48	3	5
147	1	1	1	49	3	3
150				50		2
153				51		1
156				52		1

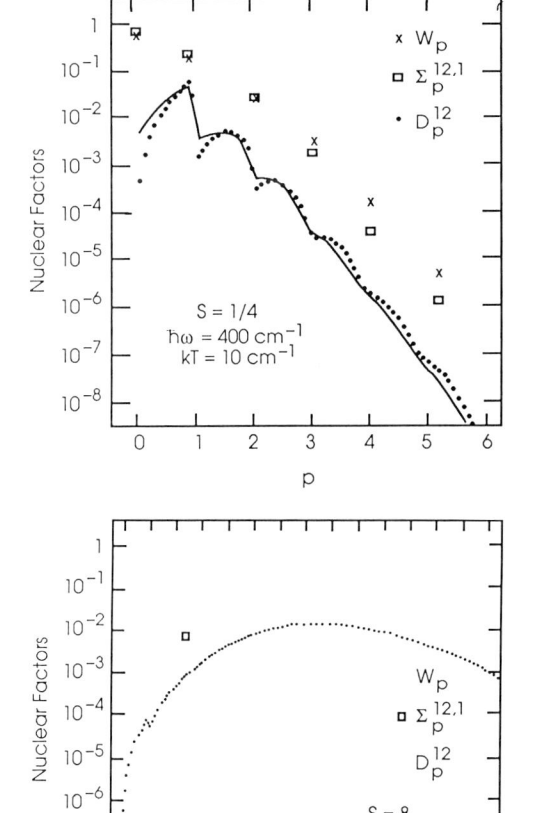

Fig. 23. Multifrequency distributions: Small offset

Fig. 24. Multifrequency distributions: Large offset

9.3 Vibrationally-Enhanced Radiative Transitions

The sums over p of the $W_{p,z}$ and of the $M_{p,z}$ distributions are obtained here, and in addition those of the $W_{p,d/dz}$ and $M_{p,d/dz}$ distributions. These former two are needed if the radiative transition probability (or f number) of a transition is seen to grow with increasing temperature. The latter two are obtained merely as an accompanying result.

The starting point is Eq. (4.87). We note first that, for both the z and d/dz operators, from Eqs. (4.70) and (4.86), the sums:

$$\sum_{i=-2}^{2} \alpha_{i,\pm} = 0 \tag{9.5}$$

Consequently, the sums:

$$\sum_{p=-\infty}^{\infty} W_{p,o_{\pm}} = \sum_{p=-\infty}^{\infty} L_p = \sum_{p=-\infty}^{\infty} \frac{1}{2}(\langle 1+m\rangle W_{p-1} + \langle m\rangle W_{p+1}) = \frac{1}{2}\langle 2m+1\rangle$$
(9.6)

and furthermore, from Eq. (5.23):

$$\sum_{p=-\infty}^{\infty} M_{p,o_{\pm}} = \sum_{p=-\infty}^{\infty} L_p = \frac{1}{2}\langle 2m+1\rangle$$
(9.7)

If therefore we deal with an absorption which is only vibrationally allowed, we expect that its absorption strength will grow as the quantity $\langle 2m+1\rangle$, using either model. This growth is not dramatic. For example, independent of S, the rate rises from its value at 0 K by an order of magnitude only when $\langle m\rangle = 4.5$, a quite high temperature. The zero-point vibrations are apparently effective in allowing significant transition probability even at 0 K.

In emission, we expect that the transition will not be totally forbidden at the upper-parabola minimum configuration, since this is not generally a configuration of high symmetry. We may anticipate encountering the operator $R_v^{-1}(1 + b_v z_v)$, where:

$$R_v = 1 + \tfrac{1}{2}|b_v|^2 \langle 2m+1\rangle_v$$
(9.8)

We might also encounter the corresponding absorption operator $R_u^{-1}(1 + b_u z_u)$, where:

$$R_u = 1 + \tfrac{1}{2}|b_u|^2 \langle 2n+1\rangle_u$$
(9.9)

These operators describe absorptions and emissions which are favored by vibrations but are not fully forbidden at the $z_u = 0$ or $z_v = a$ configurations, respectively. The total transition rate increases with temperature as R_u or R_v, respectively. We give now without proof the unequal-force-constants moments for the distributions:

$$W_{nm,z_v} = \sum_m \sum_n R_v^{-1}(1 - r_v)|\langle u_n|1 + b_v z_v|v_m\rangle|^2 r_v^m$$

$$W_{nm,z_u} = \sum_n \sum_m R_u^{-1}(1 - r_u)|\langle v_m|1 + b_u z_u|u_n\rangle|^2 r_u^n$$
(9.10)

The normalization factor is included in the weighting so that the $W_{nm,z_v \text{ or } u}$ distributions are normalized at every temperature. These moments are:

$$\langle p_u\rangle_{uv} = R_v^{-1}\left[\{S_u + k_\theta\langle 2m+1\rangle_v\} + \frac{b_v + b_v^*}{2^{1/2}}\left(\frac{\omega_u}{\omega_v}\right)^{1/2} S_u^{1/2}\langle 2m+1\rangle_v \right.$$
$$\left. + \frac{|b_v|^2}{2}\left\{S_u\langle 2m+1\rangle_v + \left(\frac{\omega_v}{\omega_u}\right) + 3k_\theta\langle 2m+1\rangle_v^2\right\}\right]$$

9.4 Comparisons of Nonradiative Rate Expressions

$$\langle(p_u - \langle p_u\rangle_{uv})^2\rangle_{uv} = R_v^{-2}\left[\left\{\left(\frac{\omega_u}{\omega_v}\right)S_u\langle 2m+1\rangle_v + 2k_\theta^2\langle 2m+1\rangle_v^2\right\}\right.$$
$$+ \frac{b_v + b_v^*}{2^{1/2}}\left\{\left(\frac{\omega_v}{\omega_u}\right)^{1/2}S_u^{1/2}\right.$$
$$+ 4k_\theta\left(\frac{\omega_u}{\omega_v}\right)^{1/2}S_u^{1/2}\langle 2m+1\rangle_v^2\right\}$$
$$+ \frac{|b_v|^2}{2}\left\{4\left(\frac{\omega_u}{\omega_v}\right)S_u\langle 2m+1\rangle_v^2 + \langle 2m+1\rangle_v\right.$$
$$+ 12k_\theta^2\langle 2m+1\rangle_v^3\right\} - \frac{(b_v + b_v^*)^2}{2}\left(\frac{\omega_u}{\omega_v}\right)S_u\langle 2m+1\rangle_v^2$$
$$\frac{b_v + b_v^*}{2^{3/2}}|b_v|^2\left(\frac{\omega_v}{\omega_u}\right)^{1/2}S_u^{1/2}\langle 2m+1\rangle_v$$
$$\pm \frac{|b_v|^4}{4}\left\{3\left(\frac{\omega_u}{\omega_v}\right)S_u\langle 2m+1\rangle_v^3\right.$$
$$+ \left(\frac{\omega_v}{\omega_u}\right)^2(\langle 2m+1\rangle_v^2 - 1) + 6k_\theta^2\langle 2m+1\rangle_v^4\right\}\right]$$

$$\langle(p_u - \langle p_u\rangle_{uv})^3\rangle_{uv} = R_v^{-3}\left[\left\{S_u + 6k_\theta\left(\frac{\omega_u}{\omega_v}\right)S_u\langle 2m+1\rangle_v^2\right.\right.$$
$$+ 4k_\theta^2\left(\frac{\omega_v}{\omega_u}\right)\langle 2m+1\rangle_v + 8k_\theta^3\langle 2m+1\rangle_v^3\right\} + \cdots\right]$$
(9.11)

where:

$$k_\theta = \frac{\cos 2\theta}{\sin^2 2\theta} = \frac{(\omega_u^2 - \omega_v^2)}{4\omega_u\omega_v} \tag{9.12}$$

The analogous $u \to v$ moments are obtained from Eq.(9.11) by the transformations:

$$u \to v, \quad v \to u, \quad k_\theta \to -k_\theta, \quad m \to n, \quad S_u^{1/2} \to -S_v^{1/2} \tag{9.13}$$

9.4 Comparisons of Nonradiative Rate Expressions

We first give descriptions of the four crossovers treated in Tables 1–4 of Chap. 1 for the z and d/dz operators in one coordinate, in the single-frequency multiple-coordinate model, and in the DDESA multiple-frequency model. We treat the z operator here primarily because of the mathematical relationship between these two operators. Nevertheless, in the real world the actual operator driving the nonradiative transition may well be some complicated function of these two operators.

Tables 12–15 treat the one-coordinate model crossovers. The $W_{p,o_\pm}$ of Eqs. (4.23), (4.69) and (4.85), (4.88), which also have the form Eq. (4.87), are used in their basic definition, Eqs. (4.22) and (4.76). They are broken down into the thermally weighted contributions from the initial-state vibrational levels indexed in the first column. The squared matrix elements themselves are given in the second and third columns. Below the tabulation, $\langle m \rangle_v$ is listed and also the $\langle m \rangle_p$ from the corresponding of the Table 1–4 set. The analogous descriptors of the transition, $\langle m \rangle_{p,o_\pm}$ are also given. These are defined by:

$$\langle m \rangle_{p,z} = \frac{1}{W_{p,z}} \sum_{m=0}^{\infty} m(1-r)r^m A_{p+m,m,z}^2$$

$$\langle m \rangle_{p,d/dz} = \frac{1}{W_{p,d/dz}} \sum_{m=0}^{\infty} m(1-r)r^m A_{p+m,m,d/dz}^2 \qquad (9.14)$$

The general trends exhibited in Tables 1–4 are also exhibited in these tables for the z and d/dz operators. Only the lowest vibrational state contributes at 0 K and the contributions from higher vibrational states grow with increasing temperature, with the strongest contributions shifting with temperature to higher states.

One can see also that there is an exact correspondence, as there should be, between $\langle m \rangle_p$ and $\langle m \rangle_{p,z}$, shown here within the roundoff error of the calculation. This correspondence comes from recognizing that the proportionality between A_{nm} and $A_{nm,z}$, from Eq. (3.49) with $n - m \to p$, namely:

$$\frac{p-S}{(2S)^{1/2}} A_{p+m,m} = A_{p+m,m,z} \qquad (9.15)$$

is a constant for a given p. The $\langle m \rangle_{p,d/dz}$ are found to be in general slightly lower.

The temperature dependence of the $W_{p,z}$ description is also identically that of the W_p description, as expected from the temperature independence of the relationship between them, Eq. (4.23). The temperature dependence of the $W_{p,d/dz}$ description is somewhat smaller. From Eq. (4.88) we expect $W_{p,d/dz}$ to be less than $W_{p,z}$ by an amount somewhat proportional to $\langle m \rangle$, whenever the curvature of the W_p distribution is positive. It is positive in the wings of the distribution, as for $p = 14, 25$, with $S = 6.25$ in Tables 12–13 and for $p = 2$ with $S = 30.25$ in Table 15, and even slightly for $p = 6$ for $S = 0.09$ in Table 14.

In Tables 16 and 17 all four crossover cases are given at a few temperatures. In Table 16, the rates themselves and also these rates normalized to unity at 0 K are given. The values of $M_{p,o_\pm}$ are not given, but from Eq. (5.6) these are linear combinations of $W_{p,o_\pm}$ and L_p, both of which are given. From Table 16, one can see that $M_{p,o_\pm}$ will have very similar temperature dependences as $W_{p,o_\pm}$ and indeed as W_p itself, since $W_{p,z}$ and W_p, as noted above, have identical temperature dependences and L_p is a linear combination of two W_p with nearby indices. Even the differences which do exist between these distributions, when viewed in the

9.4 Comparisons of Nonradiative Rate Expressions

Table 12. Z and derivative operator driven nonradiative rates: Bottom crossover

m	A^2_{14+m,z_v}	$A^2_{14+m,m,d/dz}$	$(1-r_v)r_v^m A^2_{14+m,m,z_v}$				$(1-r_v)r_v^m A^2_{14+m,m,d/dz}$			
			0 K	259 K	374 K	547 K	0 K	259 K	374 K	547 K
0	0.01477	0.01477	0.01477	0.01317	0.01160	0.00962	0.01477	0.01317	0.01160	0.00962
1	0.07537	0.05015		0.00728	0.01271	0.01712		0.00485	0.00846	0.01139
2	0.1923	0.06975		0.00201	0.00696	0.01525		0.00073	0.00253	0.00553
3	0.3197	0.03805		0.00036	0.00249	0.00885		0.00004	0.00030	0.00105
4	0.3768	0.00009		0.00005	0.00063	0.00364		0.00000	0.00000	0.00000
5	0.3166	0.06521		0.00000	0.00011	0.00107			0.00002	0.00022
6	0.1738	0.2790			0.00001	0.00020			0.00002	0.00033
7	0.0421	0.5311				0.00002			0.00001	0.00022
8	0.0009	0.6272								0.00009
9	0.0574	0.4720								0.00002
$W_{14,z_v}, W_{14,d/dz_v}$			0.01477	0.02287	0.03451	0.05577	0.01477	0.01879	0.02294	0.02847
$\langle m \rangle_v$			0.	0.122	0.273	0.536				
$\langle m \rangle_{14}$			0.	0.552	1.079	1.713				
$\langle m \rangle_{14,z_v}$			0.	0.550	1.079	1.711				
$\langle m \rangle_{14,d/dz_v}$							0.	0.342	0.641	1.093
$kT/\hbar\omega_0$			0.	0.45	0.65	0.95	0.	0.45	0.65	0.95

Table 13. Z and derivative operator driven nonradiative rates: Outside crossover

| m | A^2_{25+m,m,z_v} | \multicolumn{4}{c|}{$10^7(1-r_v)r_v^m A^2_{25+m,m,z_v}$} | \multicolumn{4}{c|}{$10^7(1-r_v)r_v^m A^2_{25+m,m,d/dz_v}$} |
		0 K	259 K	374 K	547 K	0 K	259 K	374 K	547 K
0	0.0000002761	2.761	2.462	2.17	1.8	2.761	2.462	2.17	1.8
1	0.000004143		4.003	6.99	9.4		3.737	6.52	8.8
2	0.00003203		3.354	11.60	25.4		2.910	10.06	22.0
3	0.0001699		1.928	13.21	47.0		1.545	10.59	37.7
4	0.0006941		0.853	11.58	67.0		0.629	8.53	49.4
5	0.002326		0.310	8.34	78.4		0.208	5.60	52.7
6	0.006651		0.096	5.12	78.3		0.058	3.11	47.6
7	0.01665		0.026	2.73	68.4		0.014	1.50	37.2
8	0.03723		0.006	1.32	53.4		0.003	0.64	25.7
9	0.07532		0.001	0.57	37.7		0.001	0.24	15.8
10	0.1394			0.23	24.3			0.08	8.8
11	0.2380			0.08	14.5			0.03	4.4
12	0.3769			0.03	8.0			0.01	2.0
13	0.5565			0.01	4.1				0.8
14	0.7684				2.0				0.3
15	0.9944				0.9				0.1
16	1.207				0.4				
$10^7 W_{25,z_v}$, $10^7 W_{25,d/dz_v}$		2.761	13.039	63.98	521.2	2.761	11.567	49.08	315.1
$\langle m \rangle_v$		0.	0.122	0.273	0.536				
$\langle m \rangle_{25}$		0.	1.710	3.550	6.11				
$\langle m \rangle_{25,z_v}$		0.	1.708	3.548	6.10	0.	1.576	3.224	5.39
$\langle m \rangle_{25,d/dz_v}$						0.	0.45	0.65	0.95
$kT/\hbar\omega_0$		0.	0.45	0.65	0.95				

9.4 Comparisons of Nonradiative Rate Expressions

Table 14. Z and derivative operator driven nonradiative rates: Small offset multiphonon transition

m	A^2_{6+m,m,z_v}	$A^2_{6+m,m,d/dz}$	\multicolumn{4}{c}{$10^7(1-r_v)r_v^m A^2_{6+m,m,z_v}$}	\multicolumn{4}{c}{$10^7(1-r_v)r_v^m A^2_{6+m,m,d/dz_v}$}
			0 K	259 K	374 K	547 K	0 K	259 K	374 K	547 K
0	0.0000001309	0.0000001309	1.309	1.167	1.028	0.85	1.309	1.167	1.028	0.85
1	0.0000008929	0.0000008851		0.863	1.506	2.03		0.855	1.492	2.01
2	0.000003480	0.000003419		0.364	1.260	2.76		0.358	1.238	2.71
3	0.000010174	0.000009905		0.115	0.791	2.82		0.122	0.770	2.74
4	0.00002478	0.00002391		0.030	0.414	2.39		0.029	0.399	2.31
5	0.00005312	0.00005079		0.007	0.190	1.79		0.007	0.182	1.71
6	0.0001035	0.00009805		0.001	0.080	1.22		0.001	0.075	1.15
7	0.0001873	0.0001757			0.031	0.77			0.029	0.72
8	0.0003192	0.0002968			0.011	0.46			0.011	0.43
9	0.0005182	0.0004774			0.004	0.26			0.004	0.24
10	0.0008076	0.0007370			0.001	0.14			0.001	0.13
11	0.001216	0.001099				0.07				0.07
12	0.001776	0.001590				0.04				0.03
13	0.002528	0.002241				0.02				0.02
$10^7 W_{6,z_v}, 10^7 W_{6,d/dz_v}$			1.309	2.547	5.316	15.63	1.309	2.529	5.229	15.13
$\langle m \rangle_v$			0.	0.122	0.273	0.536				
$\langle m \rangle_6$			0.	0.827	1.851	3.067				
$\langle m \rangle_{6,z_v}$			0.	0.823	1.850	3.602				
$\langle m \rangle_{6,d/dz_v}$							0.	0.816	1.830	3.557
$kT/\hbar\omega_0$			0.	0.45	0.65	0.95	0.	0.45	0.65	0.95

Table 15. Z and derivative operator driven nonradiative rates: Tunnelling crossover

m	$A^2_{n,2+n,z_u}$	$A^2_{n,2+n,d/dz}$	$10^7(1-r_u)r_u^m A^2_{n,2+n,z_u}$			$10^7(1-r_u)r_u^m A^2_{n,2+n,d/dz_u}$		
			0 K	72 K	101 K	0 K	72 K	101 K
0	0.4398-9	0.4398-9	0.00440	0.0043	0.004	0.00440	0.0043	0.004
1	0.1089-6	0.9243-7		0.0196	0.059		0.0166	0.050
2	0.000008601	0.000006009		0.0283	0.267		0.0198	0.187
3	0.00003015	0.0001657		0.0182	0.538		0.0100	0.296
4	0.0003015	0.002144		0.0059	0.547		0.0024	0.220
5	0.005329	0.01290		0.0010	0.293		0.0003	0.076
6	0.04979	0.03037		0.0001	0.082		0.0000	0.010
7	0.2433	0.01053			0.011			0.000
8	0.5669	0.03346			0.001			
	0.4593							
$10^7 W_{2,z_u}$, $10^7 W_{2,d/dz_u}$			0.00440	0.0774	1.802	0.00440	0.0534	0.843
$\langle n \rangle_u$			0.	0.019	0.061			
$\langle n \rangle_2$			0.	2.067	3.571			
$\langle n \rangle_{2,z_u}$			0.	2.067	3.572			
$\langle n \rangle_{2,d/dz_u}$						0.	1.822	3.122
$kT/\hbar\omega_0$			0.	0.25	0.35	0.	0.25	0.35

9.4 Comparisons of Nonradiative Rate Expressions

Table 16. Thermal quenching: W_p vs. M_p descriptions

Bottom crossover, $S = 6.25$, $p = 14$.

$kT/\hbar\omega$	$\langle m \rangle$	$10^2 W_{p,z}$		$10^2 W_{p,d/dz}$		$10^2 L_p$	
0.	0.	1.48	1.	1.48	1.	0.34	1.
0.45	0.122	2.29	1.55	1.88	1.27	0.56	1.62
0.65	0.273	3.45	2.34	2.29	1.56	0.90	2.63
0.95	0.536	5.58	3.78	2.85	1.93	1.66	4.81

Outside crossover, $S = 6.25$, $p = 25$.

$kT/\hbar\omega$	$\langle m \rangle$	$10^6 W_{p,z}$		$10^6 W_{p,d/dz}$		$10^6 L_p$	
0.	0.	0.276	1.	0.276	1.	0.020	1.
0.45	0.122	1.30	4.72	1.16	4.19	0.09	4.80
0.65	0.273	6.40	23.2	4.91	17.8	0.47	24.1
0.95	0.536	52.1	189.	31.5	114.	4.1	207.

Small-offset multiphonon emission, $S = 0.09$, $p = 6$.

$kT/\hbar\omega$	$\langle m \rangle$	$10^7 W_{p,z}$		$10^7 W_{p,d/dz}$		$10^7 L_p$	
0.	0.	1.31	1.	1.31	1.	0.22	1.
0.45	0.122	2.55	1.95	2.53	1.93	0.44	1.95
0.65	0.273	5.32	4.06	5.23	3.99	0.91	4.06
0.95	0.536	15.6	11.9	15.1	11.6	2.69	11.9

Inside (tunnelling) crossover, $S = 30.25$, $p = 2$.

$kT/\hbar\omega$	$\langle m \rangle$	$10^9 W_{p,z}$		$10^9 W_{p,d/dz}$		$10^9 L_p$	
0.	0.	.440	1.	.440	1.	0.0011	1.
0.25	0.0187	7.74	17.6	5.34	12.1	0.078	71.2
0.35	0.0609	180.	410.	84.	192.	3.4	3 080.

narrow temperature range in which a temperature quenching of an emission can be measured, is most frequently less than a 30% variation.

The $D_{p,o_\pm}$ distributions in Table 17 also show small probability of distinguishing experimentally among these operators. For a microsecond radiative lifetime and the outside crossover case, and matching the nonradiative rates to give $\eta \approx 1/e$ at $kT = 260$ cm^{-1}, the differences above this temperature shown here would be reflected in how the quenching curve proceeds from 40% efficiency to either 2% or to 4% for the various operators. For millisecond lifetimes, the quenching would occur at lower temperatures, and there is a chance, albeit slight, of establishing experimentally one of these operators. When one considers that the DDESA model is itself only one simple embodiment of the multiple-frequency model, it is not a very promising search.

Similar slight possibilities for finding experimental evidence favoring one of these operators exist for bottom crossovers at very low temperatures and for tunnelling crossovers at very low temperatures.

9 Representative Luminescence Centers

Table 17. Thermal quenching: W_p vs. D_p descriptions

kT, cm^{-1}	W_p	D_{3p}^3	$W_{p,z}$	$D_{3p,z}^3$	$W_{p,d/dz}$	$D_{3p,d/dz}^3$	L_p
Bottom crossover, $S = 6.25$, $p = 14$, $\hbar\omega_K = 400$ cm^{-1}								
z	3	3	2	3	2	3	2
1	3.07	0.57	1.48	2.96	1.48	2.96	0.34
180	4.76	1.23	2.29	7.09	1.88	4.91	0.56
260	7.18	2.09	3.45	12.57	2.29	6.79	0.90
380	11.61	3.68	5.58	22.92	2.85	9.82	1.66
Outside crossover, $S = 6.25$, $p = 25$, $\hbar\omega_K = 400$ cm^{-1}								
z	8	9	7	8	7	8	7
1	0.98	0.230	2.76	0.660	2.76	0.660	0.20
10	0.98	0.230	2.76	0.660	2.76	0.660	0.20
50	0.99	0.241	2.77	0.698	2.77	0.692	0.20
100	1.27	0.438	3.57	1.305	3.51	1.217	0.25
150	2.64	1.42	7.41	4.37	6.88	3.67	0.53
180	4.64	3.18	13.04	9.98	11.57	7.78	0.94
260	22.8	25.68	64.0	84.2	49.1	53.4	4.97
380	185.	331.	521.	1 126.	315.	528.	40.6
Small-offset multiphonon emission, $S = 0.09$, $p = 6$, $\hbar\omega_K = 400$ cm^{-1}								
z	9	10	7	8	7	8	7
1	0.67	0.26	1.31	0.42	1.31	0.42	0.22
180	1.31	0.60	2.55	1.04	2.53	1.02	0.44
260	2.74	1.43	5.32	2.53	5.23	2.44	0.91
380	8.07	4.96	15.63	9.09	15.13	8.54	2.69
Inside (tunnelling) crossover, $S = 30.09$, $p = 2$, $\hbar\omega_K = 200$ cm^{-1}								
z	10	11	9	10	9	10	9
1	0.33	0.0037	0.44	0.0060	0.42	0.0060	0.0011
50	5.87	3.48	7.74	5.96	1.02	2.85	0.078
70	136.7	206.0	180.	354.	2.44	110.1	3.4

Units are in 10^{-z} with z listed.

10 Experimental Studies

10.1 Eu in Oxysulfides and in Oxyhalides

10.1.1 The Energy Level Diagram and Qualitative Behavior

Figure 25 shows the single-configurational-coordinate (SCC) energy level diagram appropriate for Eu^{3+} in La_2O_2S, Y_2O_2S, and $LaOCl$. The ^7F $4f^6$ states are not shown. They are split into seven J states spanning about 6 000 cm^{-1} with J between 0 and 6. They would be placed with the zero of energy at the minimum of the $^7F_0$ parabola and with only minor offsets, with $S\hbar\omega < 100$ cm^{-1}. The first set of excited $4f^6$ states are the ^5D states, split into five J states between 0 and 4. These also have only minor offsets. The first four of these states have been identified; the $J = 4$ state and the other $4f^6$ states will be referenced only by their absorption energies in nanometers.

In each host there is an offset charge-transfer state (CTS), placed in the one diagram appropriately for the broad absorption band observed in each host.

The diagram ignores the small crystal field splittings of the various states and the degeneracies and splittings of the CTS.

Table 18 gives the energies of some of the Eu^{3+} states relative to $^7F_0$. These energies are approximate averages of the several crystal-field states for this ion.

From Fig. 25 one can expect narrow line absorptions from $^7F_0$ at 0 K and also from the thermally populated ^7F_j states at higher temperatures into the ^5D_j states. There will also be a broad band absorption into the CTS, only from $^7F_0$ at 0 K but also from the thermally populated ^7F_j states at higher temperatures. In the absence of experimental knowledge of these electric dipole matrix elements, and not having had the opportunity to develop a theoretical treatment of them in the vein of the Judd-Ofeld analysis, we take them here to be all equal. One can expect also narrow line emissions from the various ^5D states to the lower ^7F states. As might be anticipated from the diagram, the broad band emission from the CTS to the lower ^7F states are not seen because of the much faster CTS $\rightarrow$ ^5D_j feedings and the thermal inaccessability of the CTS from the lowest ^5D states.

Figure 26 shows the excitation spectrum of $(Y_{0.999}, Eu_{0.001})_2O_2S$ $^5D_0$ and $^5D_3$ emissions at 77 K. These spectra have not been corrected for the source intensity which peaked near 380 nm. These spectra show the narrow 4f → 4f absorptions into states above $^5D_3$ and the broadband ^7F → CTS at higher energies.

Figure 27 shows the emission spectra generated in this phosphor at 77 K

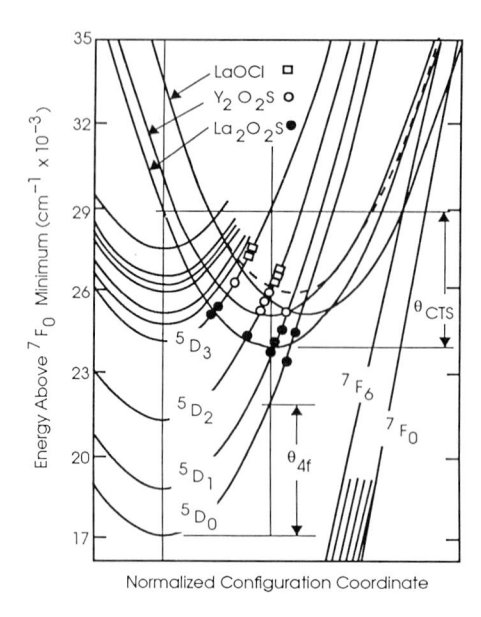

Fig. 25. SCC diagram for Eu^{3+} in several hosts

Table 18. $^7F_0 \rightarrow {}^7F_i$, ^5D_i Energies, wavenumbers

	i	Y$_2$O$_2$S	La$_2$O$_2$S	LaOCl
^{5}D	3	24 210	24 270	24 200
	2	21 360	21 410	21 400
	1	18 940	18 900	18 900
	0	17 140	17 180	17 200
^{7}F	4	2 890	2 890	2 890
	3	1 700	1 700	1 700
	2	1 190	1 190	1 190
	1	340	340	340

under two different excitation wavelengths, noted in the figure. These wavelengths are into the CTS (top curve) and into a state above $^5D_3$ which cascades directly and totally into $^5D_3$ (bottom curve). These are again spectra uncorrected for the detector spectral response, which decreased by a factor of 3 to the shorter wavelengths. The notation (i,j) is used for the transition $^5D_i \rightarrow {}^7F_j$. One important characteristic of CTS excitations is visible in Fig. 27, namely, that the $^5D_3$ emissions are excited at 397 nm. but not with the CTS, although the CTS is above the 397 nm. state. Populating the CTS leads to bypassing $^5D_3$ and the $4f^6$ levels above $^5D_3$ and to direct feeding of the lower ^5D states via CTS $\rightarrow {}^5D_2, {}^5D_1,$ and $^5D_0$ transitions. Besides confirming that indeed the lower ^5D states are fed through the CTS-^5D crossovers near the CTS minimum, this performance sug-

10.1 Eu in Oxysulfides and in Oxyhalides

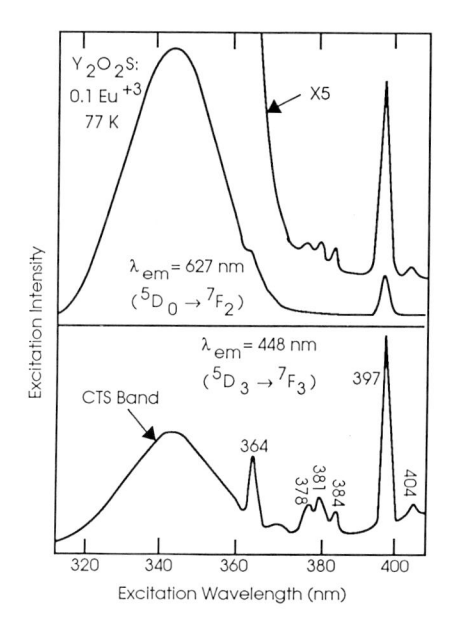

Fig. 26. Excitation spectra for $(Y_{0.999}, Eu_{0.001})_2O_2S$

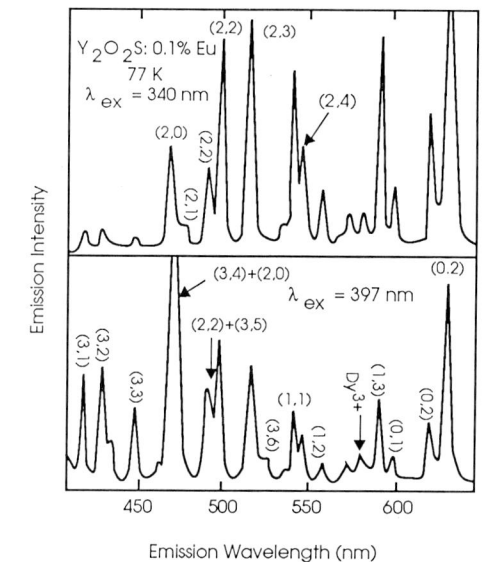

Fig. 27. Emission spectra for $(Y_{0.999}, Eu_{0.001})_2O_2S$

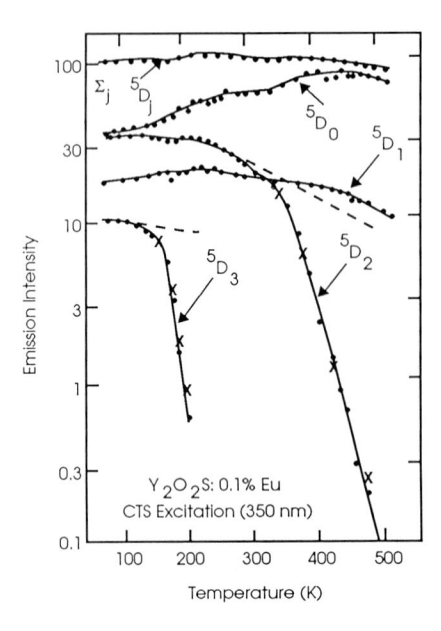

Fig. 28. Thermal quenching of $(Y_{0.999}, Eu_{0.001})_2O_2S$ emissions

gests the model of studying quantitatively the CTS feeding fractions into the ^5D manifold, namely by assessing the relative rates of these bottom crossovers through the equal-force-constants SCC-model W_p functions, or the unequal-force-constants U_{p_v} functions, or even the derivative-operator expressions in one or in many coordinates and in one or in many frequencies. Here, we restrict our attention to the Condon-operator W_p and U_{p_v} functions.

In addition, as was first reported by Wickersheim, Buchanan, and Yates [56], the ^5D_j emissions quench precipitously with increasing temperature in the then unexpected order: $^5D_3$ at the lowest temperature, then $^5D_2$, then $^5D_1$ as the temperature rises. Figure 28 shows this sequential quenching of the $(Y_{0.999}, Eu_{0.001})O_2S$ emissions. Our work established these quenchings as two-step quenchings through the higher offset CTS, i.e., as the case presented in Sect. 1.2.3.4. and pictured in Fig. 6.

The $^5D \to CTS$ quenching can have another manifestation besides this sequential precipitous disappearance of the emissions. Figure 29 shows the temperature dependences of the ^5D emissions in $(La_{0.998}, Eu_{0.002})OCl$ excited at 397 nm., i.e., excited effectively into $^5D_3$. Note the drop-off of the $^5D_2$ emissions in two steps, the first at the $^5D_3$ quenching temperature, near 270 K, and the second at the $^5D_2$ quenching temperature, near 450 K. The first drop is related not to any losses from $^5D_2$ but rather to replacing strong $^5D_3 \to {}^5D_2$ feeding via small-offset multiphonon emission with $^5D_3 \to CTS$ quenching followed by weak $CTS \to {}^5D_2$ feeding. The $CTS \to {}^5D_2$ crossover is bypassed and the dominant crossovers are nearer the CTS minimum into $^5D_1$ and $^5D_0$.

10.1 Eu in Oxysulfides and in Oxyhalides

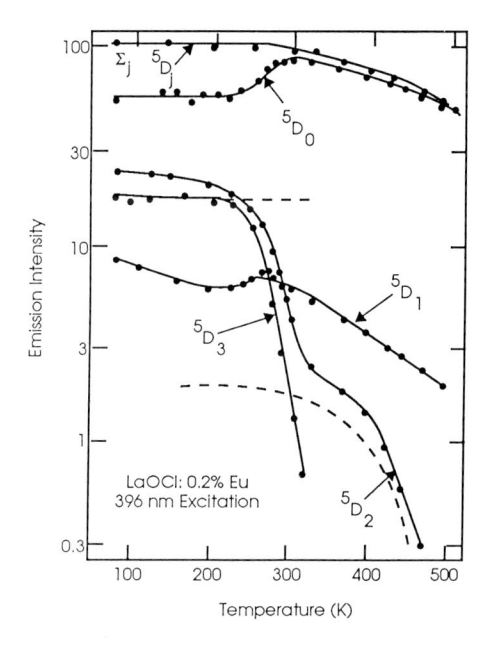

Fig. 29. Thermal quenching of $(La_{0.998}, Eu_{0.002})$ OCl emissions

10.1.2 Feeding Fractions

Excitation into a particular ^5D_j state leads to ^5D_j emissions and also, via at low concentrations a sequence of small-offset multiphonon emission transitions, to emissions originating in all lower ^5D_k states. At higher concentrations there is the added known two-center energy transfer process for populating these lower ^5D states, namely, $(^5D_j, {}^7F_0) \rightarrow (^5D_k, {}^7F_1)$, where the indices are chosen so as to conserve energy. In either case or even in any combination of them at a given temperature, there is a unique pattern of emission intensities for each ^5D_j excitation and a different unique pattern for CTS excitation. Let q_{ij} and q_{iCTS} be the quantum efficiencies for emission from ^5D_i for ^5D_j and CTS excitations, respectively. Let α_j be the quantum efficiency for transferring CTS excitation into the ^5D_j level. Then:

$$q_{iCTS} = \sum_j q_{ij} \alpha_j \qquad i, j = 0, 1, 2, 3 \tag{10.1}$$

Eq. (10.1) can be inverted to give that α_j in terms of the q_{iCTS} and q_{ij}.

For low temperatures, as at 77 K here, $\sum_i q_{ij}$ was close to unity. Thus q_{ij} was equal to the fraction of photons emitted from ^5D_i for ^5D_j excitation, and q_{iCTS} is the fraction of photons emitted from ^5D_i for CTS excitation. The inversion then gives α_j which sum to unity over j, and these are the feeding fractions.

In our model, we take:

$$\alpha_j = \frac{(2j+1) A_{5D_j, CTS} U_{p_v, j}}{\sum_{i\,unquenched} (2i+1) A_{5D_i, CTS} U_{p_v, i}} \approx \frac{(2j+1) U_{p_v, j}}{\sum_{i\,unquenched} (2i+1) U_{p_v, i}} \tag{10.2}$$

Table 19. Observed Eu^{3+} CTS $\rightarrow$ ^{5}D$_i$ feeding fractions in several hosts

Host	% Eu	T, K	α_3	α_2	α_1	α_0
La$_2$O$_2$S	0.5	77	0.02	.30	0.65	0.04
Y$_2$O$_2$S	0.1	77	0.20	.65	0.15	0.02
	0.1	295		.59	0.37	0.03
	2.0	295		.60	0.40	0.00
	2.0	495		.46	0.53	0.02
LaOCl	0.2	77	0.01	.04	0.45	0.50

Equation (10.2) uses $U_{p_U,j}$, not $U_{p_U,j,d/dz}$, nor the multicoordinate-model $M_{p_U,j,d/dz}$, nor the multiple-frequency-model $D_{p_U,j,d/dz}$. Furthermore we take all $A_{5D_i,CTS}$ equal. In this weighting by the simple degeneracy factor and equal electronic integrals, we ignore the detailed selection rules and variable transition integrals among component transitions. This latter approximation, as also for the absorption band fits of Sect. 10.1.5 below and for the quenching fits of Sect. 10.1.6 below, introduces the more serious error and probably accounts for most of the discrepancies between measured and observed feeding fractions, quenching temperatures, and temperature-dependent bandshapes. Moreover, it will be seen that the discrepancies encountered tend to point to crossovers occurring before thermal equilibrium is established: there is a consistent pattern of a few percent direct feeding at low temperature through the higher crossovers not expected from a thermal population of the CTS vibrational levels.

These discrepancies do not negate the important general conclusions reached here.

In Table 19 are listed the observed feeding fractions in these three hosts from the CTS of Eu^{3+} into the ^{5}D$_i$ states.

10.1.3 Efficiencies under Quenching Conditions

In analogy with Sect. 1.2.3.4 we list the efficiency expressions $(\eta_i)_j$, $(\eta_i)_{CTS}$, of ^{5}D$_i$ emissions for excitation into ^{5}D$_i$ or into the CTS, respectively. These are solutions of the steady state relationships among the concentrations of the various states according to:

$$\begin{bmatrix} G_{CTS} \\ G_3 \\ G_2 \\ G_1 \\ G_0 \end{bmatrix} = \begin{bmatrix} (CTS \rightarrow 3,2,1,0) & -(3 \rightarrow CTS) & -(2 \rightarrow CTS) & -(1 \rightarrow CTS) & -(0 \rightarrow CTS) \\ -(CTS \rightarrow 3) & (3 \rightarrow CTS, 2, F) & 0 & 0 & 0 \\ -(CTS \rightarrow 2) & -(3 \rightarrow 2) & (2 \rightarrow CTS, 1, F) & 0 & 0 \\ -(CTS \rightarrow 1) & 0 & -(2 \rightarrow 1) & (1 \rightarrow CTS, 0, F) & 0 \\ -(CTS \rightarrow 0) & 0 & 0 & -(1 \rightarrow 0) & (0 \rightarrow CTS, F) \end{bmatrix} \begin{bmatrix} n_{CTS} \\ n_3 \\ n_2 \\ n_1 \\ n_0 \end{bmatrix}$$
(10.3)

The solutions are obtained in the temperature region of ^{5}D$_3$ quenching by setting to zero the $(2 \rightarrow CTS)$, $(1 \rightarrow CTS)$, $(0 \rightarrow CTS)$, and all but either the G_3 or G_{CTS}

10.1 Eu in Oxysulfides and in Oxyhalides

rates. In the region of $^5D_2$ quenching, $(3 \to CTS) \gg (3 \to 2, F)$ and n_3 can be eliminated from the set of unknowns using the second equation. Similar considerations simplify Eq. (10.3) for η_2 and η_1. The results are:

$$(\eta_3)_3 = \frac{(3 \to F)n_3}{G_3} = \left[1 + \frac{(3 \to 2)}{(3 \to F)} + \frac{(CTS \to 2,1,0)}{(CTS \to 3,2,1,0)} \cdot \frac{(3 \to CTS)}{(3 \to F)}\right]^{-1}$$

$$\equiv \left[1 + \frac{(3 \to 2)}{(3 \to F)} + \beta_3(3 \to CTS)\right]^{-1},$$

$$(\eta_3)_{CTS} = \frac{(3 \to F)n_3}{G_{CTS}} = \frac{(CTS \to 3)}{(CTS \to 3,2,1,0)}(\eta_3)_3$$

$$(\eta_2)_2 = \frac{(2 \to F)n_2}{G_2} = \left[1 + \frac{(2 \to 1)}{(2 \to F)} + \frac{(CTS \to 1,0)}{(CTS \to 2,1,0)} \cdot \frac{(2 \to CTS)}{(2 \to F)}\right]^{-1}$$

$$\equiv \left[1 + \frac{(2 \to 1)}{(2 \to F)} + \beta_2(2 \to CTS)\right]^{-1},$$

$$(\eta_2)_{CTS} = \frac{(2 \to F)n_2}{G_{CTS}} = \frac{(CTS \to 2)}{(CTS \to 2,1,0)}(\eta_2)_2$$

$$(\eta_1)_1 = \frac{(1 \to F)n_1}{G_1} = \left[1 + \frac{(1 \to 0)}{(1 \to F)} + \frac{(CTS \to 0)}{(CTS \to 1,0)} \cdot \frac{(1 \to CTS)}{(1 \to F)}\right]^{-1}$$

$$\equiv \left[1 + \frac{(1 \to 0)}{(1 \to F)} + \beta_1(1 \to CTS)\right]^{-1},$$

$$(\eta_1)_{CTS} = \frac{(1 \to F)n_1}{G_{CTS}} = \frac{(CTS \to 1)}{(CTS \to 1,0)}(\eta_1)_1 \tag{10.4}$$

When our interest is in the precipitous quenching itself, we discount the weak temperature dependences due to other loss processes, e.g., the $f \to f$ transitions, and define "quenching efficiencies" $(\eta'_i)_i$ as the ratio of the efficiency to the efficiency if no $^5D_i \to CTS$ transitions were occurring. Thus:

$$(\eta'_i)_i = (\eta_i)_i \left[1 + \frac{(i \to i-1)}{(i \to f)}\right] \equiv \frac{1}{1 + \beta'_i(i \to CTS)} \equiv \frac{1}{1 + \beta'_i A_{vu} V_{p_{v_i}}}$$

$$\beta'_i = \left[\frac{(CTS \to i-1,\ldots,0)}{(CTS \to i, i-1,\ldots,0)(i \to F)}\right] \Big/ \left[1 + \frac{(i \to i-1)}{(i \to f)}\right] \tag{10.5}$$

10.1.4 Absorption Spectra at T > 0 K

Our attention will be limited to fitting the long-wavelength portion of the $^7F \to CTS$ absorption band. We then can ignore the unknown complex electronic structure of this CTS state and be content with assessing the energy only of the lowest such state.

Even so, we must take into account the absorptions from the higher ^7F_j levels.

We do so by assuming them at 340, 1190, 1700, and 2890 cm^{-1} above $^7F_0$ for $j = 1, 2, 3, 4$, respectively, as shown in Table 18, and then giving them thermal (Boltzmann) and also $(2j + 1)$ degeneracy weights. The energy placements were approximate, insofar as they correspond to 2, 7, 10, and 17 $\hbar\omega$ for the $\hbar\omega = 170$ cm^{-1} value chosen for the equal-force-constants fit below.

The absorption strength calculated for every wavelength then includes the contributions from each of the ^7F_j states, each contribution being a V_{p_v} with its p_v appropriate for its initial state energy.

Once again, the electronic dipole moment transition integral is assumed identical for all $^7F_j \to$ CTS absorptions.

10.1.5 The Fit of the Absorption Data

10.1.5.1 LaOCl

The absorption of Eu in LaOCl was described [1, 2] as fitted by a superposition of Gaussians from the thermally weighted ^7F_j states, with each Gaussian sharing common parameters A, hv_0, σ according to:

$$P_0(hv) = A \exp[-(hv - hv_0)^2/\sigma^2] \qquad \sigma = 2(kT_{eff}\theta_e^2/\theta_g)^{1/2}$$
$$kT_{eff} = E_{zp}\coth(E_{zp}/kT) \tag{10.6}$$

where kT_{eff} is an effective thermal energy which approaches the zero-point vibrational energy E_{zp} of the ground state at low temperatures and kT at high temperatures. The θ_e and θ_g are identical with our $S_v\hbar\omega_v$ and $S_u\hbar\omega_u$ of Fig. 1, respectively. In our notation, $hv_0 = \frac{1}{2}(\hbar\omega_u - \hbar\omega_v) + hv_{zp,vu} + S_v\hbar\omega_v$, i.e., the energy between the parabolas at the ground-state equilibrium configuration, and $hv = \frac{1}{2}(\hbar\omega_u - \hbar\omega_v) + hv_{zp,vu} + p_v\hbar\omega_v$.

These expressions stem from Klick and Schulman [57] in their semiclassical treatment of the luminescent center. We used this treatment in early work before we appreciated the quantum mechanical expressions set forth in this book. In the better notation of this book, these equations become, using Eq. (4.59), using the first term of the second of Eqs. (4.112), and giving the correct normalization factor:

$$P_0(hv) = \left(\frac{S_u}{2\pi S_v^2 \langle 2n+1 \rangle_u}\right)^{1/2} \exp\left[-\left(\frac{(p_v - S_v)^2 S_u}{2S_v^2 \langle 2n+1 \rangle_u}\right)\right]. \tag{10.7}$$

In reference 2 we stated that for LaOCl:Eu entering $(\theta_g \equiv S_u\hbar\omega_u) = (\theta_e \equiv S_v\hbar\omega_v) = 8200$ cm^{-1}, $hv_0 = 33400$ cm^{-1} into Eq. (10.6) provides a fit to the broad absorption band. We also suggested an alternative fit, using the parabolas with $S_u\hbar\omega_u = 7500$ cm^{-1}, $S_v\hbar\omega_v = 6800$ cm^{-1} and the same hv_0.

Because these Gaussians do fit the observed spectrum and because, using Eq. (4.59), Gaussians are approximate forms of W_p functions for large S, the W_p distributions equivalent to these Gaussians will also fit. Indeed, an adequate fit will be provided by every V_{p_v} distribution for parabola placements which maintain the hv_0 and the slope of the excited-state parabola at the ground state

10.1 Eu in Oxysulfides and in Oxyhalides

minimum configuration. There is freedom to choose the vibrational frequency as well for each such parabola placement. Explicitly, the allowed fits satisfy, using Eqs. (1.1):

$$E_{u,\theta} = S_u \hbar \omega_u \left(\frac{x}{a_\theta}\right)^2 = E_0 = S_0 \hbar \omega_0 \left(\frac{x}{a_{45}}\right)^2$$

$$E_{v,\theta} = E_{v,\min,\theta} + S_v \hbar \omega_v \left(\frac{x}{a_\theta} - 1\right)^2$$

$$\left.\frac{dE_{v,\theta}}{dx}\right|_{x=0} = -\left(\frac{2S_v \hbar \omega_v}{a_\theta}\right) = \left.\frac{dE_{v,45}}{dx}\right|_{x=0} = -\left(\frac{2S_0 \hbar \omega_0}{a_{45}}\right) \quad (10.8)$$

Having then specified a value for $S_0 \hbar \omega_0$, we have then the relationships:

$$\left(\frac{a_\theta}{a_{45}}\right) = \left(\frac{S_u \hbar \omega_u}{S_0 \hbar \omega_0}\right)^{1/2}$$

$$S_v \hbar \omega_v = \left(\frac{a_\theta}{a_{45}}\right) S_0 \hbar \omega_0 = (S_u \hbar \omega_u S_0 \hbar \omega_0)^{1/2} = \frac{S_0 \hbar \omega_0}{\tan^4 \theta}$$

$$S_u \hbar \omega_u = \frac{S_v \hbar \omega_v}{\tan^4 \theta}$$

$$E_{v,\min,\theta} = E_{v,\min,45} + S_0 \hbar \omega_0 \left[1 - \frac{1}{\tan^4 \theta}\right] \quad (10.9)$$

In Table 20, we give representative members of the set of possible fits to the LaOCl:Eu^{3+} absorption band. The θ column is totally a free choice, and the $S_v \hbar \omega_v$ and $S_u \hbar \omega_u$ columns entries for $\theta = 45°$ are unique from the original fit. Having once set these $\theta = 45°$ Franck-Condon offsets at 8200 cm^{-1}, and with a free choice of the phonon energy $\hbar \omega_v$ in the fourth column, one gets, using the second and third of Eqs. (10.9), the listed values of $S_v \hbar \omega_v$ and $S_u \hbar \omega_u$. Using Eq. (3.30), one obtains $\hbar \omega_u$, and from them S_u, S_v. Equation (3.31) then affords a_{uv} and the last of Eqs. (10.9) together with Eqs. (1.1) give the $E_{zp,vu}$ appropriate for the chosen θ.

Table 21 gives the closeness of fit obtained with these functions for the absorption strength at a set of absorption energies. We have chosen to illustrate the distributions for $\hbar \omega_v = 200$ cm^{-1} at 77 K. We give in the first column the set of absorption energies calculated, in the second column values of Eq. (4.59), namely:

$$G(p) = \left(\frac{1}{2\pi\sigma^2}\right)^{1/2} \exp\left[-\left(\frac{(p-S)^2}{2\sigma^2}\right)\right] \quad (10.10)$$

for $\sigma^2 = 41\langle 2m+1 \rangle = 42.96$, evaluated at the listed $h\nu_{abs}$ energies with the Gaussian centered for $p = S$ at 33 400 cm^{-1}. In the third, fourth, and fifth columns we give the equal-force-constants $E_{zp,vu}$, the proper equal-force constants p and the W_p function for which $G(p)$ is the Gaussian approximation. The comparison

Table 20. Representative parameters for fitting LaOCl:Eu absorptions

θ	$S_v \hbar\omega_v$	$S_u \hbar\omega_u$	$\hbar\omega_v$	$\hbar\omega_u$	S_v	S_u	a_{uv}	$E_{zp,vu}$
46.000	7 131.2	6 201.7	169.70	158.25	42.022	39.188	12.744	26 274.5
			200.00	186.51	35.656	33.251	11.739	26 275.5
			300.00	279.77	23.771	22.167	9.585	26 278.9
			400.00	373.02	17.828	16.626	8.301	26 282.3
45.167	8 011.4	7 827.1	169.70	167.74	47.209	46.663	13.702	25 389.6
			200.00	197.69	40.057	39.593	12.621	25 389.8
			300.00	296.53	26.705	26.396	10.305	25 390.4
			400.00	395.37	20.028	19.797	8.925	25 390.9
45.000	8 200.0	8 200.0	169.70	169.70	48.321	48.321	13.903	25 200.0
			200.00	200.00	41.000	41.000	12.806	25 200.0
			300.00	300.00	32.800	32.800	11.454	25 200.0
			400.00	400.00	20.500	20.500	9.055	25 200.0
44.833	8 393.1	8 590.7	169.70	171.69	49.458	50.037	14.106	25 005.9
			200.00	202.34	41.965	42.456	12.994	25 005.8
			300.00	303.51	27.977	28.304	10.610	25 005.2
			400.00	404.68	20.983	21.228	9.188	25 004.6

Table 21. Closeness of the fits to the absorption data

hv_{abs}	G_p	$E_{zp,vu}(45)$	p	W_p	$E_{vu,zp}(46)$	p_v	V_{p_v}
33 400	.0608	25 200	41	.0607	26 275	35.656	.0621
31 400	.0190		31	.0177		25.656	.0187
29 400	.00057		21	.00039		15.656	.00020

of the second and fifth columns shows that if one fits, so does the other fit the absorption band of this phosphor. The W_p function involved is for $S = 46$, i.e., for $a_{uv} = 12.806$, and is evaluated at $kT/\hbar\omega = .2676$ appropriate for $T = 77$ K. Finally, in the sixth through eighth columns we give the equivalent information for the $\theta = 46°$ case, with $a_{uv} = 12.744$ and $kT/\hbar\omega_u = .3382$. This too leads, as stated, to a reasonable fit.

We deem the various descriptions of the LaOCl:Eu absorption band in Table 20 practically indistinguishable. The selection of some best set of parameters is done below using the quenching and the feeding behaviors.

10.1.5.2 Oxysulfides

For La_2O_2S and Y_2O_2S, we have also fit the absorption bands originally using the Klick-Schulman semiclassical model equation, using Eq. (10.7) with equal force constants, with $S_u\hbar\omega_u = S_v\hbar\omega_v = 4800$ cm^{-1} for both hosts and with $hv_0 = 29900$ cm^{-1} and $= 28800$ cm^{-1} for Y_2O_2S and La_2O_2S, respectively. As in the LaOCl case, the corresponding W_p distributions, namely those listed in Table 22

10.1 Eu in Oxysulfides and in Oxyhalides

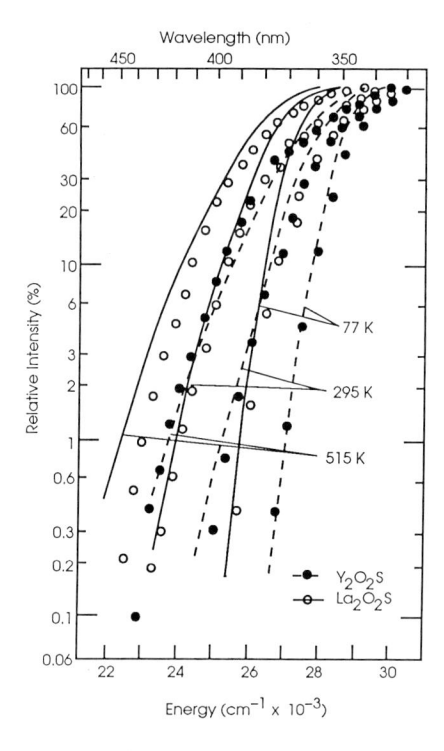

Fig. 30. Equal force constants fit of oxysulfide absorption bands

Table 22. Parameters for fitting oxysulfide: Eu $^5D \to$ CTS data

	Y$_2$O$_2$S		La$_2$O$_2$S	
u	$^7F_0$	$^7F_0$	$^7F_0$	$^7F_0$
v	CTS	CTS	CTS	CTS
*θ	44°	45°	44°	45°
*a_{uv}	7.58	10.62	8.32	10.62
S_u	14.87	28.2	17.91	28.2
S_v	13.86	28.2	16.70	28.2
*$\hbar\omega_u$	295	170	295	170
$\hbar\omega_v$	275	170	275	170
$S_v \hbar\omega_v$	3 810	4 800	4 590	4 800
*$h\nu_{zp,vu}$	26 300−5kT/2	24 900	24 700−5kT/2	23 600
*$\beta' A_{uv}$	2.5 + 8	2.3 + 7	2.5 + 8	2.3 + 7

under the columns headed by $u \to v = {}^7F_0 \to$ CTS and $\theta = 45°$, will also provide a fit. The fit is shown in Fig. 30.

Table 22 gives values for the quantities listed in the first column. The starred ones are fundamental; those not starred can be derived from the fundamental parameters. For the moment, we concentrate on the starred quantities θ, a_{uv}, $\hbar\omega_u$, $h\nu_{zp,vu}$ in the third and fifth columns, for $\theta = 45°$. From these, using Eq. (3.30), we obtain $\hbar\omega_v$. For $\theta = 45°$, $S_0 = S_u = S_v$ and then Eq. (3.31) yields S_0. At the three temperatures shown in Fig. 30 we know $\langle m \rangle$ from Eq. (4.43) and we then know the $W_p(S_0, \langle m \rangle)$ distribution from, e.g., Eq. (4.45). The proper p_j for $^7F_j \to$ CTS absorption at some energy $h\nu_{abs}$ is known from the energy balance Eq. (1.4); the $h\nu_{zp,vu}$ and the chosen $\hbar\omega_u$ are listed in Table 22; and the ^7F_j energy are listed in Table 18. The computed W_{p_j} of Eq. (4.45) is weighted by the partition function, namely, $(2j + 1)\exp\{-E(^7F_j)/kT\}/[\sum_i (2j + 1)\exp\{-E(^7F_i)/kT\}]$. The thus weighted contributions from all initial ^7F_j states are summed and plotted as the curves in Fig. 30. The points in this figure are the experimental measurements taken from Ref. [1].

Even more strongly than with the LaOCl case, we are led especially from the observed feeding fractions to an alternative fit involving unequal force constants. The alternative fit, shown in Fig. 31, is that in the second and fourth columns of Table 22, involving $\theta = 44°$.

Here it has been found necessary also to mimic anharmonic effects, i.e., expansion of the lattice with increasing temperature, by allowing the position of the CTS to decrease slightly with temperature, by $5kT/2$ units, as shown in Table 22. We will find striking evidence forcing us to adopt this variation in the energy position of the offset states when we consider ruby in Sect. 10.2. Below, we shall see that even for LaOCl we are led by the quenching data to this same magnitude for the anharmonicity.

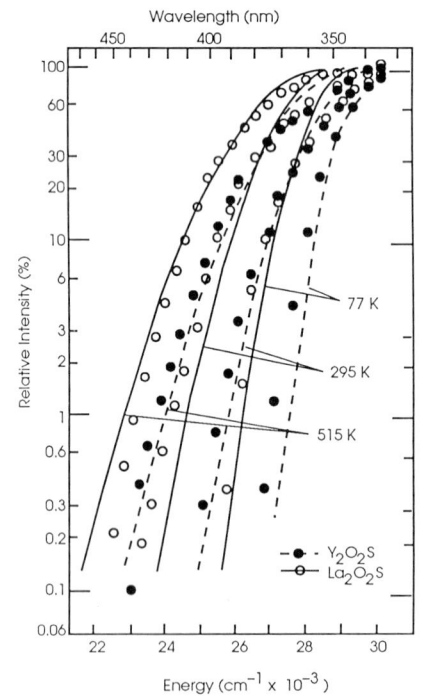

Fig. 31. 44° fit of oxysulfide absorption bands

10.1 Eu in Oxysulfides and in Oxyhalides

Once again for unequal force constants, the starred quantities in Table 22 are the fundamental ones. From them, using Eqs. (3.30), (3.31), (4.43), the numerically computed V_{p_v} distributions of Eq. (4.24), the p_j selection and the thermal and degeneracy weighting discussed in the $\theta = 45°$ fit above, we obtain the curves of Fig. 31. The points in this figure are those of Fig. 30, originally from Ref. [1].

Both $\theta = 45°$ and $\theta = 44°$ fits are deemed adequate: the selection will be made from fitting the quenching temperatures and feeding fractions in these phosphors.

10.1.6 Fitting the Quenching Data

10.1.6.1 LaOCl

Figure 29 shows that in LaOCl, the Eu^{3+} $^5D_3$ emissions quench to half efficiency near 270 K and $^5D_2$ emissions quench to that efficiency near 450 K. We now use these experimental facts to select the subset of the parameters fitting the absorption band which also fit these additional experimental facts.

We give in Table 23 additional calculated properties for several of the set of possible parameter values in Table 20. When an entry has two lines, the first is for a temperature-independent placement of the CTS, the second allows a downward adjustment by $5kT/2$ in its placement and leads to the adjusted p_j and V_{p_v} values listed on the second line of the entry. The two parameter sets which have entrees in the last three columns are deemed acceptable. These last two columns are for the Eq. (10.5) "quenching efficiencies" discounting any additional weak temperature dependences. These efficiencies are calculated using the V_{p_v} values in the fourth and sixth columns, respectively, and the prefactor $\beta'A_{vu}$ listed in the seventh column. The differences between these calculated efficiencies and the value 0.50 derived from Fig. 29 represents only a few degrees displacement of the quenching curves. We neglect this small discrepancy and ascribe it to having adopted our approximations.

None of the parameter choices in Table 23 give adequate fits for a temperature-independent placement of the CTS. Of course, upon relieving the approximation of equal electronic matrix elements, one might conceivably expect a more reasonable fit. We deem it far more probable that anharmonicity, to which we ascribe the temperature-dependent lowering of the CTS, does indeed exist and does have this effect. In our opinion, the two parameter sets pointed to in the Table are much more plausible.

The other parameter sets of this table do not yield as reasonable fits even allowing the downward shift of the CTS with temperature. For example, the next-to-last entry has entries for both invariant and downward-shifted CTS placements. The key to a successful fit is the equality of the two V_{p_v} values. This 400 cm^{-1} fit clearly is failing this key test more than the two accepted parameter sets.

10.1.6.2 Oxysulfides

Figures 32 and 33 show the fits to the La_2O_2S and Y_2O_2S quenching data for the equal-force-constants case of the third and fifth columns of Table 22 and for

Table 23. Representative parameters for fitting LaOCl:Eu quenching

θ	$\hbar\omega_v$	$p(^5D_3)$	$V_p(^5D_3)$, 270K	$p(^5D_2)$	$V_p(^5D_2)$, 450K	$\beta' A_{vu}$	$(\eta_3')_3$	$(\eta_2')_2$
45	169.7	−5.598	.5795-7	−22.481	.2374-7	5.65 + 6	.44	.56
		−2.848	.2251-6	−17.898	.1393-6			
45.16	169.7	−6.715	.4914-7	−23.598	.1905-7			
44.83	200.	−3.779	.8931-7	−18.104	.3598-7	3.82 + 6	.44	.56
		−1.440	.3379-6	−14.215	.2022-6			
44	300.	0.966	.2499-5	−8.584	.1516-6			
45	300.	−3.167	.1512-6	−12.717	.6035-7			
45.16	300.	−3.801	.1216-6	−13.351	.3568-7			
44	400.	0.734	.1094-4	−6.429	.2388-6			
44.83	400.	−1.886	.3923-6	−9.049	.8365-7			
		−0.719	.1487-5	−7.105	.4782-6			
45	400.	−2.375	.3867-6	−9.537	.6718-7			

10.1 Eu in Oxysulfides and in Oxyhalides   143

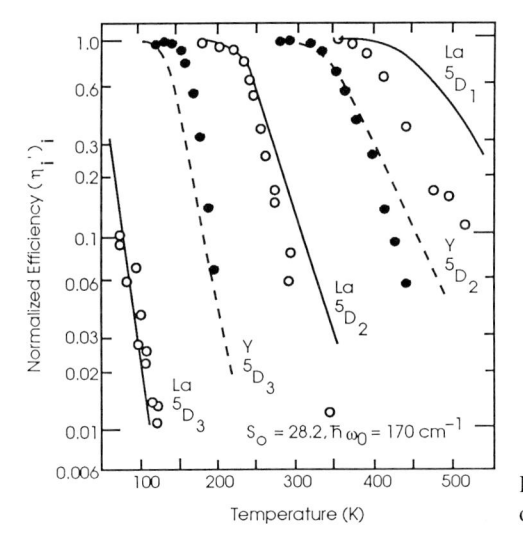

Fig. 32. Equal force constants fit of oxysulfide quenching data

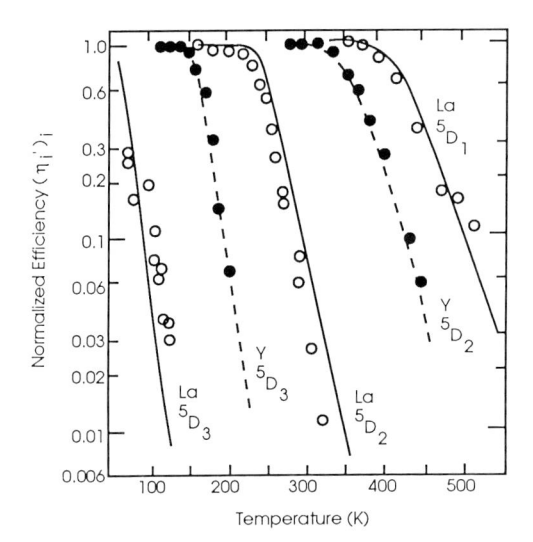

Fig. 33. 44° fit of oxysulfide quenching data

the unequal-force-constants case, involving a downward shift in the position of the CTS with temperature, of the second and fourth columns of this Table. The curves are the Eq. (10.5) "quenching efficiencies" with the V_{pv} values calculated through Eq. (4.24) and the parameters of Table 22. The points are ratios of data points taken from Figs. 4, 5, and 8 of Ref. [2]. Both fits are deemed adequate, although the unequal-force-constants fit seems slightly better.

10.1.6.3 Fitting the $(\eta_i)_i$

Finally, we return to Eqs. (10.4). In Fig. 34 $(\eta_i)_i$, the measured efficiencies of the $^5D_i \rightarrow {}^7F$ emissions in $(Y_{0.999}, Eu_{0.001})_2O_2S$ when excited into the emitting states,

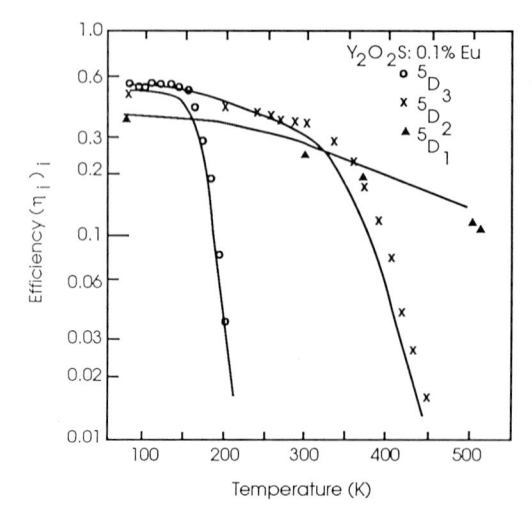

Fig. 34. Measured and calculated efficiencies in $(Y_{0.999}Eu_{0.001})_2O_2S$ are shown as a function of temperature. The data points are from Figs. 4, 5, and 6 of Ref. [2]. The plots are the calculated curves using Eqs. (10.4) with the parameters in Table 24 to describe the $4f \to 4f$ transitions and with the parameters in Table 22 to describe the $^5D \to CTS$ quenchings.

The first three columns of Table 24 all give reasonable fits to $(\eta_1)_1$. We choose the second column because we believe that $A_{uv} \approx 10^{12}$ and that $R_{uv} \approx 10^4$. The $(\eta_2)_2$ and $(\eta_3)_3$ are then calculated after a search for the S_0 which best fits each's low-temperature $4f \to 4f$ nonradiative rates.

Several points are worth making. First, Moos [27] has discussed the general forbiddenness of $^5D_1 \to {}^5D_0$ transitions relative to $^5D_2 \to {}^5D_1$ and $^5D_3 \to {}^5D_2$ transitions. This forbiddenness is seen to be a factor of ~ 3 in the case studied here.

The second point is that in visually discounting the slow $4f \to 4f$ transitions and thus obtaining experimental "quenching efficiencies" to be compared with

Table 24. Parameters for fitting $Y_2O_2S:Eu^{3+}$ $4f \to 4f$ transitions

u	$^5D_0$	$^5D_0$	$^5D_0$	$^5D_1$	$^5D_2$
v	$^5D_1$	$^5D_1$	$^5D_1$	$^5D_2$	$^5D_3$
*θ	45°	45°	45°	45°	45°
*a_{uv}	0.155	0.346	0.632	0.632	0.894
S_0	0.006	0.03	0.10	0.101	0.20
*$\hbar\omega_0$	438	435	429	435	435
$S_0\hbar\omega_0$	2.62	13.0	42.9	43.5	67.0
*$h\nu_{zp,vu}$	1 800	1 800	1 800	2 420	2 860
*A_{uv}/R_{gv}	6.6 + 10	1. + 8	1. + 6	1. + 8	1. + 8

10.1 Eu in Oxysulfides and in Oxyhalides

Eq. (10.5), no gross violence has been done, for the same description of the ^5D – CTS interactions has been used in both.

Third, an additional advantage of the unequal-force-constants fitting of the quenching performance in oxysulfide hosts, shown in Fig. 33, over the equal-force-constants counterpart in Fig. 32, is the absolute value of the $(\eta_3)_3$ data points. Because we have not followed these emission intensities below 77 K, we do not know the experimental normalization factor. Therefore we have normalized these emission intensities so as to bring them close to the fitted theoretical $(\eta_3)_3$ curves in both cases. Our belief is that the Fig. 33 efficiencies are closer to the correct values as a function of temperature. This constitutes the additional advantage tending toward establishing the unequal-force-constants description.

Fourth, we have chosen to describe the CTS – $4f^6$ interactions with phonon energies near 300 cm^{-1} and the $4f^6 - 4f^6$ interactions with above 400 cm^{-1} phonon energies. This is found to be a general trend, that the fast large-offset transitions are dominated by the lower-energy phonons while the slow small-offset transitions are dominated by the larger-energy phonons. It is to be expected that some multi-frequency model functions, such as those developed in Sect. 5, will encompass both performances simultaneously. However, it is very likely that the description thus formed will not prove particularly enlightening, in that the disjoint performance seen will be interpreted in disjoint descriptions of the corresponding single-mode A and S parameters for these transitions.

10.1.7 Fitting The Feeding Fractions

10.1.7.1 LaOCl

Table 25 gives, for several of the Table 23 cases, the 77 K U_{p_v} calculated values needed to calculate the feeding fractions according to the last form of Eq. (10.2). In all cases the α_3 was calculated extremely small.

The first and the third cases are those which gave adequate understanding of the thermal quenching data. We see here, first, that none of these parameter sets gives an adequate understanding of the feeding fractions. The difficulty is in expecting modest values for α_3 and α_2. Those cases which tend to support values of α_2 near the observed .04 weight overestimate α_1 and underestimate α_0.

We adopt the very first description, for equal force constants and ≈ 170 cm^{-1} phonon energy. We ascribe the observed α_3 and α_2 to crossovers before thermal equilibrium is established in the vibrational levels of the CTS.

It is our expectation that tenths of picoseconds are needed for establishing thermal equilibrium within the CTS and that several picoseconds might be needed to depopulate the CTS via CTS $\rightarrow$ ^5D feedings. Thus, it would be perhaps surprising if indeed 1–10% of the feedings did not show evidence of pre-thermal-equilibrium crossovers.

10.1.7.2 Oxysulfides

Also for the oxysulfides, the feeding fractions allow a final selection of a most reasonable description of the Eu^{3+} center. Table 26 shows the calculated α_j values

Table 25. Representative parameters for fitting LaOCl:Eu feeding fractions at 77 K

θ	$\hbar\omega_v$	$p(^5D_2)$	$U_p(^5D_2)$	$p(^5D_1)$	$U_p(^5D_1)$	$p(^5D_0)$	$U_p(^5D_0)$	α_3	α_2	α_1	α_0
desired											
45	169.7	22.481	.0000375	37.006	.01657	47.378	.05795	.01	.04	.45	.50
45.16	169.7	23.874	.0001712	38.570	.03050	49.063	.05204	.00	.00	.47	.52
44.83	200.	17.894	.0000208	30.077	.01058	38.775	.05289	.00	.01	.63	.36
44	300.	8.005	.0000002	15.667	.00345	21.138	.00674	.00	.00	.37	.62
45	300.	12.717	.0008726	20.933	.03841	26.800	.07592	.00	.00	.61	.39
45.16	300.	13.507	.00230	21.820	.05556	27.756	.07289	.00	.02	.59	.38
44	400.	5.995	.0008581	11.742	.00157	15.845	.01438	.00	.05	.66	.29
44.83	400.	8.944	.001452	15.036	.03761	19.385	.08219	.00	.18	.20	.61
45	400.	9.537	.003258	15.700	.05398	20.100	.08803	.00	.04	.56	.41
								.00	.06	.61	.33

10.2 Oxysulfides: Other Rare Earths

Table 26. CTS → ^5D_i feeding fractions in oxysulfides: Eu

Host	% Eu	T, K	α_3	α_2	α_1	α_0	
La$_2$O$_2$S	0.5	77	0.02	.30	0.65	0.04	obs.
		77	0.00	0.31	0.63	0.06	calc. for $\theta = 44°$
		295		0.32	0.57	0.11	calc. for $\theta = 44°$
		77	0.00	0.02	0.91	0.07	calc. for $\theta = 45°$
		295		0.21	0.67	0.12	calc. for $\theta = 45°$
Y$_2$O$_2$S	0.1	77	0.20	.65	0.15	0.02	obs.
	0.1	295		.59	0.37	0.03	obs.
	2.0	295		.60	0.40	0.00	obs.
	2.0	495		.46	0.53	0.02	obs.
		77	0.13	0.83	0.04	0.00	calc. for $\theta = 44°$
		295		0.85	0.14	0.01	calc. for $\theta = 44°$
		77	1.-6	0.61	0.38	0.003	calc. for $\theta = 45°$
		295		0.61	0.37	0.02	calc. for $\theta = 45°$

at several temperatures for the two descriptions of this center in these oxysulfide hosts in Table 22.

Once again, the crucial data are the large observed α_3 for (Y$_{0.999}$, Eu$_{0.001}$)O$_2$S at 77 K and the large observed α_2 for (La$_{0.995}$, Eu$_{0.005}$)O$_2$S. Here, the $\theta = 44°$ descriptions are clearly preferable to the $\theta = 45°$ descriptions. The remaining discrepancies are, for (La$_{0.995}$, Eu$_{0.005}$)O$_2$S, indicative of pre-thermal-equilibrium crossovers into $^5D_3$. For (Y$_{0.999}$, Eu$_{0.001}$)O$_2$S, a slightly larger $\theta \approx 44.2°$ might afford a slightly better fit but there would still be some 10% feeding into $^5D_3$ to ascribe to pre-thermal-equilibrium crossovers. Of the two descriptions considered, however, the $\theta = 44°$ parameters clearly are the better choice.

10.2 Oxysulfides: Other Rare Earths

10.2.1 La$_2$O$_2$S:Tm^{3+}

10.2.1.1 The Experimental Behavior

The energy level diagram for Tm^{3+} in a La$_2$O$_2$S host lattice [20] is shown in Fig. 35. The $4f^{12}$ states are placed as reported in Ref. [58], ignoring crystal field splittings.

Strong host-lattice absorption prevents direct observation via typical absorption or excitation spectra of any level above 35 500 cm^{-1}. Below 35 000 cm^{-1}, however, excitation and emission transitions are seen at energies, listed in Table 27, which are consistent with the diagram. The spectra supporting the energy values listed in Table 27 are in Figs. 36–38.

Figure 38 shows the nonlinear two-photon excitation (TPE) spectrum of

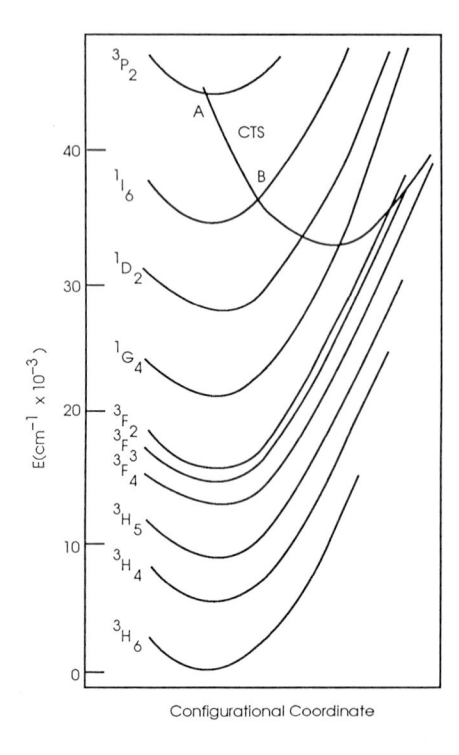

Fig. 35. Energy level diagram for La$_2$O$_2$S: Tm^{3+}

Table 27. Optical transition energies, cm^{-1}, in La$_2$O$_2$S:Tm^{3+}

Transition[a]	Excitation			Emission		
$^1I_6 \rightarrow {}^3H_6$	34 000	34 500		...[b]		
$^1I_6 \rightarrow {}^3H_4$				28 325		
$^1I_6 \rightarrow {}^3H_5$				25 650	25 750	
$^1I_6 \rightarrow {}^3F_4$				21 325	21 475	
$^1D_2 \rightarrow {}^3H_6$	27 500	27 625	27 850	27 350	27 500	
$^1D_2 \rightarrow {}^3H_4$				21 850	21 925	22 050
$^1D_2 \rightarrow {}^3H_5$				18 975	19 175	19 300
$^1D_2 \rightarrow {}^3F_4$				14 875	14 975	15 025
$^1G_4 \rightarrow {}^3H_6$	20 800	20 950	21 025	20 600	20 800	
$^1G_4 \rightarrow {}^3H_4$				15 175	15 225	15 350

[a] Observed crystal field states have not been included in this assignment.

[b] $^1I_6$ emissions have been observed only for direct excitation into $^1I_6$ itself; after excitation into the host lattice absorption band, the $^1I_6$ level is poorly fed. The resolution used here was too low to allow recording the $^1I_6 \rightarrow {}^3H_6$ emission.

10.2 Oxysulfides: Other Rare Earths

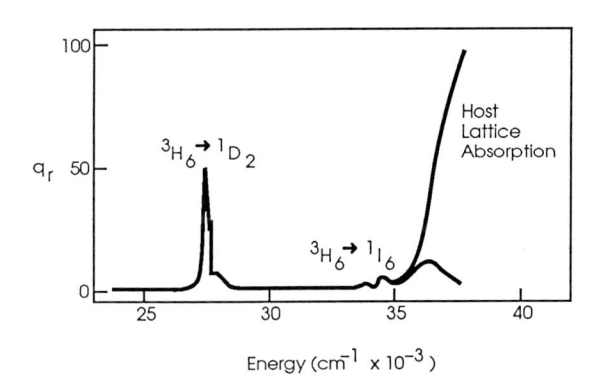

Fig. 36. Excitation spectrum for $La_2O_2S:Tm^{3+}$ $^1D_2 \to {}^3H_4$ emission at 21 930 cm^{-1} at 5 K

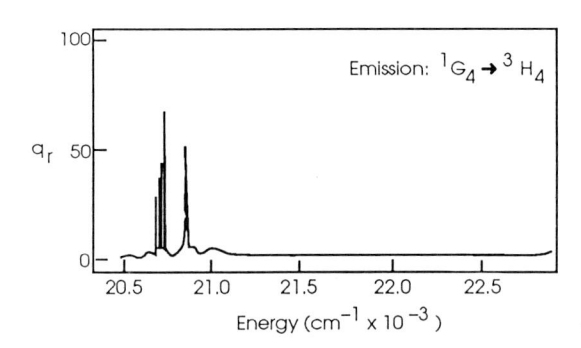

Fig. 37. Excitation spectrum for $La_2O_2S: Tm^{3+}$ $^1G_4 \to {}^3H_4$ emission at 15 380 cm^{-1} at 5 K

$^1D_2 \to {}^3H_6$ emission at 27 400 cm^{-1}. Two spectra are shown, the weaker one with a 50% transparency filter, the stronger one without such a filter in the excitation beam. The excitation was the output of a tunable dye laser (Molectron DL 200) pumped with a nitrogen laser (Molectron UV 14). The dyes used for output between 18 500 and 24 000 cm^{-1} were Coumarin 120 (Exciton), 7D4MC (Molectron), and Coumarin 485 (Exciton). The intensities of the output from the various dyes was measured using Lumogen T rot GG (BASF) as a standard.

The ratio of these two curves is a factor of 4 and the excitation is therefore two-photon nonlinear absorption into states at twice the energy of the scale.

In comparing Figs. 37 and 38, we see three features of importance. First, the sharp lines between 20 500 and 21 000 cm^{-1} have too much overlap not to ascribe the Fig. 38 sharp lines to excited state absorption from $^1G_4$ into the broad-band state. Second, the broad band itself, here ascribed to the CTS, is reached by the nonlinear two-photon absorption of the exciting laser beam without any intermediate absorption. Third, the sharp line near 22 000 cm^{-1} is also a nonlinear two-photon absorption into a small-offset Tm^{3+} $4f^{12}$ state, here ascribed to $^3P_2$. The nonlinear absorption in both cases is made evident by the factor of four in the intensity of the two curves of Fig. 38, which are themselves related by a factor of two in the excitation intensity.

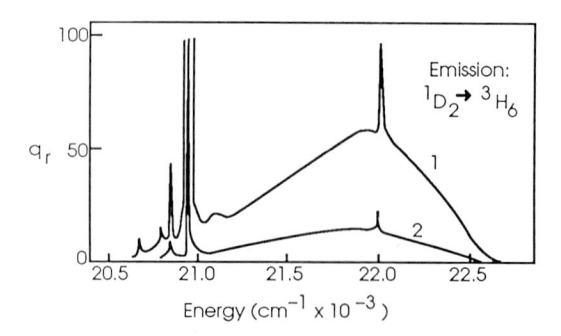

Fig. 38. 2-Photon excitation spectrum of La$_2$O$_2$S:Tm^{3+} ^{1}D$_2$ → ^{3}H$_6$ 5 K 27 400 cm^{-1} emission

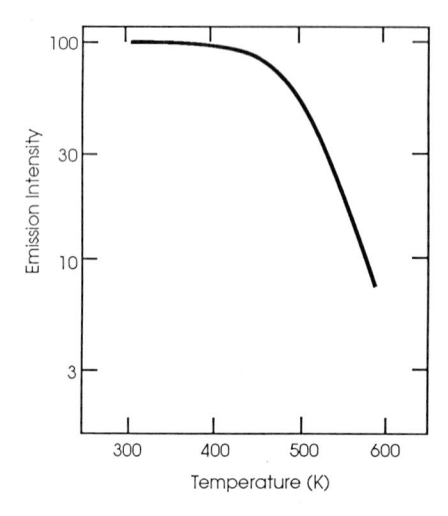

Fig. 39. Thermal quenching of La$_2$O$_2$S: Tm^{3+} ^{1}D$_2$ → ^{3}H$_4$ emission at 457 nm

Figure 39 shows the quenching of the Tm^{3+} ^{1}D$_2$ → ^{3}H$_4$ emissions at 457 nm. under ^{3}H$_6$ → ^{1}D$_2$ excitation at 363 nm.. The quenching is precipitous near 500 K. In Ref. [5] it was stated that what were described as ^{3}P$_0$ emissions and are better described as ^{1}I$_6$ emissions quench precipitously at ∼ 40 K. There, indeed, these two quenchings were stated to follow the quenching behavior,

$$I(T)/I_0 = [1 + A \exp(-E/kT)]^{-1} \quad (10.11)$$

with A ∼ 10^7 and $E = 6000$ cm^{-1} and 400 cm^{-1}, for ^{1}D$_2$ and ^{1}I$_6$ quenchings, respectively.

10.2.1.2 Fitting with W$_p$ Functions

The first point to determine is that the broad band absorption shown in Fig. 38 cannot explain the observed sequential quenching of ^{1}I$_6$ at 40 K and of ^{1}D$_2$ at 500 K.

The best chance, which fails, is to explain this absorption curve as indeed two bands, the first at 42 500 cm^{-1} with a width at half height equal to twice the

10.2 Oxysulfides: Other Rare Earths

two-photon width at quarter height, namely 2200 cm^{-1}, and the second at 43 700 cm^{-1} with a width at half height 2400 cm^{-1}. Using the relationships:

$$\sigma^2 = \frac{(\Delta_{1/2})^2}{8 \ln 2} = S\langle 2m + 1\rangle(\hbar\omega)^2 \approx S(\hbar\omega)^2 \tag{10.12}$$

noting that $S\hbar\omega$ will be larger the smaller $\hbar\omega$ is, and finally adopting the smallest plausible $\hbar\omega$, namely 100 cm^{-1}, we are led to $S = 87.28$, $S\hbar\omega = 8728$ cm^{-1}. Since $h\nu_{zp,vu}$ for $^1I_6$ is 34 000 cm^{-1}, and even allowing for anharmonicity by introducing a downward shift by $(5/2)kT$ in the CTS placement, we find, through the relations:

$$E_{zp,vu,CTS}(T) = 42\,500 \text{ cm}^{-1} - S\hbar\omega + \tfrac{1}{2}\hbar\omega - \tfrac{5}{2}kT$$

$$p = (E_{zp,vu,CTS}(T) - E_{zp,vu,I})/\hbar\omega \tag{10.13}$$

that $p = 2.65$ at 50 K and that at this temperature $W_p = 1.30 \times 10^{-15}$. No reasonable value for the electronic integral will give a nonradiative rate that will compete against a millisecond time constant radiative process. That is, no quenching of this level is predicted near 40 K. Furthermore, no variant of this cited parameter set, aimed at describing a state with its absorption maximum near 42 500 cm^{-1}, both gave nonradiative rates which with reasonable electronic integrals compete against emission and simultaneously gave a sharp increase of this nonradiative rate near 40 K.

There is a level placement which will indeed explain the observed sequential quenchings. We take $S = 16$, $\hbar\omega = 200$ cm^{-1}, $E_{zp,vu,CTS} = 34\,200$ cm^{-1}, which, with $E_{zp,vu,I} = 34\,000$ cm^{-1}, gives from Eqs. (10.13) $p_I = -1$, $\Delta_{1/2} = 1884$. We ascribe values of radiative lifetimes of the order 10^{-4} and $10^{-3.5}$ to $^1I_6$ and to $^1D_2$, respectively. We allow a downward shift in the CTS placement according to:

$$E_{zp,vu,CTS}(T) = E_{zp,vu,CTS}(0\,K) - 2.8\,kT \tag{10.14}$$

and we take $A_{vu} = 10^{12.5}$ for both $^1I_6 \to$ CTS and $^1D_2 \to$ CTS transitions. Table 28 gives the quenching efficiency of $^1I_6$ emissions predicted and Table 29 gives those for $^1D_2$ emissions.

In Tables 28 and 29 we do indeed find $^1I_6$ quenching at ~ 40 K and $^1D_2$ quenching at ~ 500 K.

The question now arises whether anything in the absorption spectra gives evidence of such a state. From Eq. (10.12) we estimate the width at half height for this band to be near 1 900 cm^{-1}. A band shaped as the one hand-sketched into Fig. 36 would have its peak absorption at $\sim 37\,000$ cm^{-1}, near 34 200 +

Table 28. Quenching of $^1I_6$

T, K	10	35	40	45	50
W_{-1}	.572-18	.496-9	.145-8	.351-8	.758-8
$N = 10^{12.5}W_{-1}$	1.81-6	1.57 + 3	4.59 + 3	1.11 + 4	2.40 + 4
$\eta = R/(R + N)$	.98	.86	.68	.47	.29

Table 29. Quenching of $^1D_2$ Tm^{3+} emissions in La_2O_2S through the CTS

T, K	300	350	400	450	500	550	600
$E_{zp,vu,CTS}(T)$	33 616	33 519	33 422	33 324	33 227	33 130	33 032
$E_{zp,vu,D}$	27 500	27 500	27 500	27 500	27 500	27 500	27 500
$p_{CTS,D}$	−30.558	−30.094	−29.608	−29.121	−28.635	−28.148	−27.662
$W_{|p_{CTS,D|}}$	.00570	.00575	.00804	.01050	.01295	.01536	.01773
r^p	1.862-13	1.796-11	5.615-10	8.174-9	6.961-8	4.018-7	1.730-6
$W_{p_{CTS,D}}$	1.062-15	1.032-13	4.513-12	8.579-11	9.014-10	6.171-9	3.069-8
$N = 10^{12.5} W_{p_{CTS,D}}$	.00335	.3263	14.271	271.3	2850.	19 514.	97 050.
η	1.0	1.0	.996	.92	.53	.14	.03

10.2 Oxysulfides: Other Rare Earths

$3200 = 37\,400\,\mathrm{cm}^{-1}$, would have its width at half height at $1900\,\mathrm{cm}^{-1}$, and would account for some of the gradual rise of the host lattice absorption in this spectral region, i.e., the very weak shoulder on the onset of the host-lattice absorption curve in Fig. 36 near this energy.

The demand, because of the $^1I_6$ and $^1D_2$ quenchings, for placing an offset state at this energy with this $S\hbar\omega$ offset is in our opinion strong. The circumstantial evidence would be strengthened if the amplitude of this shoulder should be found to depend on the Tm^{3+} concentration. It would also be convincing to demonstrate that when $^1I_6$ emissions quench, they are replaced with $^1D_2$ emissions through two-step $^1I_6 \to CTS \to {}^1D_2$ processes.

It is to be expected that two-photon absorptions might have their transition selection rules so that some states are not coupled well to the ground state. The assignment of the states uncovered by the two photon experiments is an aim of research in itself. Their lifetime against decay into the host-lattice electronic states would be of great interest.

10.2.2 $Y_2O_2S:Yb^{3+}$

10.2.2.1 The Experimental Behavior

The experimental behavior [59, 21, 60] of Yb^{3+} in $Y_2O_2S:Yb^{3+}$ is understandable in terms of the energy level diagram shown in the insert in Fig. 40. In this figure, the two $4f^{13}$ $^2F_{5/2}$ and $^2F_{7/2}$ states are labeled u and u'. The state labeled v is the offset CTS of Yb^{3+}.

Also shown in this diagram are computed optical spectra: the narrow line seen in both emission and absorption, two broad band emissions labeled $v \to u'$ and $v \to u$, and the broad band absorption labeled $u \to v$. These computed spectra are in agreement with the positions and shapes at 300 K, as observed both by

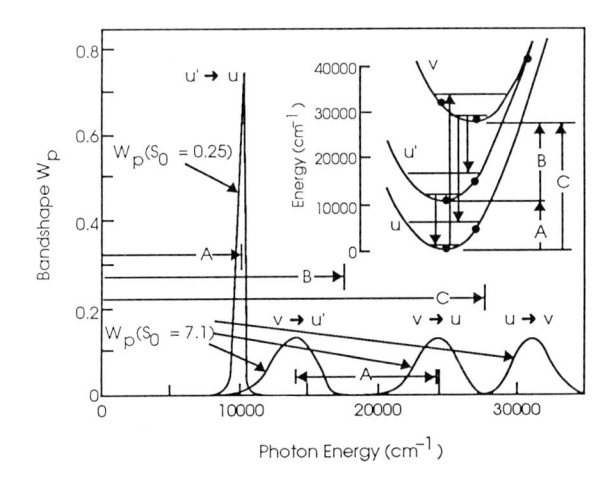

Fig. 40. Optical transitions of $Y_2O_2S:Yb^{3+}$ and the SCC energy diagram

Nakazawa [60] and by us. Higher CTS are seen in the absorption spectrum but are ignored in Fig. 40. The zero-phonon energies appropriate to the transitions labeled A, B, and C in the energy diagram are shown as corresponding vectors along the energy scale.

When excitation is into the CTS at $\sim 30\,000$–$35\,000\,\mathrm{cm}^{-1}$ at low temperatures, both broad bands $v \to u'$ and $v \to u$ as well as the narrow line $u' \to u$ are seen in emission. At low Yb^{3+} concentrations, the ratio of the $v \to u'$ and $u' \to u$ integrated intensities is unity. One is seeing the sequence of two emissions $v \to u' \to u$ competing with the direct $v \to u$ emission.

As the temperature is raised, nonradiative $v \to u'$ transitions take over from both broad-band transitions. The $u' \to u$ intensity grows with temperature and the $v \to u'$ intensity decreases with temperature.

At higher Yb^{3+} concentrations, an additional depopulation of the CTS is observed at low temperatures, due to the energy transfer $v, u \to u', u'$. Then both broad bands are decreased in intensity and the line emission is increased, even at low temperatures.

Figure 41 shows schematically the nonradiative $v \to u'$ transitions which become dominant at high temperature and the energy transfer $v, u \to u', u'$ transitions which become dominant at high Yb^{3+} concentrations. Note especially that the energy transfer occurs between a large-offset center and a small-offset center, and has a large energy mismatch, that is, creates many phonons. Nevertheless this transfer can dominate over emission.

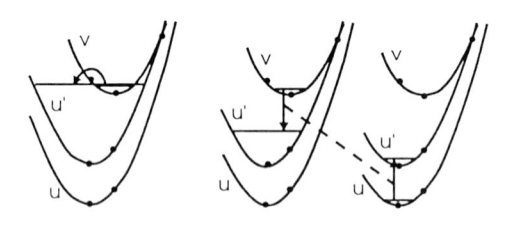

Fig. 41. Nonradiative and energy transfer transitions in $\mathrm{Y_2O_2S:Yb^{3+}}$ and the SCC energy diagram

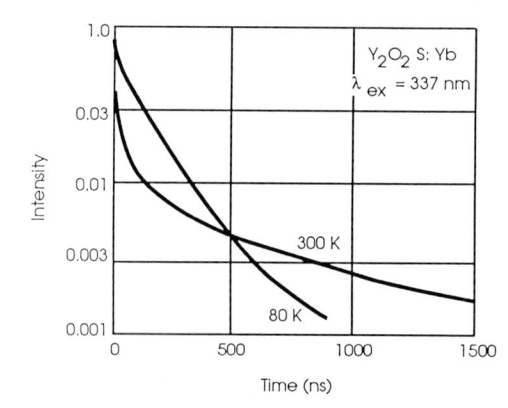

Fig. 42. Decay curves for Yb^{3+} CTS emissions in $\mathrm{Y_2O_2S}$

10.2 Oxysulfides: Other Rare Earths

The CTS emissions are found to quench near 250 K through $v \to u'$ nonradiative transitions. Also, at 77 K, the quantum efficiencies of the $v \to u$ and $v \to u'$ emissions are near unity for $(Y_{.9995}Yb_{.0005})_2O_2S$ and ~ 0.75 for $(Y_{.99}Yb_{.01})_2O_2S$.

The decay curves of these two compositions were given by Nakazawa [60] as shown in Fig. 42.

Note that these decay curves are distinctly nonexponential. We shall return to this later in this Chapter. At any rate, the radiative decay time for these emissions is of the order of 0.4 μs.

10.2.2.2 Fitting the Optical Spectra with SCC Model Functions

For the narrow-line transition $u \to u'$ and its inverse, we use equal force constants with $S = 0.25$ and $\hbar\omega = 500\,\text{cm}^{-1}$.

We give in Table 30 both Nakazawa's and our fitting parameters for the broad absorption and emission bands. Nakazawa used the semiclassical model of Eq. (10.6) and we have translated his parameters into the notation of this book as in the corresponding Eqs. (10.6)–(10.9). The required parameters are those starred. From them, using Eq. (3.30) we determined θ, then S_u, S_v, and $\hbar\omega_u$, and from them, using Eq. (3.31), a_{uv}. The zero-point energy was chosen to fit the absorption and emission energy placements. The second moments in absorption were calculated using the second of Eq. (4.112) truncated to the first term.

The $\theta = 45°$, $S_0 = 7.1$, $\hbar\omega = 500\,\text{cm}^{-1}$ fit of the second column was our best equal-force-constants fit of the data. The three other columns are our unequal-force-constants fits with the same sum of first moments and products of second moments over absorption and emission, through Eqs. (9.1) and (9.2). Our finding was that the $\theta = 44°$ fit was closest among these coarse-grid selections for the 300 K lineshapes. All five would be acceptable to us, but we would have some reluctance to admit θ below 43° since such phonon inequalities have not been needed in any other case.

The agreement in the second moments between Nakazawa's and our 44° fit at 300 K demonstrates the reproduceability of the properties of Yb^{3+} in this host. One can see in his spectra the downward shift in the position of the CTS with increasing temperature, and this should be introduced in his placement of the $h\nu_{zp,vu,CTS}$. We see some preference for his 300 cm^{-1} phonon energy in the observed temperature dependence of the linewidth; but, as will be made clear most dramatically in our discussion of ruby in Sect. 10.3, the phonon energy which explains the optical band shapes can be found different from and smaller than that which explains quenchings.

10.2.2.3 Fitting the Quenchings with SCC Model Functions

Our interest is in simultaneously understanding the optical bands, the $v \to u'$ quenching at ~ 250 K, and the low-temperature v, $u \to u'$, u' energy-transfer quenching at high Yb^{3+} concentrations. To this end, we have explored the descriptions of the $v \to u'$ quenching transitions in Table 30. We describe the energy transfer rate as in Eq. (6.6), namely, the combination of the large-offset $v \to u'$ and the small-offset $u \to u'$ distributions for generating the phonons

Table 30. Fitting the $4f^{13}$-CTS optical bands of $Y_2O_2S:Yb^{3+}$

	Nakazawa	Ours				
θ	43.584	45°	44°	43°	42°	
a_{uv}	7.054	5.329	5.318	5.283	5.228	
$\hbar\omega_u$, cm^{-1}	300.	500.	519.02	541.42	567.58	
S_u	13.	7.1	7.316	7.467	7.548	
$*S_u\hbar\omega_u$	3900	3550	3797	4043	4284	
$*\hbar\omega_v$	271.7	500.	484.02	470.81	460.15	
S_v	11.778	7.1	6.823	6.493	6.119	
$*S_v\hbar\omega_v$	3200	3550	3302	3057	2816	
$h\nu_{zp,vu,CTS}$	29800	29000-6.5 kT	29266-6.5 kT	30021-6.5 kT	30522-6.5 kT	
$\langle 2n+1 \rangle_{80\,K}$	1.048	1.000248	1.000177	1.000118	1.000074	
$\langle(p-\langle p\rangle)^2\rangle_v$ (80 K)	11.179	7.102	6.363	5.646	4.961	
σ^2, cm^{-1} (80 K)	825240	1775440	1490720	1251503	1050471	
$\langle 2n+1 \rangle_{300\,K}$	1.6219	1.2000	1.1810	1.1610	1.1407	
$\langle(p-\langle p\rangle)^2\rangle_v$ (300 K)	17.301	8.520	7.514	6.555	5.659	
σ^2, cm^{-1} (300 K)	1557100	2129940	1760412	1453097	1198197	

10.2 Oxysulfides: Other Rare Earths

needed for the energy offset in the $v, u \to u', u'$ transition. For equal force constants, this combined distribution is, through Eq. (6.5), itself a W_p function with its S the sum of the two component S's. For unequal force constants, this sum must be computed numerically.

We take anharmonicity and subsequent lattice expansion into account by lowering the CTS proportional to kT as shown. We adopt those values of $A_{vu'}$ and A_{xfer} which in our opinion are at the upper limit of plausibility so that the nonradiative rates are as large as possible. The rates are then described, for equal force constants, by $p = 37.6 - .013kT$, $S = 7.1$, $A_{u'v} = 10^{14}$ and, using Eq. (6.5), $p = 17.2 - .013\,kT$, $S = 7.35$, $A_{xfer} = 10^8$, respectively, for the $v \to u'$ nonradiative rate and the $v, u \to u', u'$ energy transfer rate.

At other θ values, all parameters are determined by Eqs. (9.1) and (9.2) from the equal-force-constants set in Table 30. In Table 31 we give some of the output of this Eqs. (9.1), (9.2) arithmetic for selecting the p values in the calculation of the $v \to u'$ rate at some representative temperatures.

In Fig. 43 we give the nonradiative rates thus calculated as a function of temperature. The dotted curves for each θ are the calculated $v \to u'$ rates. The full curves give the sum of the nonradiative $v \to u'$ and energy-transfer $v, u \to u', u'$ rates. These latter rates must compete near 250 K with the observed radiative rate of $\approx 2.5 \times 10^6$. It is seen that the $\theta = 44°$ fit, however good it is for the optical spectra, cannot afford a fit to this quenching. One needs some value near $\theta = 42°$. Secondly, the energy-tranfer rate at 77 K must be a quarter of the radiative rate, or near 0.6×10^6 in order to have the observed $\eta = 0.75$ for the 1% concentration samples at this temperature. This requirement is also met by the $\theta = 42°$ choice.

Table 31. Selecting the parameters for fitting the Y_2O_2S: Yb^{3+} quenchings

θ	45°	44°	43°	42°
a_{uv}	5.329	5.318	5.283	5.228
$\hbar\omega_u$, cm^{-1}	500.	519.02	541.42	567.58
$\hbar\omega_v$	500.	484.02	470.81	460.15
$h\nu_{zp,vu}(^2F_{7/2})$	10 000	10 000	10 000	10 000
$h\nu_{zp,vu,CTS}$	29 000-6.5 kT	29 266-6.5 kT	30 021-6.5 kT	30 522-6.5 kT
$\Delta E(80\ K)$	18 838	18 904	19 660	20 160
$p_U(80\ K)$	37.277	36.423	36.311	32.106
$kT/\hbar\omega_v(80\ K)$	.11120	.11487	.11810	.12083
$\Delta E(200\ K)$	18 097	18 362	18 892	19 619
$p_U(200\ K)$	36.193	35.378	34.893	34.565
$kT/\hbar\omega_v(200\ K)$	.27800	.27818	.29524	.30208
$\Delta E(250\ K)$	17 871	18 137	18 893	19 393
$p_U(250\ K)$	35.741	34.944	34.895	34.167
$kT/\hbar\omega_v(250\ K)$	.34750	.35898	.36905	.37760
$\Delta E(300\ K)$	17 645	17 910	18 666	19 167
$p_U(300\ K)$	35.289	34.507	34.476	33.769
$kT/\hbar\omega_v(300\ K)$	.41700	.43077	.44286	.45312

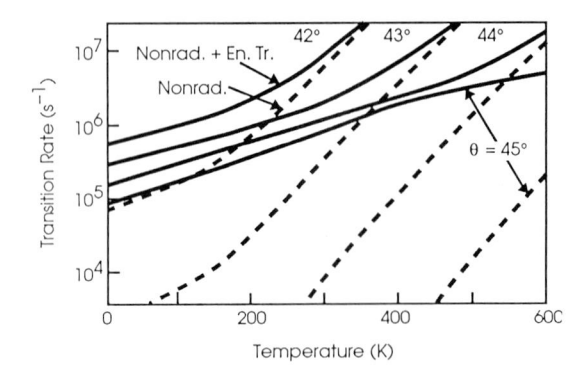

Fig. 43. Nonradiative rates vs. temperature in $Y_2O_2S:Yb^{3+}$

We are thus in the position that we can understand the optical bandshapes by using $\theta = 43.6°$ and $\hbar\omega_u = 300\,cm^{-1}$ but are forced, against our better judgment as to plausible values for the Manneback angle, to $\theta = 42°$ and $\hbar\omega_u = 576.6\,cm^{-1}$ for the energy transfer behavior and the thermal quenching behavior.

The defeat is, however, a lead into an effect beyond the SCC model. The nonexponential decays shown by Nakazawa were interpreted correctly by him as evidence of CTS breakup into a free hole and a deeply trapped electron as Yb^{2+}. If the dissociated CTS are dominantly nonradiatively recombined, then this rate must be added to the two nonradiative rates considered thus far to assess the competition between radiation and loss processes. AT 300 K, as much as 90% of the emission betrays the influence of this process, and thus the calculated total nonradiative rates shown in Fig. 43 must be raised by even an order of magnitude. Then the calculated nonradiative rates for all θ values become large enough to explain the observed quenchings. One may well find that $\theta = 43.6°$, with $A_{u'v}$ adjusted downwards by somewhat less than one order towards more plausible values, and even perhaps with $\hbar\omega_u$ smaller than the $500\,cm^{-1}$ value used here, will afford a fit to the observed quenchings.

Such dissociation phenomena will be discussed in detail in Chap. 11 for $Y_2O_2S:Eu^{3+}$. We do not have the detailed experimental information required to assess this dissociation with finality in the Yb^{3+} phosphors here.

10.3 Alkali and Alkaline Earth Halides:Sm

10.3.1 The Model and Expectations

The Sm^{2+} ion is a $4f^6$ ion and has energy levels analogous to Eu^{3+} Here, however, the offset state is not the CTS as for Eu^{3+} but the $4f^5 5d$ state of Sm^{2+}. The proper energy level diagram for BaClF host was given by Gâcon [61] and coworkers and is reproduced in Fig. 44.

The diagrams for KCl and $BaCl_2$ hosts were given by us [15] and are reproduced in Fig. 45.

10.3 Alkali and Alkaline Earth Halides: Sm

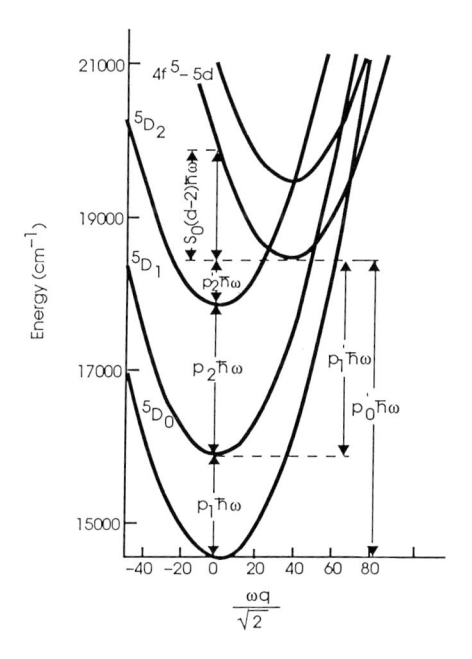

Fig. 44. Energy level diagram for Sm^{2+} in BaClF

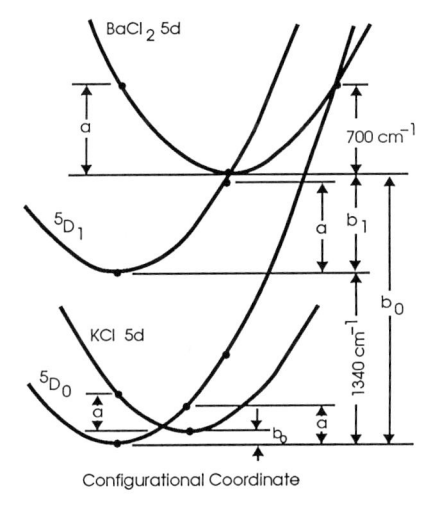

Fig. 45. Energy level diagram for Sm^{2+} in $BaCl_2$ and KCl

In Fig. 45 the energies are coded a, b_0, and b_1 and these are listed in Table 32.

The energies of the various states, given with the energy of the $^7F_0$ parabola minimum at the zero of the energy scale, are given in Table 33 for KCl and $BaCl_2$. These are also the zero-phonon energies of the transitions from $^7F_0$ into these states for the equal force constants model.

From the relationship of Fig. 44 to the corresponding Eu^{3+} diagram of Fig. 25, one can see that in BaClF $^5D_2$ should quench via $^5D_2 \rightarrow 4f^5 5d \rightarrow {}^5D_1$ two-step quenchings at a relatively low temperature, and that $^5D_1$ should quench

Table 32. Sm^{2+} $4f^55d \to {}^5D_j$ energies, in cm^{-1}, for the SCC diagram in Fig. 45

code	Energy	KCl	BaCl$_2$
a	$S\hbar\omega$	300	700
b_0	$p_0\hbar\omega$	120	2 100
b_1	$p_1\hbar\omega$	—	760

Table 33. Sm^{2+} energy levels

State	KCl	BaCl$_2$
$4f^55d$	—	16 630
${}^5D_1$	15 870	15 900
$4f^55d$	14 675	—
${}^5D_0$	14 530	14 560
${}^7F_5$	3 140	3 130
${}^7F_4$	2 320	2 270
${}^7F_3$	1 530	1 490
${}^7F_2$	810	820
${}^7F_1$	280	290
${}^7F_0$	0	0

via ${}^5D_1 \to 4f^55d \to {}^5D_0$ two-step quenchings at higher temperatures. Only in the latter case is there any possibility at all of seeing broad-band $4f^55d \to {}^7F$ emissions. Even here, however, one would well expect that, at temperatures at which $4f^55d$ is thermally accessible to ${}^5D_1$, the $4f^55d \to {}^5D_0$ nonradiative rate would compete very strongly against the $4f^55d \to {}^7F$ emission rate. Gâcon sees no broad band emissions in this host.

From Fig. 45 one expects thermal quenching of ${}^5D_1 \to {}^7F$ emissions and their replacement with ${}^5D_0 \to {}^7F$ emissions through ${}^5D_1 \to 4f^55d \to {}^5D_0$ two-step

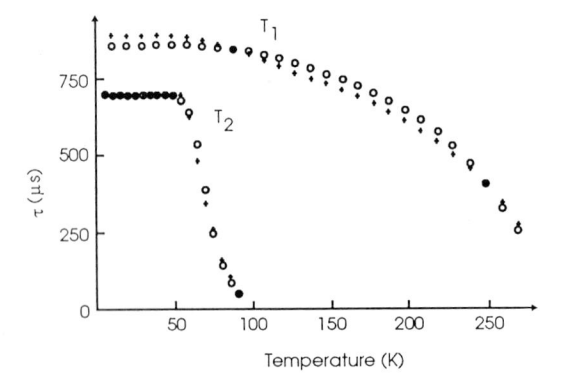

Fig. 46. Temperature dependence of Sm^{2+} ${}^5D_2$ and ${}^5D_1$ emissions in BaClF

10.3 Alkali and Alkaline Earth Halides: Sm

quenchings at some modest temperature in $BaCl_2$ and even at 0 K in KCl. In addition one expects replacing the slow $^5D_0 \rightarrow \,^7F$ emissions with the faster $4f^5 5d \rightarrow \,^7F$ emissions at some low temperature in KCl and at some higher temperature in $BaCl_2$. One can also expect that at low temperatures in $BaCl_2$, excitation into the $4f^5 5d$ state would lead dominantly to $^5D_1$ feeding and not to $^5D_0$ feeding.

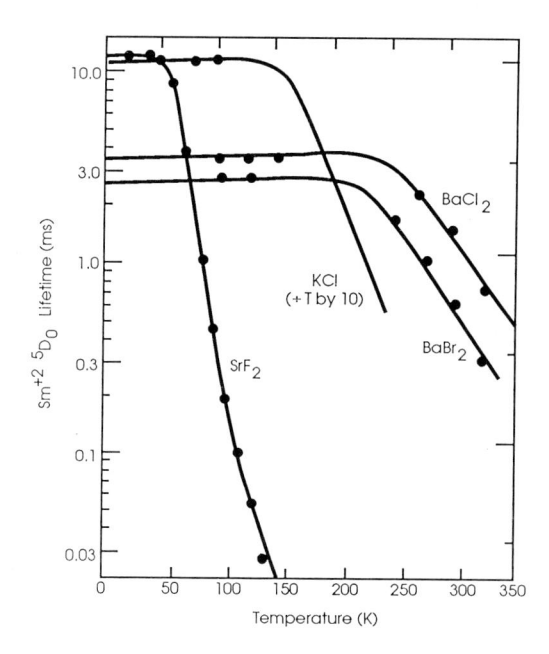

Fig. 47. Temperature dependence of Sm^{2+} $^5D_0$ emissions in several hosts

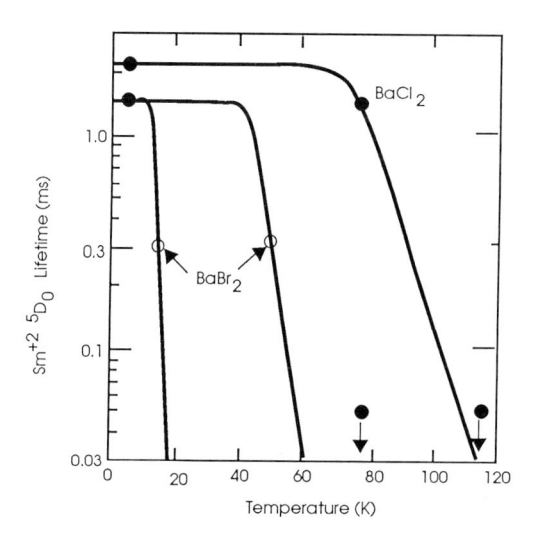

Fig. 48. Temperature dependence of Sm^{2+} $^5D_1$ emissions in several hosts

162   10 Experimental Studies

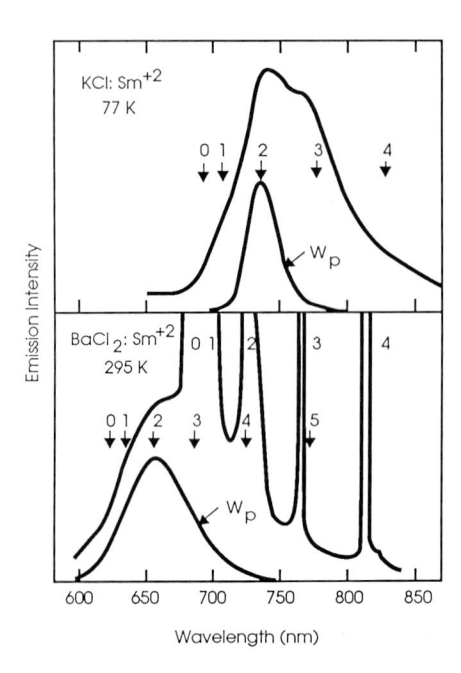

Fig. 49. Emission spectra of Sm^{2+} $^5D_0$ in KCl and $BaCl_2$

Gâcon shows the temperature dependence of the lifetime of the $^5D_2$ and $^5D_1$ emissions. His figure is reproduced here in Fig. 46. The filled points are his experimental data.

We have studied [15] the $^5D_0$ lifetime of Sm^{2+} in KCl, SrF_2, $BaBr_2$, and $BaCl_2$, given in Fig. 47. The data here were taken from Fong and coworkers [62, 63] and shown as points in the figure. The $^5D_1$ lifetimes in $BaBr_2$ and $BaCl_2$ are taken from the same source and are shown as points in Fig. 48.

We will confine our remarks to the KCl and $BaCl_2$ host data. Figure 49 gives the emission bands seen in KCl and in $BaCl_2$ at the cited temperatures. These are also from the same Fong references.

We have also studied the feeding rates from the $4f^55d$ state into the ^5D states. The $f^5d \rightarrow {}^5D_0$ rate is $\approx 10^{10}$ s^{-1} near 77 K. The $f^5d \rightarrow {}^5D_1$ rate is at least two orders faster. These facts again are implicit in the cited references to Fong and coworkers and will be demonstrated in the discussion below.

10.3.2 The Fitting Parameters

To fit the Sm^{2+} experience Gâcon and coworkers [61] and we [15] used the parameters given in Table 34.

The open circles in Fig. 46 are Gâcon's fit. The governing equations are:

$$\frac{1}{\tau_1} = P_1^r + P_{10} + \frac{P_{d0}P_{1d}}{P_{d1} + P_{d0}} \sim P_1^r + P_{10} + P_{d0}\exp(-p_{d1}\hbar\omega_{d1}/kT)$$

10.3 Alkali and Alkaline Earth Halides: Sm

Table 34. Parameters for fitting the Sm^{2+} experience

BaFCl

S_{10} 0.05	p_{10} 6	$\hbar\omega_{10}$ 223	A_{10} 7.56 + 12	S_{d0} 7.13	p_{d1} 19	$\hbar\omega_{d0} = \hbar\omega_{d1}$ 205	A_{d0} 0.80 + 12
S_{21} 0.25	p_{21} 9	$\hbar\omega_{21}$ 216	A_{21} 6.70 + 12	S_{d2} 7.20	p_{d2} 3	$\hbar\omega_{d2}$ 203	A_{d2} 0.67 + 10
P_1^r, s^{-1} 1012	P_2^r, s^{-1} 1381						

BaCl$_2$

S_{10} 3.5	$\hbar\omega$ 200	p_{d0} 10.35	$p_{d0}\hbar\omega$ 2070	A_0' 2.07 + 10	$\tau_{0,R}$(msec) 3.5	$E_{zp,vu}(^5D_0)$, cm^{-1} 14560
S_{21} 3.5	$\hbar\omega$ 200	p_{d1} 3.7	$p_{d1}\hbar\omega$ 760	A_1' 7.30 + 10	$\tau_{1,R}$(msec) 2.15	$E_{zp,vu}(^5D_1)$, cm^{-1} 15900	$E_{zp,vu}(4f^55d)$, cm^{-1} 16630

KCl

S_{10} 2.1	$\hbar\omega$ 144	p_{d0} 1.0	$p_{d0}\hbar\omega$ 145	A_0' 1.45 + 9	$\tau_{0,R}$(msec) 10.8	$E_{zp,vu}(^5D_0)$, cm^{-1} 14530	$E_{zp,vu}(4f^55d)$, cm^{-1} 14675

$$\frac{1}{\tau_2} = P_2^r + P_{21} + \frac{P_{2d}(P_{d1} + P_{d0})}{P_{d2} + P_{d1} + P_{d0}} \sim P_2^r + P_{21}$$

$$+ \frac{P_{d2}P_{d1}}{P_{d2} + P_{d1}} \exp(-p_{d2}\hbar\omega_{d2}/kT) \tag{10.15}$$

We have relabeled Gâcon's parameters in the first section of Table 34 for their use in Eq. (10.15). The equality relationships in Eq. (10.15) are Eq. (1.19) with Eq. (1.13) introduced for $\eta_{gu}|_{G_u}$. The approximate forms in Eq. (10.15) use the inequality $P_{d1} > P_{d0}$, the relationship giving the upward rate in terms of the downward rate $P_{id} = P_{di}\exp(-p_{di}\hbar\omega/kT)$ (where p_{di} is nonnegative), and the model equations $P_{ij} = A_{ij}W_{p_{ij}}(S_{ij}, \hbar\omega_{ij}, T)$. The validity of Gâcon's approximation $P_{d1} > P_{d2}$, which allows replacing the pre-exponential fraction in the $1/\tau_2$ expression with p_{d2}, is unclear. However, the general fit of Eq. (10.15) to the data of Fig. 46 is uncontested.

The full curves in Fig. 47 are of the expression:

$$\tau_J = \frac{\tau_{J,R}}{1 + \tau_{J,R}A'_J\exp(-p_J\hbar\omega/kT)} \tag{10.16}$$

with the parameters in the second and third sections of Table 34 for $J = 0$. The full curve in Fig. 48 is also Eq. (10.16) with the Table 34 parameters for $J = 1$. (We have also modeled the SrF_2 and $BaBr_2$ behavior but do not address these here.) Eq. (10.16) is Eq. (1.19) with Eq. (1.23) substituted for $\eta_{gu}|_{G_u}$, with the identity $\tau_{J,R} = 1/(u \to g)$, and with the reasonable approximation in the neighborhood of the quenching $A'_J = A_J W_{p_J}(S_J, \langle m \rangle) \approx \text{const}$.

The A' value in Table 34 is the $v \to g$ nonradiative rate and is near 10^{10}. This fit to the quenching is the cited implicit determination of the $4f^55d \to {}^5D_0$ feeding rate in Fong's data.

The broad bands of Fig. 49 are the superposition of individual components from the $4f^55d$ band into the several ${}^7F$ states. The positions of these component transitions are indicated by arrows in the figure, and the component into ${}^7D_2$ is drawn as calculated from the W_p distribution with the parameters in Table 34. The observed shape is very probably a superposition of such W_p distributions placed at the indicated energies, but a definitive demonstration of this awaits a theory of the relative amplitudes of these components.

The $4f^55d \to {}^5D$ feedings in $BaCl_2$ are also understandable on the basis of the W_p distribution for $S = 3.5$, $\hbar\omega = 200\,\text{cm}^{-1}$. The $4f^55d \to {}^5D_1$ transition creates ~ 3.7 phonons and the $4f^55d \to {}^5D_0$ transition creates ~ 10.4 phonons. The W_p value at 0 K for $p = 3.7$ is ~ 0.19 while that for $p = 10.4$ is ~ 0.001. This ratio of ~ 200 is in the neighborhood of that estimated from the spectral data at 4 K and at 77 K supplied by Fong and coworkers [63]. They show at 4 K the 7308 ${}^5D_0 \to {}^7F_2$ line about one order weaker than the ${}^5D_1 \to {}^7F_4$ lines, and about one order stronger at 77 K, where ${}^5D_1$ is about half quenched. The ratio at each temperature is proportional to the ratio of the populations n_0/n_1. At 77 K this ratio is about 1 and one can infer then that at 4 K the ratio is about 0.01. Thus ${}^5D_1$ is fed at least two orders faster than ${}^5D_0$ from the $4f^55d$ state.

10.4 Ruby

Figure 50 gives the SCC energy diagram for ruby. There are small-offset ^2E, $^2T_1$, and $^2T_2$ states and large-offset $^4T_2$ and $^4T_1$ states relative to the ground $^4A_2$ state.

The model parameters we have used to describe transitions between these states are given in Table 35. Also given in this table are calculated rates for all downward transitions at 0 K. The rates at higher temperatures would be larger than those listed.

In Table 35 one sees that excitation into $^4T_1$ leads in less than 10^{-12} sec. to ^2E dominantly through $^2T_1$, excitation into $^4T_2$ excitation also reaches ^2E this quickly, and that, while there is only weak excitation into the small-offset states, these too lead quickly to ^2E.

It is experimentally manifest that in ruby the demand is unavoidable for a downward shift in the offset energies with increasing temperature. Figure 51 shows emission spectra of ruby at 295 K, where one sees dominantly the $^2E \rightarrow {}^4A_2$ line emission, and at 663 K, where one sees this emission at the same energy.

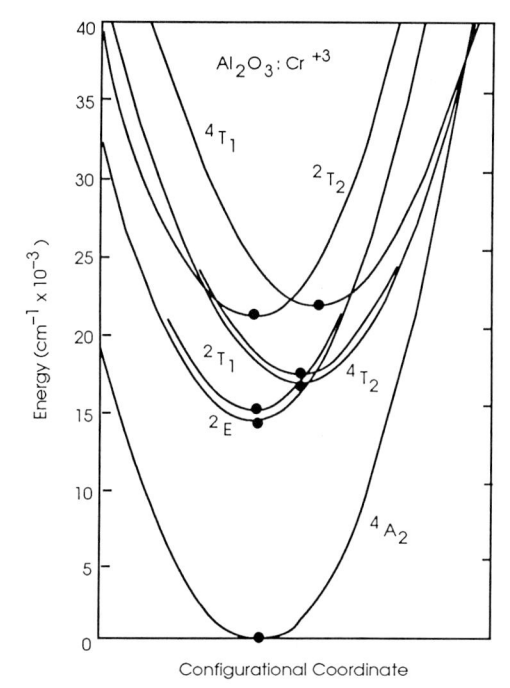

Fig. 50. SCC diagram for Al_2O_3:Cr^{3+}

Table 35. Parameters and nonradiative downward rates in ruby

Transition		θ	a_{uv}	$S_f\hbar\omega_f$ (cm^{-1})	$\hbar\omega_f$ (cm^{-1})	$h\nu_{zp,if}$ (cm^{-1})	$10^{14}U_{pv}$ or $10^{14}V_{pv}$ (sec^{-1})
$^4T_1$	$\to\ ^2T_2$	44°	5.230	3 790	536	610	8.4 + 11
	$\to\ ^4T_2\pi$	45°	1.297	350	833	4 300	4.7 + 9
	$\to\ ^4T_2\sigma$	45°	1.297	350	833	4 830	8.6 + 8
	$\to\ ^2T_1$	44°	5.230	3 790	536	6 640	2.4 + 12
	$\to\ ^2E$	44°	5.230	3 790	536	7 180	1.3 + 12
	$\to\ ^4A_2$	44°	4.051	3 790	893	21 600	2.7 + 4
$^2T_2$	$\to\ ^4T_2\pi$	44°	3.526	1 500	500	3 690	1.3 + 12
	$\to\ ^4T_2\sigma$	44°	3.526	1 500	500	4 220	3.8 + 11
	$\to\ ^2T_1$	45°	0.849	161	893	6 030	2.5 + 5
	$\to\ ^2E$	45°	0.849	161	893	6 570	2.6 + 4
	$\to\ ^4A_2$	45°	0.849	161	893	20 990	2.1 − 27
$^4T_2\sigma$	$\to\ ^2T_1$	44°	3.526	1 725	536	1 810	2.0 + 13
	$\to\ ^2E$	44°	3.526	1 725	536	2 350	1.5 + 13
	$\to\ ^4A_2$	44°	2.731	1 725	893	16 770	2.3 + 3
$^2T_1$	$\to\ ^2E$	45°	1.095	161	536	540	2.2 + 13
	$\to\ ^4A_2$	45°	0.849	161	893	14 960	1.6 − 13
^2E	$\to\ ^4A_2$	45°	0.849	161	893	14 420	2.5 − 12

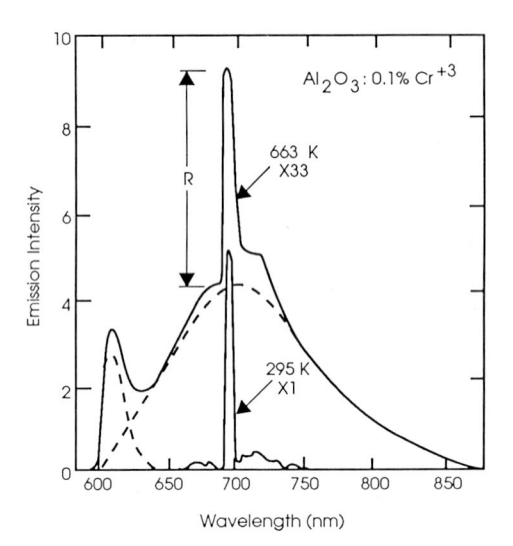

Fig. 51. Emission spectra of ruby at two temperatures

It is a factor of 30 weaker, but, more importantly, it is positioned at the maximum of the stronger $^4T_2 \to\ ^4A_2$ broad band. There is no way that the SCC diagram of Fig. 50, which is correct for the 4 K placement of the broad $^4A_2 \to\ ^4T_2$ absorption band and the $^2E \to\ ^4A_2$ line emission, can allow the maximum of this absorption to be at the narrow-line energy at 663 K unless the offset state moves down in

10.4 Ruby

energy with increasing temperature. (The narrow band near 620 nm is due to filter overlap between emission and detection filters and should be ignored.)

McClure [64] has shown that all of these large-offset states move downward with respect to the small-offset states with increasing temperature, by approximately the same amount, namely, by $\sim -1.4\,\text{kT}$. The $^2E \rightarrow {}^4A_2$ line emission energy placement has only a tenth this temperature sensitivity and all placements of the small-offset states are treated here as independent of temperature.

This shift is also in the temperature-dependent optical band positions. Fig. 52 shows fits to the absorption spectra into $^4T_2\,\sigma$ and into $^4T_1\,\sigma$ and also to the emission spectrum for $^4T_2\,\sigma \rightarrow {}^4A_2$. The data is again from McClure [64]. The fit of the SCC model shown in Fig. 52 uses the parameters of Table 36.

While these parameters look different from those in Table 35 for the same

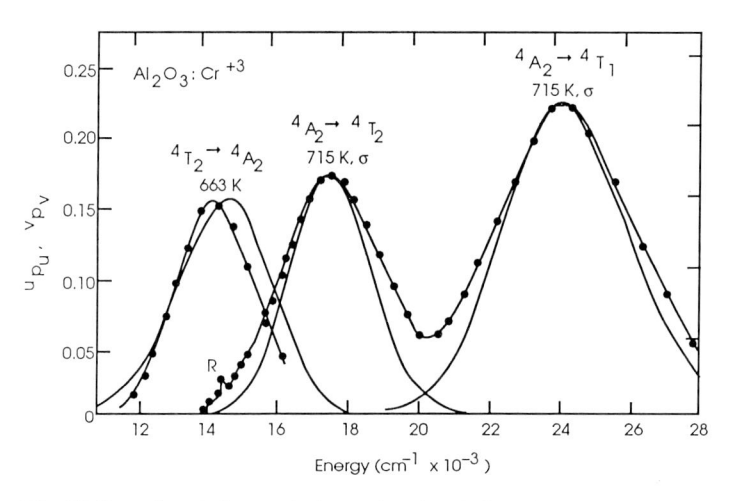

Fig. 52. Broad emission and absorption bands in ruby

Table 36. Parameters for fitting the optical bands in ruby

θ	a_{uv}	$\hbar\omega_v$ (cm^{-1})	$\hbar\omega_u$ (cm^{-1})	S_v	S_u	$S_v\hbar\omega_v$ (cm^{-1})	$S_u\hbar\omega_u$ (cm^{-1})	$h\nu_{zp,vu}$ (cm^{-1})
$^4A_2 \rightarrow {}^4T_2$								
44°	3.526	500	536	3	3.217	1500	1725	16770−1.4 kT
$^4A_2 \rightarrow {}^4T_1$								
44°	5.230	500	536	6.6	7.069	3300	3790	21600−1.4 kT
$^4A_2 \rightarrow {}^2E$								
45°	1.265	400	400	0.4	0.4	160	160	14420

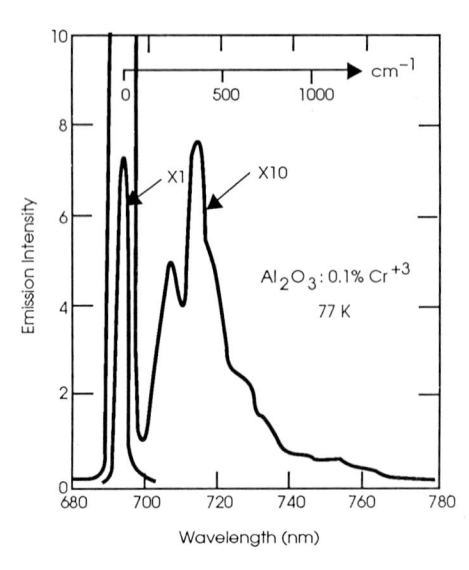

Fig. 53. Emission spectrum of ruby at 77 K

transitions, they are related by having the same parabola placement, i.e., they share the same $S\hbar\omega$ values and the same θ and differ only in the phonon energy invoked.

In Fig. 52 the V_{p_v} distributions for the $^4T_1$ band are shown multiplied by 1.91 to allow for the slightly stronger absorption strength of this transition.

The small offset states in Fig. 50 are shown inaccurately at zero offset. Their actual offset parameter S can be determined as the ratio of the intensities of the one-phonon and zero-phonon sidebands at low temperatures. Figure 53 shows the 77 K emission spectrum of ruby. The ratio under consideration is ~ 0.4 and the first moment $S\hbar\omega$ is $\sim 160\,\mathrm{cm}^{-1}$. The two prominent phonon energies in this spectrum are 280 and 430 cm^{-1}, and the 200 cm^{-1} phonons so prominent in the broad-band absorption in Fig. 54 below are not especially evident here. Thus our description of this narrow-line transition is as in the third section of Table 36.

It is of next importance to see that certain features of the spectra cannot be understood simultaneously with one phonon energy. A striking example is in the fine structure seen in absorption at 2 K. Figure 54 shows Brossel and Margerie's [65] $^4A_2 \rightarrow\ ^4T_2$ absorption spectrum at this temperature. A strong progression of 200 cm^{-1} phonons is seen at the low energy side. If we fit the spectrum with 200 cm^{-1} phonons, however, the width at half height is significantly too small. We will show later that $\sim 800\,\mathrm{cm}^{-1}$ phonons are needed to understand the thermal quenching of this emission. The fit to the spectrum with 833 cm^{-1} phonons gives a significantly larger width at half height than observed. The spectrum is best understood with 500 cm^{-1} phonons, as adopted in Table 36.

The need for $\sim 800\,\mathrm{cm}^{-1}$ phonons to understand the quenching is illustrated in Fig. 55. This figure shows data taken by us as solid points and data from Kisliuk and Moore [24] as open points. In the model, η_{total} quenches above 450 K

10.4 Ruby

because of $^4T_2 \to {}^4A_2$ crossovers. Figure 55 shows the quenching curve predicted using the description in Table 35 for this crossover. It also shows three alternative quenching curves, for using the larger-offset but higher-energy crossover from $^4T_1$, for using $\theta = 44.5°$ rather than $\theta = 44°$, and for using 500 cm^{-1} rather than the 833 cm^{-1} adopted in Table 35. All three alternative fits give dramatically inferior understanding of the observed quenching.

Figure 56 shows the thermal quenching of the ruby $^2E \to {}^4A_2$ narrow-line emission and the buildup of the $^4T_2 \to {}^4A_2$ broad-band emission. Once again, the points are our measurements and the open points those of Kisliuk and Moore [24]. The ordinate is either emission ratio or lifetime in msec as shown.

The curves in Fig. 56 require a more detailed analysis. We consider the three-level system $u = {}^2E$, $v = {}^4T_2 \sigma$, and $g = {}^4A_2$. We take the descriptions of the transitions between these states as in Table 35. We take the offset parabola downshift to be -1.4 kT. The emission rates are 140 sec^{-1} for $^2E \to {}^4A_2 (u \to g)$

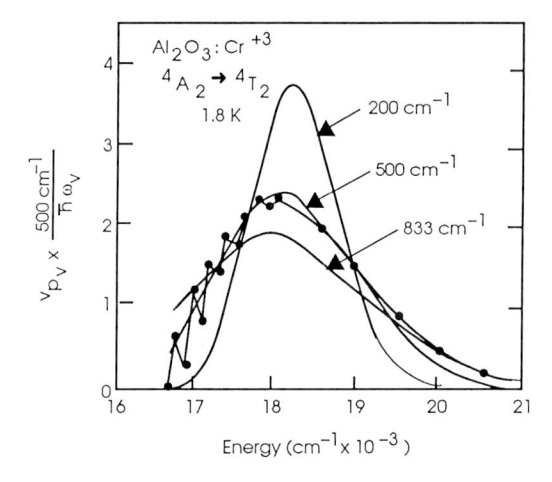

Fig. 54. Absorption spectrum into $^4T_2 \sigma$ of ruby at 2 K

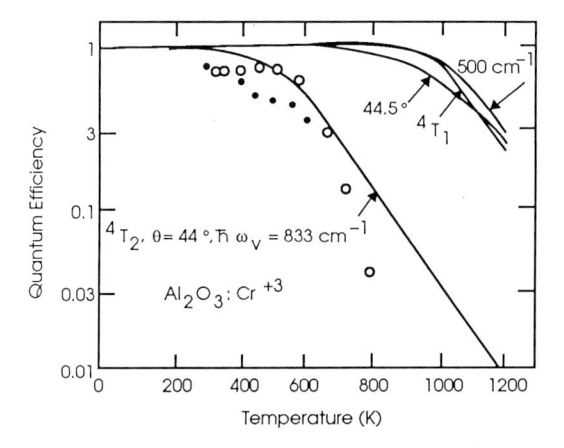

Fig. 55. Thermal quenching of ruby total emission

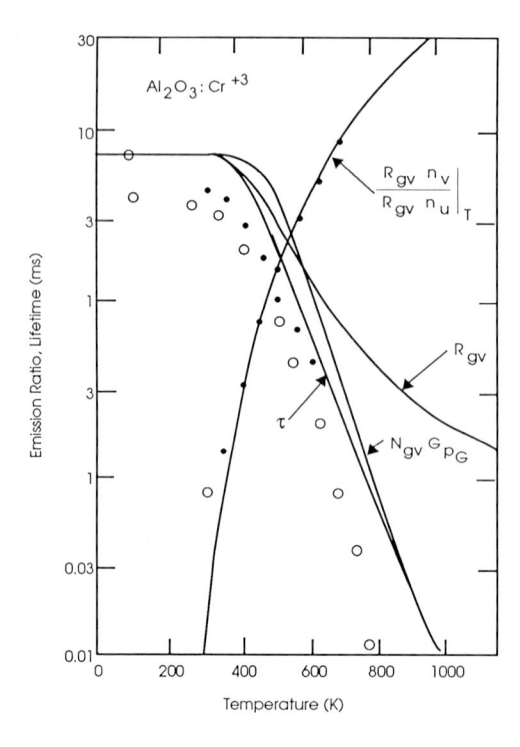

Fig. 56. Quenching of the ruby lifetime and buildup of broad-band to narrow-line emission ratio

radiation and 23 000 sec^{-1} for $^4T_2 \to {}^4A_2(v \to g)$ broad-band emission. The degeneracy of ^2E is 4 and of $^4T_2$ is 8.

In thermal equilibrium the population ratio between the u and v states is:

$$\left.\frac{n_v}{n_u}\right|_T = \frac{8}{4} e^{-E_{zp,vu}/kT} = 2e^{1.4} e^{-2350 \text{ cm}^{-1}/kT} \tag{10.17}$$

where a factor $[1 - \exp(-\hbar\omega_u/kT)]/[1 - \exp(-\hbar\omega_v/kT)]$ near unity is ignored. The ratio of emission intensities is then:

$$\left.\frac{R_{gv}n_v}{R_{gu}n_u}\right|_T = \frac{23\,000}{140}\left.\frac{n_v}{n_u}\right|_T = 1330\, e^{-2350 \text{ cm}^{-1}/kT} \tag{10.18}$$

This function is plotted in Fig. 56 as the thus-labeled curve and both the points of Kisliuk and Moore and our points follow the curve.

Excitation into v either feeds into u or goes radiatively or nonradiatively to g. Thus, labeling the quantum efficiency for emission from state i with excitation into state j $(\eta_i)_j$, we have for the relationship of the total efficiencies:

$$(\eta_v + \eta_u)_v = \frac{A_{uv}U_{p_U}(\eta_v + \eta_u)_u + R_{gv}}{A_{uv}U_{p_U} + R_{gv} + A_{gv}G_{p_G}}. \tag{10.19}$$

In Eq. (10.19), $A_{uv}U_{p_U}$ is $\approx 10^{13}$, while R_{gv} and $A_{gv}G_{p_G}$ are $\approx 10^4$ and $(\eta_v + \eta_u)_v$ and $(\eta_v + \eta_u)_u$ differ only by $\approx 10^{-9}$. Misu [25] and Morgenshtern and coworkers

10.4 Ruby

[66] have reported $(\eta_v + \eta_u)_v \approx 0.7$ and $(\eta_v + \eta_u)_u \approx 1.0$ at room temperature. Such a 30% difference cannot be explained in this model. If indeed there is such a difference, some factor not in this model, perhaps site multiplicity with the excitations into u and v reaching inequivalent and non-communicating Cr^{3+} ions, must be invoked.

We now examine the T_c at which thermal equilibrium is reached. For excitation at rates G_u into u and G_v into v, the steady state equations are:

$$(R_{gu} + A_{vu}V_{pv})n_u - A_{uv}U_{pu}n_v = G_u$$
$$-A_{vu}V_{pv}n_u + (A_{uv}U_{pu} + R_{gv} + A_{gv}G_{p_G})n_v = G_v \quad (10.20)$$

Thus, for excitation into u ($G_v = 0$) and for excitation into v ($G_u = 0$) we have for the ratio of the populations:

$$\left.\frac{n_v}{n_u}\right|_{G_u} = \frac{A_{vu}V_{pv}}{A_{uv}U_{pu} + R_{gv} + A_{gv}G_{p_G}} \approx \frac{A_{vu}V_{pv}}{A_{uv}U_{pu}} = \left.\frac{n_v}{n_u}\right|_T$$

$$\left.\frac{n_v}{n_u}\right|_{G_v} = \frac{R_{gu} + N_{vu}V_{pv}}{A_{uv}U_{pu}} = \frac{R_{gu}}{A_{uv}U_{pu}} + \left.\frac{n_v}{n_u}\right|_T \quad (10.21)$$

where we have recognized and used the thermal equilibrium requirement:

$$A_{vu}V_{pv}n_u = A_{uv}U_{pu}n_v \quad (10.22)$$

Thus, for excitation into u, u and v are always in thermal equilibrium and, for excitation into v, they are in thermal equilibrium above a temperature T_c defined by:

$$R_{gu}n_u = A_{uv}U_{pu}n_v \quad (10.23)$$

For the rates $A_{uv}U_{pu} = 1.5 \times 10^{13}$ sec^{-1} and $R_{gu} = 140$ we calculate $T_c = 123$ K. This estimate is not very sensitive to the rates assumed: a factor of ten change in $A_{uv}U_{pu}$ would change the T_c only by 10°. We can now use this knowledge that u and v are in thermal equilibrium to simplify the derivations and the expressions for τ under the three hypotheses examined in Fig. 56.

The curve in Fig. 56 labeled R_{gv} is the temperature-dependent lifetime if the slow narrow-line $^2E \to {}^4A_2$ emissions were replaced at high temperatures by the fast broad-band $^4T_2 \to {}^4A_2$ emissions without any losses. Under these conditions the expression for the overall lifetime is:

$$\tau \equiv \frac{n_{total}}{W_{total}} = \frac{n_u + n_v}{R_{gu}n_u + R_{gv}n_v} = \frac{1 + \left.\frac{n_v}{n_u}\right|_T}{R_{gu}\left[1 + \left.\frac{R_{gv}n_v}{R_{gu}n_u}\right|_T\right]}$$

$$= \frac{1 + 8.110\exp(-2350\,\text{cm}^{-1}/kT)}{140[1 + 1330\exp(-2350\,\text{cm}^{-1}/kT)]} \quad (10.24)$$

The curve labeled $N_{gv}G_{p_G}$ is the lifetime if the $^4T_2 \to {}^4A_2$ emissions were strictly forbidden, but $^2E \to {}^4A_2$ emissions could be replaced with $^4T_2 \to {}^4A_2$ nonradiative transitions. Under this hypothesis the expression for the overall lifetime is:

$$\tau \equiv \frac{n_{total}}{W_{total}} = \frac{n_u + n_v}{R_{gu}n_u + A_{gv}G_{p_G}n_v} = \frac{1 + \left.\frac{n_v}{n_u}\right|_T}{R_{gu}\left[1 + \left(\frac{A_{gv}G_{p_G}}{R_{gu}}\right)\left.\frac{n_v}{n_u}\right|_T\right]} \quad (10.25)$$

This $N_{gv}G_{p_G}$ curve in Fig. 56 is Eq. (10.25) with $n_v/n_u|_T$ given by Eq. (10.17), $R_{gu} = 140$ sec^{-1}, and with $A_{gv} = 10^{14}$ sec^{-1} and G_{p_G} calculated for $\theta = 44°$, $a_{uv} = 2.731$, and $p_G = 18.77$.

The curve labeled τ is for the hypothesis actually adopted, that at low temperatures $^2E \to {}^4A_2$ emissions predominate, and that they are replaced above 300 K at first without loss with $^4T_2 \to {}^4A_2$ broad-band emissions, and, above 400 K, partially also with $^4T_2 \to {}^4A_2$ nonradiative transitions. The expression for the lifetime is now:

$$\tau \equiv \frac{n_{total}}{W_{total}} = \frac{n_u + n_v}{R_{gu}n_u + R_{gv}n_v + A_{gv}G_{p_G}n_v}$$

$$= \frac{1 + \left.\frac{n_v}{n_u}\right|_T}{R_{gu}\left[1 + \left.\frac{R_{gv}n_v}{R_{gu}n_u}\right|_T + \left.\frac{A_{gv}G_{p_G}}{R_{gu}}\frac{n_v}{n_u}\right|_T\right]}$$

$$= \frac{1 + \left.\frac{n_v}{n_u}\right|_T}{R_{gu}\left[1 + \left.\frac{R_{gv}n_v}{R_{gu}n_u}\right|_T\left(1 + \frac{A_{gv}G_{p_G}}{R_{gv}}\right)\right]} \quad (10.26)$$

This τ curve in Fig. 56 is Eq. (10.26) with $R_{gv}n_v/R_{gu}n_u|_T$ given by Eq. (10.18) and all other quantities defined as in Eq. (10.17). The fit is deemed adequate to support the hypothesis tested.

There are other effects noticeable in the data, particularly some very-low-temperature (below 350 K) inefficiency in the narrow-line $^2E \to {}^4A_2$ emission. The lifetimes are known very well, having been measured using photon counting over three orders of magnitude. The change in lifetime with temperature is established; but it cannot be explained within the SCC model and is probably due to energy-transfer transitions between Cr^{3+} ions and some unknown "killer" center.

11 Effects Beyond the Model: Oxysulfide: Eu Storage and Loss Processes

11.1 The Need for Enhancement of the Model

In this chapter, we shall describe storage and loss performances which are clearly beyond any of the models considered thus far. Figure 57 shows the phosphorescence seen in $(Y_{0.999}Eu_{0.001})_2O_2S$. While not one transition in Fig. 25 has a time constant greater than a few milliseconds, $^5D_0 \rightarrow {}^7F$ phosphorescence is seen lasting for five minutes. The ordinate is the ratio of the phosphorescence to the steady-state emission intensity as percent.

Figure 58 shows the amplitude of this $^5D_0$ phosphorescence 15 seconds after excitation into $^5D_1$, $^5D_2$, and the CTS, as a percent of the steady-state emission intensity, at the excitation temperature of the abscissa. Before each point was taken for this figure, the phosphor was annealed at 500 K. Also shown in this figure as a dotted line is the thermal quenching behavior of the $^5D_2$ emissions. What one sees here is that only when $^5D_2$ is quenched to the CTS does $^5D_2$ excitation cause phosphorescence. A similar conclusion is true also about $^5D_1$ excitations, i.e., only when $^5D_1$ is quenched does this excitation lead to phosphorescence.

There is even longer excitation-energy storage in these phosphors. Figure 59 shows the thermally stimulated $^5D_0$ glow in $(Y_{0.999}Eu_{0.001})_2O_2S$ following 10 minute excitations into the labeled states at the labeled temperatures. The glow intensity is given as a percent of the steady state $^5D_0$ emission intensity. The heating rate was 6°/minute. An analogous performance is seen in $La_2O_2S:Eu^{3+}$. (Some long-term phosphorescence is observed in $La_2O_2S:Yb^{3+}$: see Sect. 10.2.2.)

Such behavior has led us to the diagram of energy storage-loss processes shown in Fig. 60, to the SCC diagram of Fig. 64 of Sect. 11.4, and to the careful set of measurements and the fitting to model equations given in this chapter.

The key concept is that the CTS can be thought of as a exciton trapped at the Eu^{3+} impurity, i.e., as $Eu^{2+} - S^{1-}$, as an extra electron on the Eu^{3+} giving Eu^{2+} coupled with a hole on the nearest-neighbor S^{2-} ion giving S^{1-}. With some thermal activation energy, the hole can wander away from the Eu^{2+} and can be trapped elsewhere. This is the state that can live for months. When the trapped hole is released, either thermally or optically, it can find the Eu^{2+}, reform the CTS, effect CTS $\rightarrow {}^5D$ feedings, and give $^5D \rightarrow {}^7F$ line emissions.

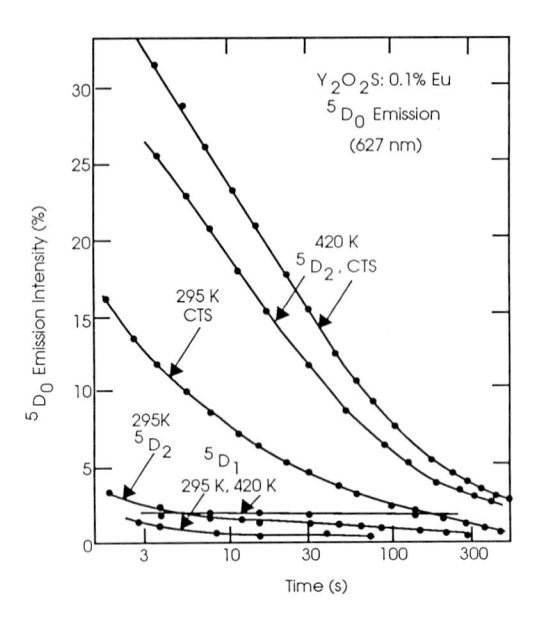

174   11 Effects Beyond the Model: Oxysulfide: Eu Storage and Loss Processes

Fig. 57. $^5D_0$ Phosphorescence in $Y_2O_2S:Eu^{3+}$

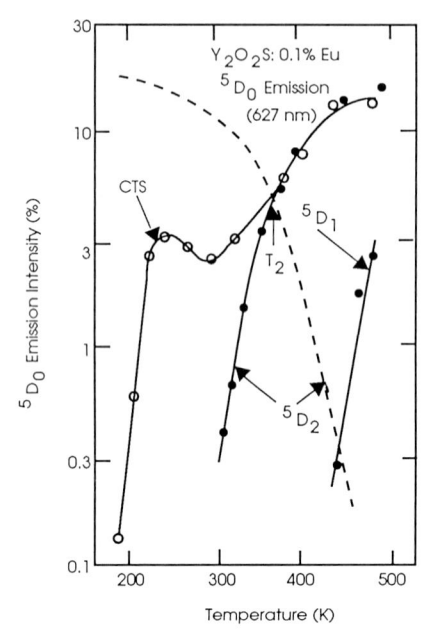

Fig. 58. Magnitude of $^5D_0$ phosphorescence in $Y_2O_2S:Eu^{3+}$

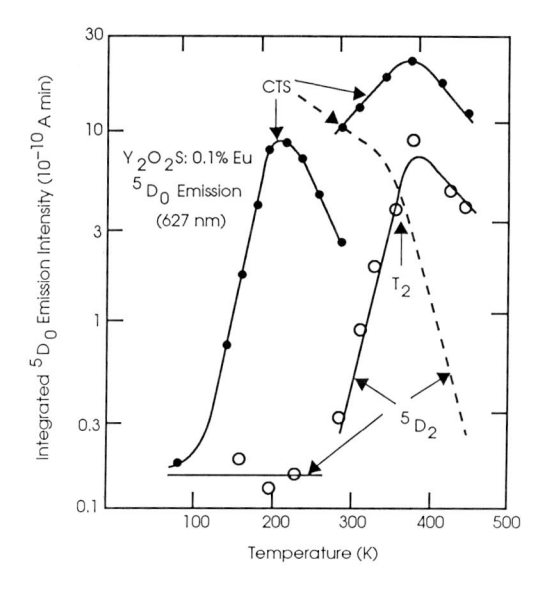

Fig. 59. Magnitude of $^5D_0$ thermal glow in $Y_2O_2S:Eu^{3+}$

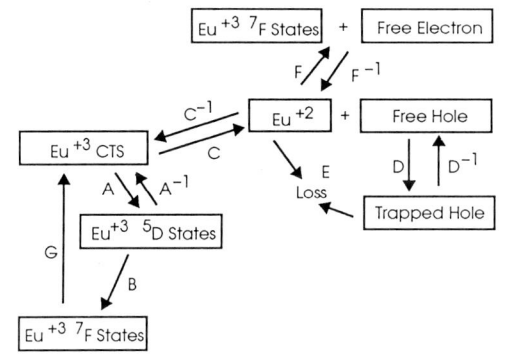

Fig. 60. Diagram of storage-loss processes in oxysulfide:Eu phosphors

11.2 Synopsis of the Experiments to Probe the Model

In oxysulfide:Eu phosphors, the $^5D \rightarrow {}^7F$ emission intensity in response to a long pulse of CTS excitation has the behavior shown in Fig. 61. The emission builds up rapidly, over a few ^5D lifetimes, to an initial value B_0. The temperature dependence of B_0 is taken up in Sect. 11.4.

As excitation proceeds, B slowly increases to a higher level and tends to an asymptotic value B_∞. The dependence of this value on excitation intensity, i.e., the nonlinearity of the quantum efficiency with excitation intensity, is taken up in Sect. 11.6. The rate of increase, i.e., the rise time, is discussed in Sect. 11.7.

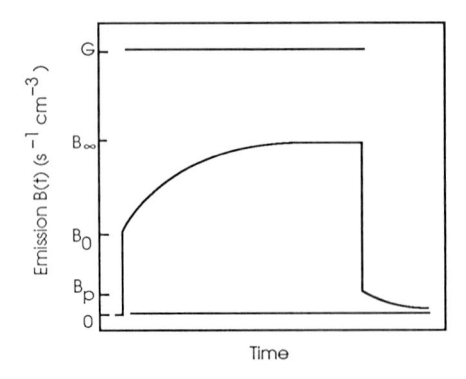

Fig. 61. Long-term time dependence of emission intensity in oxysulfides: Eu

When excitation is removed, then the emission falls rapidly to a level B_P. The dependence of B_P on excitation intensity is taken up in Sect. 11.9. During the decay, B slowly falls to zero, but this fall will not be studied closely here.

In Fig. 61 and in later figures showing the experimental data excitation G and emission $B(t)$ are shown in units such that B_∞/G is the steady-state quantum efficiency for all $^5D \rightarrow {}^7F$ emissions. G itself is given in units of 10^{-10} amps., a unit which we have calibrated to correspond to a $^5D_0 \rightarrow {}^7F_2$ emission strength of $\sim 2 \times 10^9$ photons per second. The sample volume excited was $\sim 10^{-3}$ cm^3.

The B_0/G depend upon sample history and are smallest when the phosphor has been previously exhausted of storage by heating or by infrared (IR) radiation prior to excitation. This smallest B_0/G is independent of Eu concentration and of excitation intensity. It decreases with increasing temperature from near unity at low temperature to about 0.1 at 500 K. At high temperatures, particularly at high Eu concentrations, B_∞/G increases with G, from near B_0/G at low excitation intensities to nearer unity B_∞/G at high excitation intensities.

The time constant of the slow rise from B_0 to B_∞ decreases sublinearly with excitation intensity G. (By "sublinear", we mean as G^α where $\alpha < 1$ and by "superlinear" we mean with $\alpha > 1$, where α need not be a constant.) We take up this behavior in Sect. 11.8.

The B_P is asymmetric with the buildup in the sense that $B_P < B_\infty - B_0$ and B_P increases sublinearly with G.

11.3 The Model Equations: Notation

The storage and loss processes in the model are diagrammed in Fig. 60. Process G is $^7F \rightarrow {}^5D$ excitation. Process A is the CTS $\rightarrow {}^5D$ feeding, and A^{-1} is the inverse $^5D \rightarrow$ CTS quenching. Process B is the $^5D \rightarrow {}^7F$ emission. Process C is

11.3 The Model Equations: Notation

the thermal dissociation of the CTS into Eu^{2+} and a free hole. Process C^{-1} is the inverse capture of a free hole by a Eu^{2+}. Process D is the trapping of free holes; D^{-1} is the inverse freeing of the trapped holes, either thermally or by IR radiation. Process E is the nonradiative recombination of Eu^{2+} and a trapped hole, which is prominant when the Eu concentration is high. Process F is the dissociation of Eu^{2+} into Eu^{3+} and a free electron; F^{-1} is the inverse capture of a free electron by Eu^{3+} to form Eu^{2+}. Luminescence under bandgap excitation proceeds by the sequence F^{-1}, C^{-1}, A, B.

Following dissociation of the CTS, the free hole is quickly either trapped (D) or recombined at an available Eu^{2+} (C^{-1}). The free hole concentration is always small, and the charge balance is between Eu^{2+} and trapped holes. Loss and storage are complementary processes. At high temperatures and high Eu concentrations, the extra electron trapped as Eu^{2+} can wander from Eu to Eu until it finds a trapped hole and recombines with it. At low concentrations and low temperatures, the hopping between Eu sites is slow and the trapped holes can remain trapped. These trapped holes can be released either by heating the phosphor or by IR irradiation, can find the Eu^{2+} and reform the CTS, and can then lead to the $^5D \to {}^7F$ line emissions; the sequence is D^{-1}, C^{-1}, A, B. The IR must be chosen to free the holes without simultaneously freeing the electron from Eu^{2+}.

Formulas for these processes are given in Table 37. Notations used in this table are given in Table 38.

Table 37. Rates of processes diagrammed in Fig. 60

Process	Rate[a]
$^7F \to$ CTS excitation	G
CTS $\to {}^5D$ feeding	$A = ap_0$
$^5D \to$ CTS quenching	$A^{-1} = a^{-1}n_D$
$^5D \to {}^7F$ emission	$B = bn_D$
CTS dissociation	$C = \sigma_{p0} v_T \bar{p}_0 p_0$
Free-hole capture by Eu^{2+}	$C^{-1} = \sigma_{p0} v_T n_0 p$
Trapping of free holes[b]	$D = \Sigma_i D_i = \Sigma_i \sigma_{pi} v_T n_i p$
Detrapping of holes	$D^{-1} = \Sigma_i D_i^{-1} = \Sigma_i \sigma_{pi} v_T \bar{p}_i p_i$
Loss[b]	$E = \Sigma_i E_i = \Sigma_i e_i p_i n_0$
Eu^{2+} dissociation	$F = \sigma_{n0} v_T \bar{n}_0 n_0$
Free-electron capture by Eu^{3+}	$F^{-1} = \sigma_{n0} v_T N_{Eu} n$

[a] Steady-state quantities are designated by adding the subscript ∞. The quantities A^{-1}, a^{-1}, C^{-1}, D^{-1} and F^{-1} are rates of inverse processes and are not numerical inverses of the corresponding A, a, C, D, and F rates.

[b] The index i in the rates D, D^{-1}, and E run over hole traps of all types i. The subscripts $1i$ and $2i$ are used to differentiate between traps that are shallow and deep.

Table 38. Notation used in Table 37

Quantity	Symbol
[CTS]	p_0
[Eu^{2+}]	n_0
[Hole traps of type i]	N_i
[Holes in traps of type i]	p_i
[Empty traps of type i]	n_i
ΣN_i	N_{trap}
[Free holes]	p
[Free electrons]	n
Density of states, valence band	N_v
Density of states, conduction band	N_e
$N_v \exp(-E_{p0}/kT)$	$\bar{p}_0$
$N_v \exp(-E_{pi}/kT)$	$\bar{p}_i$
$N_e \exp(-E_{n0}/kT)$	$\bar{n}_0$
[Eu sites]	N_{Eu}
[Populated ^5D states]	n_D
Hole binding energy to Eu^{2+} to form CTS	E_{p0}
Hole binding energy to trap i	E_{pi}
Electron binding energy to Eu^{3+} to form Eu^{2+}	E_{n0}
Hole-capture crosssection for Eu^{2+}	σ_{p0}
Hole-capture crosssection for trap i	σ_{pi}
Electron-capture crosssection for Eu^{3+}	σ_{n0}
Thermal velocity of free carriers	v_T
Rate constant for CTS $\rightarrow$ ^5D feeding	a
Rate constant for $^5D \rightarrow$ CTS thermal quenching	a^{-1}
Rate constant for $^5D \rightarrow$ ^7F emission	b
Rate constant for nonradiative recombination of Eu^{2+} and holes in traps of type i	e_i

[ ] means concentration.

11.4 CTS Dissociation: The B_0/G Behavior

The CTS population relaxes via CTS $\rightarrow$ ^5D feeding at the rate ap_0 and via CTS dissociation at the thermally activated rate $\sigma_{p0} v_T \bar{p}_0 p_0$. We define the fraction of CTS excitations which feed the ^5D states and the fraction which ionizes, at any time during the response to excitation sketched in Fig. 61, from Fig. 60 and Table 37, as, respectively:

$$f_{feed} = \frac{A}{A+C} = [1 + (\sigma_{p0} v_T N_v/a)e^{-E_{p0}/kT}]^{-1}$$

$$f_{ionize} = 1 - f_{feed} \tag{11.1}$$

We give our attention first to B_0.

11.4 CTS Dissociation: The B_0/G Behavior

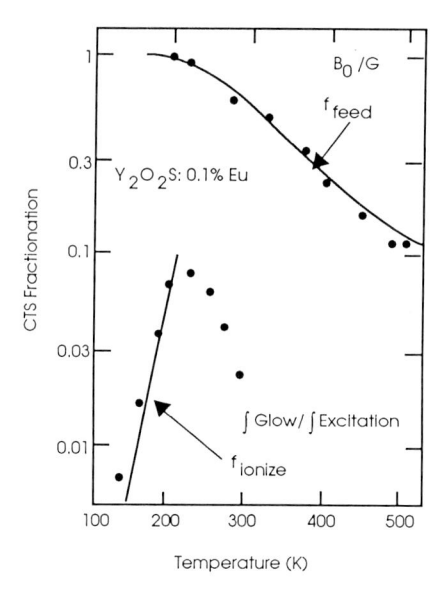

Fig. 62. Temperature-dependent fractionation of the CTS in $(Y_{0.999}Eu_{0.001})_2O_2S$

From the Fig. 60 diagram, $A = B + A^{-1}$ and $A + C = G + A^{-1} + C^{-1}$. When no ^5D states are quenching, $A^{-1} = 0$. Also, at the initial excitation of a phosphor previously exhausted of storage, $C^{-1} = 0$. Under these conditions, $A = B = B_0$, $A + C = G$, and

$$f_{feed} = \frac{A}{A+C} = \frac{B_0}{G} \tag{11.2}$$

Thus, except for the complications due to A^{-1} to be discussed below, the initial quantum efficiency B_0/G is equal to f_{feed}, and the initial storage efficiency C_0/G is equal to f_{ionize}.

The measured B_0/G and normalized storage data are given for Y_2O_2S and La_2O_2S in Figs. 62 and 63.

In these figures, the points along the solid lines are the measured normalized storage (the rising curves) and the measured B_0/G (the falling curves). Each B_0/G data point was taken after a thermal anneal at 500 K to exhaust the phosphor of storage. The data points in Fig. 62 for normalized storage and the open squares in Fig. 63 for this quantity show the areas under the 250 K glow peak (including phosphorescence) measured after annealing at 500 K, cooling to the abscissa temperature, and exciting for a fixed time into the CTS. The crosses in Fig. 63 add the estimated temporary storage released during the excitation period itself.

The solid lines in these figures are the functions f_{feed} and f_{ionize} according to Eqs. (11.1) using the parameters in columns 2 and 3 of Table 39.

One sees a general agreement between the measured direct feeding and measured storage and the values calculated using Eqs. (11.1).

An equally good fit is obtained if one takes into account non-zero values for the rate of process A^{-1}.

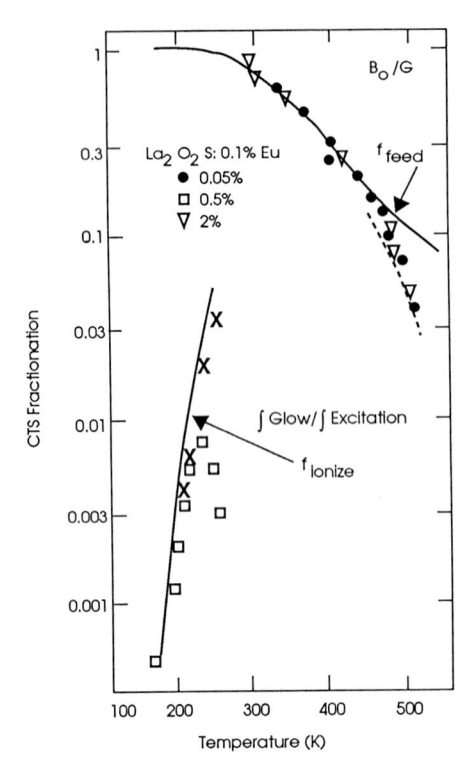

Fig. 63. Temperature-dependent fractionation of the CTS in $La_2O_2S:Eu^{3+}$

Table 39. Parameters describing CTS dissociation in oxysulfides

Host	$\sigma_{p0}v_T N_v/a$	$E_{p0}(\text{cm}^{-1})$	$\sigma_{p0}v_T N_v/a'$	$E_{p0}(\text{cm}^{-1})$
Y_2O_2S	230	1 200	45	900
La_2O_2S	900	1 700	75	1 300

Since the Eu^{3+} ^5D states quench so precipitously, at any temperature they can be grouped into (upper) quenched states and (lower) unquenched states. For the quenched states, the CTS $\rightleftarrows$ ^5D_j inverse A and A^{-1} processes cancel one another and can be ignored. For the unquenched states, the A^{-1} processes are zero, and the analysis above utilizing $A^{-1} = 0$ is again applicable. It is then useful to introduce modified CTS feeding and ionizing fractions f'_{feed} and f'_{ionize} which are based on the unquenched ^5D_j states alone:

$$f'_{feed} = \frac{A_{unquenched}}{A_{unquenched} + C} = [1 + (\sigma_{p0}v_T N_v/a')e^{-E_{p0}/kT}]^{-1}$$

$$f'_{ionize} = 1 - f'_{feed} \tag{11.3}$$

where a' is the constant in the $a'p_0$ partial CTS feeding rate into the unquenched ^5D_j states. The a' is a constant with stepped decreases with increasing temperature. We crudely approximate a' as:

11.5 The SCC Model for Understanding Storage-Loss Processes 181

$$a'(T) = \frac{j_q + 1}{4} a \qquad (11.4)$$

where j_q is the j of the highest unquenched ^5D_j state. As successively lower ^5D_j states quench, the a' decreases by $a/4$.

As in the analysis above for the initial excitation of a phosphor with $A^{-1} = C^{-1} = 0$, $A_{unquenched} = B = B_0$, $A_{unquenched} + C = G$, and the parallel result $B_0/G = f'_{feed}$ obtains. Equations (11.3) and (11.4) leads to closely the same solid curves as in Figs. 63 and 64 using the parameters in columns 4 and 5 of Table 39. These values obtained using these equations are thought to be closer to those which might ultimately be calculable from models of such parameters.

When, as in the La_2O_2S host near 500 K, $^5D_0$ itself is quenching, then this quenching must be taken explicitly into account. The expression for f'_{feed} in $B_0/G = f'_{feed}$ becomes:

$$f'_{feed} = \{1 + [(\sigma_{p0} v_T N_v/a') e^{-E_{p0}/kT}][1 + (a'/b') e^{-E'/kT}]\}^{-1} \qquad (11.5)$$

where a' is the partial CTS feeding rate into $^5D_0$, b' is the $^5D_0 \to {}^7F$ emission rate, and E' is the $^5D_0 \to$ CTS activation energy.

This equation is derived by the sequence:

$$\frac{B_0}{G} = \frac{1}{1 + \dfrac{C}{B}} = \frac{1}{1 + \dfrac{C}{A} \cdot \dfrac{A}{B}} = \frac{1}{1 + \dfrac{C}{A}\left(\dfrac{B + A^{-1}}{B}\right)} = \frac{1}{1 + \dfrac{C}{A}\left(1 + \dfrac{A^{-1}}{B}\right)}$$

$$= \frac{1}{1 + \dfrac{C}{A}\left(1 + \dfrac{A}{B} e^{-E'/kT}\right)}. \qquad (11.6)$$

Introducing the expressions in Table 37 for these process rates, one obtains Eq. (11.5). The last step in Eq. (11.6) uses the ratio of inverse rates encountered previously in, e.g., Sect. 4.12.

Equation (11.5) is plotted in Fig. 63 as the dotted curve and fits better the observed f'_{feed} data points above 500 K where $^5D_0$ is in fact quenching. Thus, the general understanding of the CTS dissociation into Eu^{2+} and a free hole embraces all the feeding and storage effects measured.

11.5 The SCC Model for Understanding Storage-Loss Processes in Oxysulfide:Eu Phosphors

We give in Fig. 64 an SCC model for the states involved in the processes listed in Table 37. The ^7F, ^5D, and CTS have been placed previously in such a diagram, Fig. 25. We show here only selected members of these states. The CTS minimum lies 24 000 cm^{-1} above the $^7F_0$ minimum in La_2O_2S and 25 100 cm^{-1} above in Y_2O_2S. The state named $^7F_0 + e(\infty) + h(\infty)$ is the state reached by absorbing bandgap radiation. This bandgap energy is $\sim 35 000$ cm^{-1} for La_2O_2S and $\sim 37 000$ cm^{-1} for Y_2O_2S. Because the excitation is far away from the Eu center, the environment near the Eu is unaffected and this state is shown with zero offset.

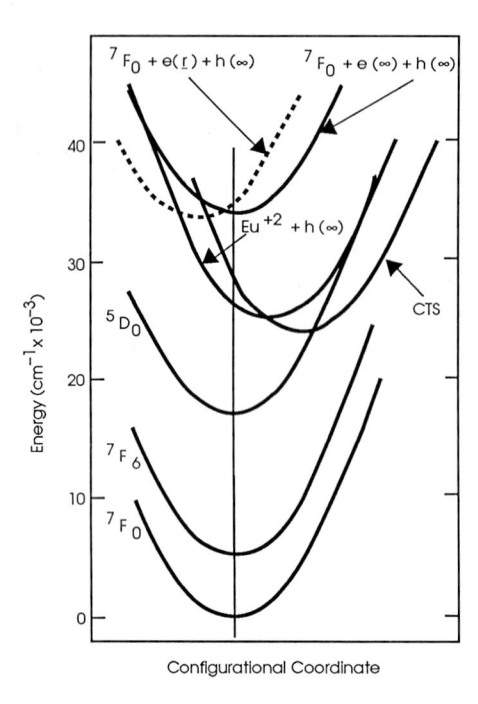

Fig. 64. SCC model for understanding storage-loss processes in oxysulfide:Eu phosphors

The state labeled $Eu^{2+} + h(\infty)$ is the state reached by process C. It is 1300 cm^{-1} or 900 cm^{-1} above the CTS, respectively, for La_2O_2S or Y_2O_2S, from the entries in column 5 of Table 39. The difference, $35\,000 - (24\,000 + 1300) = 9700$ cm^{-1} or $37\,000 - (25\,000 + 900) = 11\,000$ cm^{-1}, respectively, is the stabilization energy for capturing a free electron by Eu^{3+} to form Eu^{2+}, process F^{-1} of Table 37. This is the first direct measurement of this quantity in this host known to us.

Also shown in this diagram as a dashed curve is a state where the electron is in the conduction band near the Eu impurity. It is clear that the attraction of the Eu^{3+} for an electron is a very short-ranged interaction. Whether there is a bound state or not is speculative at this time; it is shown, however. Its presence would be demonstrated by low (4 K)-temperature measurements of the σ_{n0}, which should be $\sim 10^{-15}$ cm^{-2} at room temperature where this state is thermally emptied but would probably be larger at low temperatures where this state is stable against thermal ionization.

11.6 The Steady-State Efficiency and its Dependence on Excitation Intensity: B_∞/G

11.6.1 The Observed Behavior

Figures 65 and 66 show the rise curves of $La_2O_2S:Eu^{3+}$ at Eu concentrations 0.0005 and 0.005, respectively. Similar curves were found for 0.02 Eu in this host and for $Y_2O_2S:Eu^{3+}$ with Eu concentration 0.001.

11.6 The Steady-State Efficiency

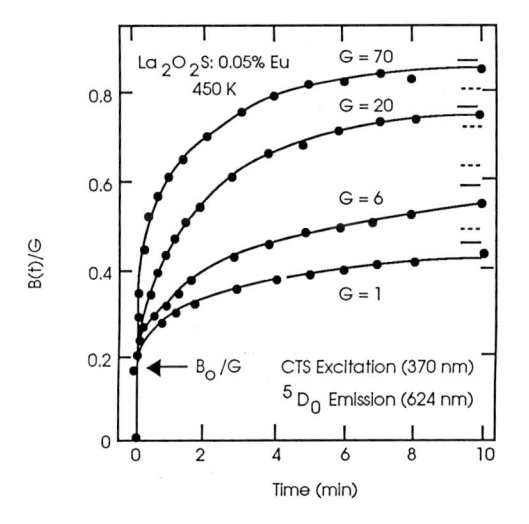

Fig. 65. Rise curves of $^5D_0 \to {}^7F_2$ emission in $(La_{0.9995}Eu_{0.0005})_2O_2S$

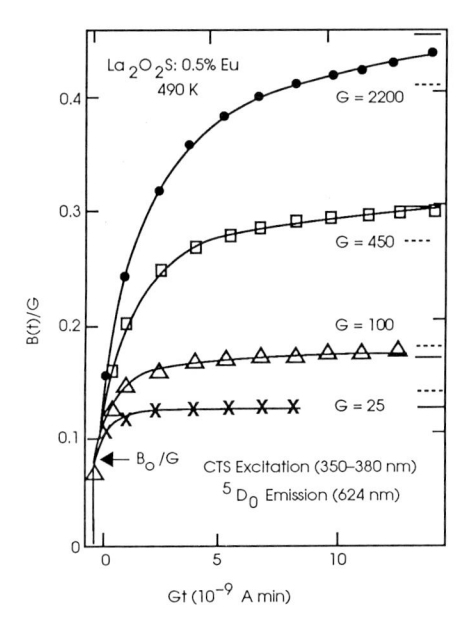

Fig. 66. Rise curves of $^5D_0 \to {}^7F_2$ emission in $(La_{0.995}Eu_{0.005})_2O_2S$

The solid curves connect the points which are the measured values at the indicated excitation intensities. The solid lines at the right edge of the graph give the measured steady-state efficiencies at these excitation intensities, towards which the points are tending. The dotted lines are our calculated values of these steady-state efficiencies as developed in the following Sects. 11.6.2 and 11.6.3.

In both these figures, the values of the initial quantum efficiency B_0/G are seen independent of the excitation intensity at the value indicated in each figure. In both figures at all temperatures, the quantum efficiency rises slowly with time and ultimately approaches a constant value, lower than unity, which increases sublinearly with excitation intensity.

11.6.2 The Equation for B_∞/G

We assume here (1) that the Eu^{2+} concentration n_0 is charge balanced by the trapped-hole charge, (2) that traps can be divided into two types, shallow traps (type 1) kept mostly empty by detrapping ($D_{1i} \approx D_{1i}^{-1}$) and deeper traps (type 2) kept mostly empty by loss ($D_{2i} \approx E_{2i}$), and (3) that most of the loss occurs through the deeper traps ($E \approx E_2 = \Sigma_i E_{2i}$).

Under these assumptions, the relation between B_∞ and G is best stated through a parametric variable $\hat{n}_{0\infty}$ which increases sublinearly with excitation intensity:

$$G = \frac{1}{2}\beta(1 + \gamma)\frac{\hat{n}_{0\infty}^3 + \hat{n}_{0\infty}^2}{1 + \gamma\hat{n}_{0\infty}}$$

$$\frac{B_\infty}{G} = f'_{feed} + \left(\frac{\hat{n}_{0\infty}}{1 + \hat{n}_{0\infty}}\right)f'_{ionize} \qquad (11.7)$$

Equation (11.7) and its parameters will be derived in Sect. 11.6.3. At any given temperature, β, γ, f'_{feed}, and f'_{ionize} are constants. The $\hat{n}_{0\infty}$ parametric variable will be shown to be the ratio of the probability that a free hole will generate emission to the probability that it will be trapped in deep traps and lost. The $\hat{n}_{0\infty}$ will be found proportional to the Eu^{2+} concentration; it can therefore be regarded as a scaled measure of this concentration.

Figure 67 shows the general behavior of Eq. (11.7). This figure gives the B_∞/G calculated as a function of G/β for two values of γ, namely, 0 (deep traps only) and 1 (both shallow and deep traps). (These characterizations will become clear in the derivation.) Each point is fixed by assigning a value to $\hat{n}_{0\infty}$ and then calculating the ($B_\infty/G, G/\beta$) couple which satisfies Eq. (11.7). The curves show a gradual rise in quantum efficiency with excitation intensity, which is most pronounced when f'_{feed} is small, i.e., when most of the excitations into the CTS cause ionization and trapping.

The dotted line segments at the right edge of Figs. 65 and 66 are calculated values obtained from Eq. (11.7) with the parameters given in Table 40. Figure 68 shows the nonlinear efficiency observed in Y_2O_2S with the two listed Eu concentrations and at the three listed temperatures. This figure also shows the fits obtained using Eq. (11.7) with the parameter sets shown both in the figure and also in Table 40. The β is given in units of 10^{-9} amp in Fig. 68 without this unit being cited and is given in units of 10^{-10} amp in Table 40 as cited there. Thus the nonlinear quantum efficiency with excitation intensity is also understandable with the model of the breakup of the CTS into a Eu^{2+} and a free hole.

11.6 The Steady-State Efficiency

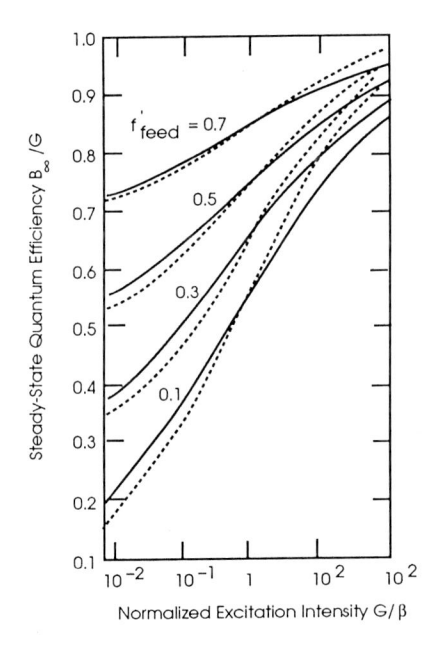

Fig. 67. B_∞/G vs. G/β: Calculated nonlinear efficiency

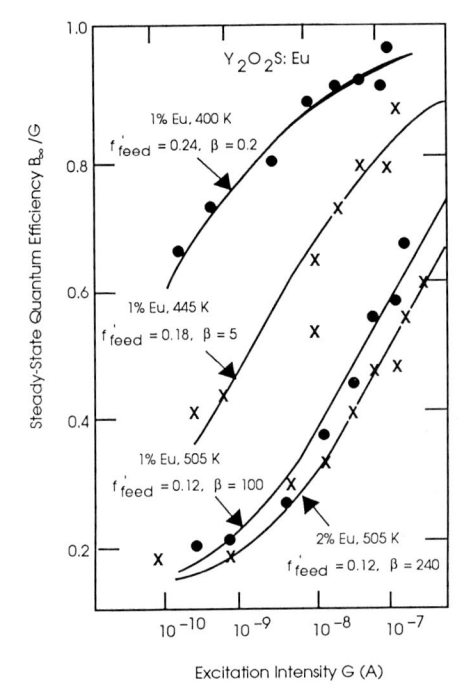

Fig. 68. B_∞/G vs. G for two $Y_2O_2S:Eu^{3+}$ phosphors

Table 40. Parameters describing nonlinear efficiencies in oxysulfides: Eu^{3+}

Host	T	Eu conc.	$f_{feed} = B_0/G$	β, $(10^{-10}$ amp.)	γ
La_2O_2S	450 K	0.0005	0.18	3.6	0
	490 K	0.005	0.085	120.	0
Y_2O_2S	400 K	0.01	0.24	2.	0
	445 K	0.01	0.18	50.	0
	505 K	0.01	0.12	1 000.	0
	505 K	0.02	0.12	2 400.	0

11.6.3 Derivation of Nonlinear Efficiency Expression

In the steady state the rates satisfy:

$$G = B_\infty + E_\infty, \quad C_\infty = C_\infty^{-1} + E_\infty, \quad D_\infty = D_\infty^{-1} + E_\infty \quad (11.8)$$

Since:

$$B/C = \frac{f'_{feed}}{f'_{ionize}} \quad (11.9)$$

it follows that:

$$G = \left(\frac{B}{C}\right)C_\infty + E_\infty = \frac{f'_{feed}}{f'_{ionize}}(C_\infty^{-1} + E_\infty) + E_\infty = \left(f'_{feed}\frac{C_\infty^{-1}}{E_\infty} + 1\right)\frac{E_\infty}{f'_{ionize}} \quad (11.10)$$

We now define:

$$\hat{n}_0 = \frac{f'_{feed}C^{-1}}{D_2} = f'_{feed}\left(\frac{\sigma_{p0}n_0}{\langle\sigma_{p2i}\rangle N_{2\,trap}}\right) \quad (11.11)$$

where $\langle f \rangle$ is the average of f:

$$\langle f \rangle = \frac{\Sigma_i f_i N_i}{\Sigma_i N_i}. \quad (11.12)$$

According to Eq. (11.11), $\hat{n}_{0\infty}$ has the attributes cited in the discussion after Eq. (11.7).
From Eqs. (11.10) and (11.11) one finds that:

$$G = \left(\frac{\hat{n}_{0\infty} + 1}{f'_{ionize}}\right)E_\infty \quad (11.13)$$

and:

$$\frac{B_\infty}{G} = 1 - \frac{E_\infty}{G} = \frac{\hat{n}_{0\infty} + f'_{feed}}{1 + \hat{n}_{0\infty}} = f'_{feed} + \frac{\hat{n}_{0\infty}}{1 + \hat{n}_{0\infty}}f'_{ionize} \quad (11.14)$$

This is the second of Eqs. (11.7). When $\hat{n}_{0\infty} = 1$, B_∞/G has the value $f'_{feed} + \frac{1}{2}f'_{ionize}$,

11.6 The Steady-State Efficiency

midway between its limiting values f'_{feed} for $\hat{n}_{0\infty} \ll 1$ and $f'_{feed} + f'_{ionize} = 1$ for $\hat{n}_{0\infty} \gg 1$.

The first of Eqs. (11.7) is derived under the approximations that the trap-emptying processes D^{-1} and E keep the traps mostly empty of holes:

$$N_i = n_i + p_i \approx n_i \tag{11.15}$$

that the traps can be divided into two types: shallow traps (1) kept mostly empty by detrapping:

$$D_{1i\infty} = D_{1i\infty}^{-1} + E_{1i\infty} \approx D_{1i\infty}^{-1} \tag{11.16}$$

and deep traps (2) kept mostly empty by loss:

$$D_{2i\infty} = D_{2i\infty}^{-1} + E_{2i\infty} \approx E_{2i\infty} \tag{11.17}$$

and that most losses occur from the deep traps:

$$E_\infty = E_{1\infty} + E_{2\infty} \approx E_{2\infty} = \Sigma_i E_{2i\infty} \tag{11.18}$$

Then:

$$E_\infty = E_{2\infty} = D_{2\infty} = \Sigma_i D_{2i\infty} = \Sigma_i \sigma_{p2i} v_T N_{2i} p_\infty = \langle \sigma_{p2i} \rangle v_T N_{2\,trap} p_\infty \tag{11.19}$$

Equations (11.13) and (11.19) are the starting equations in the derivation of the first of Eq. (11.7). We need an expression for p_∞:

$$p_\infty = \frac{n_{0\infty}^2}{n_{0\infty} N_{1\,trap} \langle 1/\bar{p}_{1i} \rangle + \langle \sigma_{p2i}/e_{2i} \rangle v_T N_{2\,trap}} \tag{11.20}$$

This expression is obtained from the charge balance equation:

$$n_0 = \Sigma_i p_{1i} + \Sigma_i p_{2i} \tag{11.21}$$

by the following sequence of relationships. First, we use the definitions of type 1 and type 2 traps according to Eqs. (11.16) and (11.17), namely:

$$D_{1i\infty} = D_{1i\infty}^{-1} \qquad D_{2i\infty} = E_{2i\infty} \tag{11.22}$$

Using the expressions for these rates in Table 37 one gets:

$$\sigma_{p1i} v_T n_{1i\infty} p_\infty = \sigma_{p1i} v_T \bar{p}_{1i} p_{1i\infty}$$

$$\sigma_{p2i} v_T n_{2i\infty} p_\infty = e_{2i} p_{2i\infty} n_{0\infty} \tag{11.23}$$

Introducing Eqs. (11.23) into the charge balance Eq. (11.21) gives:

$$n_{0\infty} = \sum_i \left(\frac{\sigma_{p2i} v_T n_{2i\infty} p_\infty}{e_{2i} n_{0\infty}} \right) + \sum_i \frac{n_{1i\infty} p_\infty}{\bar{p}_{1i}} \tag{11.24}$$

Multiplying through by $n_{0\infty}$, solving for p_∞, and using the definition of averages according to Eq. (11.12) gives the desired expression for p_∞, namely, Eq. (11.20).

We want this equation, however, not with the parameter $n_{0\infty}$ but with the parameter $\hat{n}_{0\infty}$. To this end we solve Eq. (11.11) for $n_{0\infty}$ in terms of $\hat{n}_{0\infty}$, introduce this expression into Eq. (11.20), and obtain:

$$p_\infty = \frac{\hat{n}_{0\infty}^2 \langle \sigma_{p2i} \rangle^2 N_{2\,trap}}{\hat{n}_{0\infty} N_{1\,trap} \langle 1/\bar{p}_{1i} \rangle \sigma_{p0} f'_{feed} \langle \sigma_{p2i} \rangle + \langle \sigma_{p2i}/e_{2i} \rangle v_T \sigma_{p0}^2 f'^2_{feed}}$$

$$= \frac{\hat{n}_{0\infty}^2 \langle \sigma_{p2i} \rangle^2 N_{2\,trap}}{v_T \langle \sigma_{p2i}/e_{2i} \rangle \sigma_{p0}^2 f'^2_{feed} \left[1 + \dfrac{\hat{n}_{0\infty} N_{1\,trap} \langle 1/\bar{p}_{1i} \rangle \langle \sigma_{p2i} \rangle}{\langle \sigma_{p2i}/e_{2i} \rangle \sigma_{p0} f'_{feed} v_T}\right]} \qquad (11.25)$$

We define:

$$\gamma = \frac{\langle 1/\bar{p}_{1i} \rangle N_{1\,trap} \langle \sigma_{p2i} \rangle}{f'_{feed} \sigma_{p0} v_T \langle \sigma_{p2i}/e_{2i} \rangle} \qquad (11.26)$$

so that the bracketed term in the denominator in Eq. (11.25) is $1 + \gamma \hat{n}_{0\infty}$. Thus, finally:

$$p_\infty = \frac{\hat{n}_{0\infty}^2 \langle \sigma_{p2i} \rangle^2 N_{2\,trap}}{v_T \langle \sigma_{p2i}/e_{2i} \rangle \sigma_{p0}^2 f'^2_{feed} (1 + \gamma \hat{n}_{0\infty})} \qquad (11.27)$$

Returning now to Eq.(11.13), introducing Eq. (11.19) for E_∞, then introducing Eq. (11.27) for p_∞, we obtain:

$$G = \hat{n}_{0\infty}^2 \frac{(1 + \hat{n}_{0\infty})}{(1 + \gamma \hat{n}_{0\infty})} \left[\frac{\langle \sigma_{p2i} \rangle^3 N_{2\,trap}^2}{\langle \sigma_{p2i}/e_{2i} \rangle \sigma_{p0}^2 f'^2_{feed} f'_{ionize}}\right] \qquad (11.28)$$

By defining a constant parameter judiciously, namely:

$$\frac{\beta(1 + \gamma)}{2} = \left[\frac{\langle \sigma_{p2i} \rangle^3 N_{2\,trap}^2}{\langle \sigma_{p2i}/e_{2i} \rangle \sigma_{p0}^2 f'^2_{feed} f'_{ionize}}\right] \qquad (11.29)$$

one can rewrite Eq. (11.28) as the first of Eqs. (11.7). This choice for the constant makes:

$$\beta = \frac{2\langle \sigma_{p2i} \rangle^3 N_{2\,trap}^2}{(1 + \gamma) f'_{ionize} f'^2_{feed} \sigma_{p0}^2 \langle \sigma_{p2i}/e_{2i} \rangle} \qquad (11.30)$$

The main advantage in the choice Eq. (11.29) for the constant is that, in Eq. (11.7), when $\hat{n}_{0\infty} = 1$, $G = \beta$, i.e., that β is the G which makes $\hat{n}_{0\infty} = 1$ and leads to the efficiency midway between the zero-intensity efficiency and the infinite-intensity efficiency, namely, $f'_{feed} + \tfrac{1}{2} f'_{ionize}$.

When no shallow traps are present, $\gamma = 0$ and the two Eqs. (11.7) can be combined to give:

$$\frac{G}{\beta} = \frac{1}{2} f'_{ionize} \frac{(B_\infty/G - f'_{feed})^2}{(1 - B_\infty/G)^3} \qquad (11.31)$$

Equation (11.31) is derived by solving the second of Eq. (11.7) for $\hat{n}_{0\infty}$ and then calculating $(1 + \hat{n}_{0\infty})$:

$$\hat{n}_{0\infty} = \frac{\dfrac{B_\infty}{G} - f'_{feed}}{1 - \dfrac{B_\infty}{G}}$$

$$1 + \hat{n}_{0\infty} = \frac{f'_{ionize}}{1 - \frac{B_\infty}{G}} \tag{11.32}$$

Introducing Eq. (11.32) into the $\gamma = 0$ form of the first of Eq. (11.7) gives Eq. (11.31).

11.7 The $n_{0\infty}$ Achieved

Larach and Faughnan of RCA Laboratories looked for but did not find an EPR signal due to the Eu^{2+} created by CTS excitation. We give two consistent estimates of the $n_{0\infty} = Eu^{2+}$ concentration achieved under the most favorable conditions investigated.

The time-integrated phosphorescence and glow curves is one measure of the number of Eu^{2+} centers produced by the excitation. The component of the steady-state emission intensity equal to $B_\infty - B_0$ is a second measure of this number. Both these measures, the first from the data of Figs. 57–59, the second from the data of Figs. 65, 66, and 68, agree and give $\sim 10^{13}$ detected photons, or $\sim 10^{16}$ cm^{-3} Eu^{2+} in the 10^{-3} cm^3 sample volume irradiated.

11.8 The Rise Time

In the next two sections the discussion of the rise time and of the decay is specifically aimed at the excitation-intensity dependence of these characteristics. A full understanding of the rise curve and of the decay curve would require elucidation of the trap depth distribution and of the capture crosssections of these traps. It would also force a determination of the nature of the loss process, i.e., whether it is dominated by surface recombinations or not, and of the rate constants. We confine ourselves here to a discussion based on the processes of Table 37 and the rates in Table 38.

The rise time is roughly the time needed to create $n_{0\infty}$ Eu^{2+} centers at the rate allowed by the CTS dissociation rate $C = f'_{ionize} G$ while it is opposed by the CTS reformation rate C^{-1}. This rate C^{-1} is initially zero when all traps are empty and becomes equal to the rate C at the "end" of the rise. The average rate is taken as $f'_{ionize} G/2$. (Of course the real performance is the somewhat exponential rise which reaches its "end" at $t = \infty$.) Under this picture, the rise time is:

$$\Delta t_{rise} = \frac{n_{0\infty}}{\left(\frac{f'_{ionize} G}{2}\right)} = \frac{2\langle\sigma_{p2i}\rangle N_{2\,trap}}{\sigma_{p0}} \frac{\hat{n}_{0\infty}}{f'_{feed} f'_{ionize} G}. \tag{11.33}$$

This formula shows that the rise time Δt_{rise} should decrease with increasing G as $G^{1/2}$ to $G^{2/3}$. The exact dependence can be predicted by solving Eq. (11.7) for $\hat{n}_{0\infty}$ and introducing this value into Eq. (11.32).

190   11 Effects Beyond the Model: Oxysulfide: Eu Storage and Loss Processes

In Fig. 65 $B(t)/G$ is plotted against t and the rise transient is shown decreasing with increasing G. In Fig. 66 $B(t)/G$ is plotted against Gt and the rise transient is shown increasing with increasing G. We have plotted these observed curves against $f'_{feed}f'_{ionize}Gt/\hat{n}_{0\infty}$ as suggested by Equation (11.33) and have found the rise transient then roughly independent of G.

11.9 The Assymetry Between Phosphorescence and Build-Up

Slow phosphorescence B_P and slow buildup $(B_\infty - B_0)$ were found to be asymmetric, in the sense that the initial slow phosphorescence was smaller than the total buildup $(B_\infty - B_0)$. The phosphorescence was also found to be sublinear with excitation intensity G. Both these behaviors can be understood with the model of this chapter.

The B_P sublinearity is understandable because phosphorescence is a consequence of detrapping D^{-1} via the sequence D^{-1}, C^{-1}, A, B. In the model here, the D^{-1} is proportional to $\Sigma_i p_i$, which varies sublinearly with G because it is itself equal to the Eu^{2+} concentration n_0 which varies sublinearly with G. The exact dependence will be discussed in the next section.

The relationship between initial phosphorescence and total buildup is elucidated by considering the relations between rates in Table 37 at steady state:

$$G_\infty + A_\infty^{-1} + C_\infty^{-1} = A_\infty + C_\infty$$

$$A_\infty = B_\infty + A_\infty^{-1}$$

$$C_\infty + D_\infty^{-1} = C_\infty^{-1} + D_\infty$$

$$D_\infty = D_\infty^{-1} + E_\infty \tag{11.34}$$

These relations are for the steady state concentrations p_0 of the CTS, n_D the Eu^{3+} ^5D states, p the free holes, and $\Sigma_i p_i$ the trapped holes, respectively. We need also Eq. (11.2) for B_0, namely:

$$B_0 = f'_{feed}(G + C^{-1}) = f'_{feed}G \tag{11.35}$$

since at the onset of excitation of a phosphor previously exhausted of storage $C^{-1} = 0$. Finally, we need the equations describing the initial phosphorescence, which are identical to Eq. (11.34) with the subscript ∞ replaced by the subscript P, with the notation added (t) at $t = 0$, and with G set to zero.

The development then proceeds as follows:

$$\frac{B_P(0)}{B_\infty - B_0} = \frac{A_P(0) - A_P^{-1}(0)}{A_\infty - A_\infty^{-1} - f'_{feed}G} = \frac{C_P^{-1}(0) - C_P(0)}{(G + C_\infty^{-1} - C_\infty) - f'_{feed}G}$$

$$= \frac{C_P^{-1}(0) - C_P(0)}{(C_\infty^{-1} - C_\infty) + f'_{ionize}G} = \frac{D_P^{-1}(0) - D_P(0)}{D_\infty^{-1} - D_\infty + f'_{ionize}G}$$

$$= \frac{D_\infty^{-1} - D_\infty}{D_\infty^{-1} - D_\infty + f'_{ionize}G} \tag{11.36}$$

11.10 An Expression for Phosphorescence

In explanation, since detrapping D^{-1} and excitation G generate free holes at the rates D^{-1} and $f'_{ionize}G$, respectively, the free hole concentration p can be regarded as a superposition of component concentrations p_G and p_D proportional to G and D^{-1}, respectively. When excitation is removed, G and p_G decrease rapidly to zero, but D^{-1} and p_D and also n_0 maintain their steady-state values D_∞^{-1}, $p_{D\infty}$ and $n_{0\infty}$.

We do not have a global picture of the various terms in the expression Eq. (11.36). Whatever value the numerator takes, however, it is clear that the ratio of phosphorescence to total buildup will decrease with increasing excitation intensity G, in agreement with the observations.

11.10 An Expression for Phosphorescence

The expression for the phosphorescence to be obtained here also uses the relations of the previous section for the onset of phosphorescence:

$$D_P^{-1}(0) + C_P(0) = C_P^{-1}(0) + D_P(0)$$
$$D_P^{-1}(0) = B_P(0) + D_P(0) \tag{11.37}$$

and the relation, true of Eq. (11.2) when $G = 0$ at the onset of phosphorescence, using Eq. (11.11):

$$B_P(0) = f'_{feed} C_P^{-1}(0) = f'_{feed} C_\infty^{-1} = \hat{n}_{0\infty} D_{2\infty} \tag{11.38}$$

We write:

$$B_P(t) = \frac{B_P(t) D_P^{-1}(t)}{D_P^{-1}(t)} = \frac{D_P^{-1}(t)}{\left[\dfrac{B_P(t) + D_P(t)}{B_P(t)}\right]} = \frac{D_P^{-1}(t)}{\left[1 + \dfrac{D_P(t)}{B_P(t)}\right]} \tag{11.39}$$

Now we introduce the two types of traps, shallow and deep, observe the equalities Eq. (11.22) defining these types, restrict the losses to the deep traps, and find:

$$B_P(t) = \frac{D_{1P}^{-1}(t) + D_{2P}^{-1}(t)}{\left[1 + \dfrac{D_{1P}(t) + D_{2P}(t)}{B_P(t)}\right]} = \frac{D_{2P}(t)\left[\dfrac{D_{1P}^{-1}(t) + D_{2P}^{-1}(t)}{D_{2P}(t)}\right]}{1 + \left[\dfrac{D_{1P}(t)}{D_{2P}(t)} + 1\right]\dfrac{D_{2P}(t)}{B_P(t)}} \tag{11.40}$$

We define the time-independent quantities:

$$\delta = \frac{D_{1P}(t)}{D_{2P}(t)} = \frac{\langle \sigma_{p1i} \rangle N_{1\,trap}}{\langle \sigma_{p2i} \rangle N_{2\,trap}}$$

$$\varepsilon = \frac{\hat{n}_0(t) D_{2P}^{-1}(t)}{D_{2P}(t)} = \frac{\hat{n}_0(t) D_{2P}^{-1}(t)}{E_{2P}(t)} = \hat{n}_0(t) \left[\frac{\sum_i^{N_{2trap}} \sigma_{p2i} \bar{p}_{2i} v_T p_{2i}}{\sum_i^{N_{2trap}} e_{2i} n_0 p_{2i}}\right]$$

$$= \left[\frac{v_T \langle \sigma_{p2i} \bar{p}_{2i} \rangle}{n_0(t) \langle e_{2i} \rangle}\right]\left[\frac{f'_{feed} \sigma_{p0} n_0(t)}{\langle \sigma_{p2i} \rangle N_{2\,trap}}\right] = \frac{f'_{feed} \sigma_{p0} v_T \langle \sigma_{p2i} \bar{p}_{2i} \rangle}{N_{2\,trap} \langle e_{2i} \rangle \langle \sigma_{p2i} \rangle} \tag{11.41}$$

11 Effects Beyond the Model: Oxysulfide:Eu Storage and Loss Processes

where the second step of the equation for ε uses the second of Eqs. (11.22) and where:

$$\hat{n}_0(t) \equiv \frac{B_p(t)}{D_2(t)} = \left[\frac{f'_{feed}\sigma_{p0}}{\langle\sigma_{p2i}\rangle N_{2\,trap}}\right] n_0(t) \tag{11.42}$$

Next we note that during the early stages of phosphorescence:

$$\hat{n}_0(t) = \hat{n}_{0\infty}\alpha_0(t)$$
$$D_1^{-1}(t) = D_1^{-1}(0)\alpha_1(t) = D_1(0)\alpha_1(t) = D_{1\infty}\alpha_1(t) = \delta D_{2\infty}\alpha_1(t)$$
$$D_2^{-1}(t) = D_2^{-1}(0)\alpha_2(t) \tag{11.43}$$

where the $\alpha_0(t)$, $\alpha_1(t)$, and $\alpha_2(t)$ decrease slowly from unity, and obtain:

$$B_p(t) = \left[\frac{\delta\alpha_1(t) + \left(\dfrac{\varepsilon}{\hat{n}_{0\infty}}\right)\alpha_2(t)}{1 + \left(\dfrac{\delta + 1}{\hat{n}_{0\infty}\alpha_0(t)}\right)}\right] D_{2\infty} \tag{11.44}$$

We have applied Eq. (11.44) to the phosphorescence data of $(Y_{0.99}Eu_{0.01})_2O_2S$ in Table 41.

In Table 41, we have fits to the data for deep traps only ($\delta = 0$) and for

Table 41. Phosphorescence in Y_2O_2S: 0.01 Eu^{3+}

	Equation (11.44) parameters				$G(10^{-10}$ A)		
	α_0	δ	$\delta\alpha_1$	$\varepsilon\alpha_2$	40	300	1 150
G/β					7.0	53	200
$\hat{n}_{0\infty}$					2.1	4.5	7.0
$(B_\infty - B_0)/G = [\hat{n}_{0\infty}/(1 + \hat{n}_{0\infty})]f'_{ionize}$					0.53	0.64	0.68
		$B_p(2$ sec$)/G$					
Observed					0.045	0.030	0.021
Fitted	1	0	0	1.00	0.081	0.026	0.012
	1	1/3	0.24	0.11	0.045	0.029	0.021
	1	2/3	0.24	0.17	0.045	0.029	0.021
		$B_p(15$ sec$)/G$					
Observed					0.0039	0.0025	0.0011
Fitted	2/3	0	0	0.080	0.0056	0.0019	0.0009
	2/3	1/3	0.011	0.042	0.0040	0.0020	0.0013
	2/3	2/3	0.011	0.050	0.0040	0.0020	0.0013
	1/3	0	0	0.110	0.0047	0.0019	0.0010
	1/3	1/3	0.011	0.075	0.0040	0.0021	0.0013
	1/3	2/3	0.011	0.090	0.0040	0.0021	0.0014

11.10 An Expression for Phosphorescence

both shallow and deep traps. We see that for deep traps only we overestimated the variation in $B_P(2\text{ sec})$ with G by about a factor of three over the measured range. We obtained a better fit with both types of traps, and find that $\delta \sim 1/3$ to $2/3$, that $\alpha_1(t)\delta$ decreased from near 0.24 at 2 sec to 0.011 at 15 sec, that $\alpha_0(t)$ decreased from near unity at 2 sec to near $1/3$–$2/3$ at 15 sec, and that $\alpha_2(t)\varepsilon$ decreased from near 0.14 at 2 sec. to near 0.06 at 15 sec. Detrapping the shallow traps dominated the 2 sec phosphorescence, whereas both deep and shallow traps participated about equally at 15 sec.

Other forms for $\alpha_0(t)$ and $\alpha_1(t)$ also lead to equally good fits, e.g., for decreases of the bimolecular form $(1 + \alpha \hat{n}_{0\infty} t)^{-1}$. As stated, we do not claim finality with our treatment of these transient behaviors.

12 The Exponential Energy-Gap "Law" for Small-Offset Cases

Riseberg, Moos, and coworkers [27] have studied $4f \to 4f$ nonradiative transitions of 3+ rare-earth ions in several hosts. They used for the nonradiative rate the expression:

$$W_{p,M} = A_{RM}\varepsilon^p \langle 1+m \rangle^p \tag{12.1}$$

where A_{RM} is a constant, $p\hbar\omega_0$ is the energy gap between the initial and final electronic state, and $\langle m \rangle$ is, as in our notation, Planck's thermal occupancy measure of the temperature. The temperature dependence in Eq. (12.1) was attributed to Kiel's [29] formalism, where it is linked to the creation of lattice phonons according to the squared matrix element $\langle v_{m+1} | Q | v_m \rangle$. The 0 K value of Eq. (12.1) is the exponential energy gap law $A_{RN}\varepsilon^p$ and was offered simply as an empirical fit. The different hosts were all found to give straight line semilog fits, plotting the nonradiative rate for various electronic states of various activators versus p. Each host exhibited its own different slope and its own A_{RM}.

In the following two equations, approximate forms for these rates are given in our notation. These rates are to be understood in terms of one or another of the expressions shown, with approximate forms indicated by $\sim$ and with the $T = 0$ K forms following the arrows. The E in the subscripts refers to the Einstein model.

$$R_{nr} = A_{nr} W_p(S, \langle m \rangle)$$

$$R_{nr,d/dz} = A_{nr,d/dz} W_{p,d/dz}(S, \langle m \rangle) \to A_{nr,d/dz} \frac{(p-S)^2}{2S} W_p(S, \langle m \rangle) + \cdots$$

$$\sim A_{nr,d/dz} \frac{p^2}{2S} W_p(S, \langle m \rangle)$$

$$R_{nr,d/dz,E} = A_{nr,d/dz,E} M_{p,d/dz}(S, \langle m \rangle) \tag{12.2}$$

These expressions, through Eqs. (4.58), (4.88), (4.23), (4.86), and (5.6)–(5.7), allow the additional approximate forms:

$$R_{nr} \approx A_{nr} e^{-S\langle 2m+1 \rangle} \frac{(S\langle 1+m \rangle)^p}{p!} \to A_{nr} e^{-S} \frac{S^p}{p!}$$

$$R_{nr,d/dz} \to A_{nr,d/dz} \{\tfrac{1}{2} W_{p-1} + S[\tfrac{1}{2} W_{p-2} - W_{p-1} + \tfrac{1}{2} W_p]\}$$

$$R_{nr,d/dz,E} \to A_{nr,d/dz,E} \{\tfrac{1}{2} W_{p-1} + S(1-\gamma)[\tfrac{1}{2} W_{p-2} - W_{p-1} + \tfrac{1}{2} W_p]\} \tag{12.3}$$

12 The Exponential Energy-Gap "Law" for Small-Offset Cases

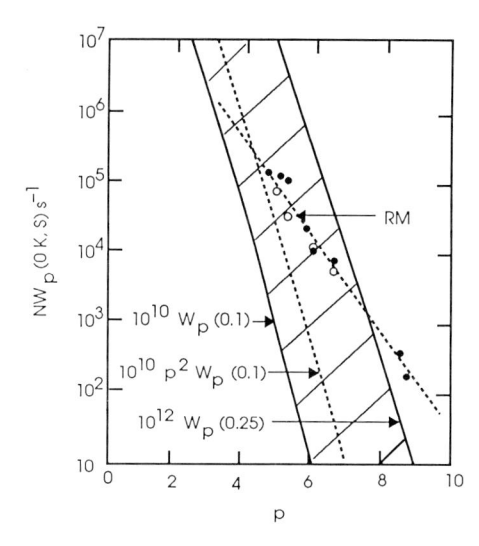

Fig. 69. Observed and fitted low-temperature rare earth $+3\ 4f \to 4f$ non-radiative rates

Figure 69 shows the Riseberg-Moos energy gap law applied to LaF_3 (filled circles) and to Y_2O_3 hosts. We do not dispute the practical utility of the correlations cited. They have been widely used to estimate nonradiative rates of transitions between rare earth energy levels. However, S values at times close to unity and A_{RM} values near 10^7 are invoked.

Yamada, Shionoya, and Kushida [67] have given a similar treatment for energy transfer rates in Y_2O_2S. Their fit is given in Fig. 70. It requires an S near 2. Again, we have no quarrel with the practical utility of this fit.

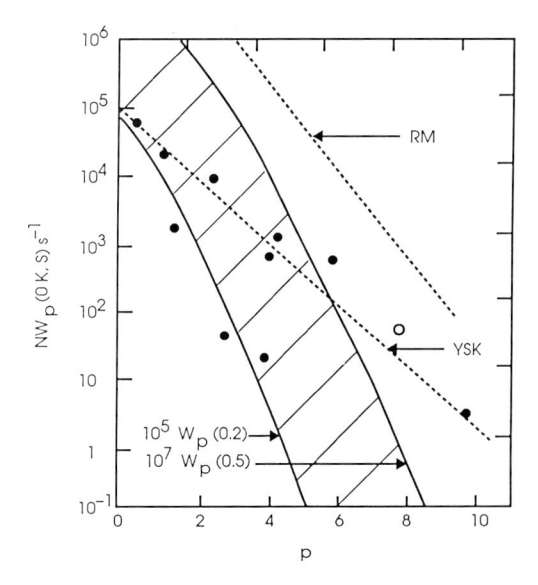

Fig. 70. Observed and fitted low-temperature rare earth energy transfer rates

Our insistence is that a more fundamental understanding of this correlation must be based on the fact that the W_p functions at 0 K are functions of two arguments, namely p and S, and that the Riseberg-Moos correlation implies that all these transitions have a common value for S in a given host. Indeed, van Dijk [68] (see also Jortner and Englman [69] and Englman as quoted by van Dijk) derives the energy gap law using the Einstein model, by pointing directly to the near constancy of S, within a factor of two. He gets, in a form reminiscent of the third of Eqs. (12.3), the rate, his Eq. (9):

$$k_{NR} = \beta_{el} \exp[-(\Delta E_0 - 2\hbar\omega_{max})\alpha] \tag{12.4}$$

and he illustrates this expression for $S = 0.25, 0.5,$ and 1.0.

We will take up the Riseberg-Moos expression first. The required connection to fit the RM expression and the last form of the first of Eqs. (12.3) to the same data over the range from p_1 to $p_1 + \Delta p$ is:

$$S = \hat{p}\varepsilon \qquad A_{RM} = A_{nr} \qquad \hat{p} \equiv \left[\frac{(p_1 + \Delta p)!}{p_1!}\right]^{1/\Delta p} \tag{12.5}$$

We give in Table 42 our connections between the Riseberg-Moos ε^p and our $S^p/p!$ expressions for all hosts treated by Moos. These S values are in fact uniformly unacceptably too large, because S is also the ratio of the one-phonon sideband and zero-phonon intensities.

We have no direct measures of the offsets between excited states. However, even for $LaCl_3$ and $LaBr_3$ hosts, the S value is a factor of two or three larger than expected, while for the other hosts S values no larger than 0.2–0.3 are expected.

In Fig. 69 we also give graphs of W_p functions with S parameters needed to span the same data points covered by the Riseberg-Moos expression. The A_{nr} values are $10^{10} - 10^{12}$. The S values do indeed span only a factor of 2–5, from 0.10–0.25; but this variation is crucial. What is needed, in our opinion, is a theory which links in each host the A_{nr} and S (or, better, the $S\hbar\omega$) values to the energy gap, i.e., to the p values, covering the observed data points as pictured in Figs. 69 and 70. Such a small increase of S with energy gap might well be expected because a transition with a larger energy gap implies a larger redistribution of the charge cloud of the ion and a larger readjustment of the environment to the

Table 42. Connection between Riseberg-Moos ε and our S

Host	ε	$\hat{p}$	S
$LaCl_3$	0.037	5	0.185
$LaBr_3$	0.037	5	0.185
LaF_3	0.14	6.5	0.91
SrF_2	0.20	7	1.4
Y_2O_3	0.12	6	0.72

12 The Exponential Energy-Gap "Law" for Small-Offset Cases

transition. It is not implausible that a larger energy gap might also compel a slightly larger electronic integral A_{nr}.

Probably a simple correlation even within one host is too much to expect. The need to reflect different levels of forbiddenness for the transition, depending on the spin, for example, might complicate the theoretical output. Moreover, the inequality among hosts in the rate of any specific transition must be built into this correlation. Nevertheless the need for such a correlation remains even if it costs simplicity.

Now we attend to the derivative operator expressions of van Dijk and of Englman and Jortner. In Fig. 69 we give a curve for the derivative operator in an adequate approximate form, as is evident from the second of Eqs. (12.2). The invocation of the derivative operator does not afford an understanding of the observed empirical fit within the context of the Riseberg-Moos exponential energy-gap law, that is, with the significantly smaller slope of the RM expression. The slope of the derivative expression curve is barely distinguishable from that for W_p functions.

In conclusion, one must in our opinion give attention to the implied correlation between the energy gap, the values for the offset parameter S, and the values for the electronic integral A_{nr}. One can, as usual, have motivations in physics to invoke the derivative operator expression, but one will be hard put to find experimental evidence forcing abandonment of the Condon-operator expression using the W_p functions themselves.

13 Conclusions

The W_p function is the distribution describing the optical band shapes, whether broad or nearly confined to the zero-phonon line, in two models: the single-configurational-coordinate model and the Einstein-Huang-Rhys-Pekar single-frequency multiple-coordinate model.

This function is also useful for describing the nonradiative rate between two electronic states. In this use, it is multiplied by "the electronic integral", treated as a parameter to be obtained from fitting experimental data on quenching, and of the order of $10^{11} - 10^{14}$.

The W_p function has a very useful recursion relationship:

$$S\langle m\rangle W_{p+1} + pW_p - S\langle 1+m\rangle W_{p-1} = 0. \tag{13.1}$$

This recursion formula is useful not only to calculate values for the distribution, but also to obtain the algebraic identity between the one- and the many-coordinate model W_p functions and to obtain algebraic formulas for the z and d/dz operators in the single-frequency model, in both one and in many dimensions. These algebras are based on the Manneback recursion formulas for the matrix elements of the overlap matrix A_{nm}.

Descriptions are given also of distributions related to W_p for the three operators considered, for the case of unequal force constants. Here the work must be numerical, because there is no natural grouping of the thermal weights into a simple discrete set of allowed energies. The numerical handling adopted is such that at 0 K the weights become the proper set of A_{n0}^2 or A_{m0}^2 squared overlap integrals, in emission and in absorption, respectively.

Descriptions are given also, for all these operators, for a multiple-frequency model in which all frequencies are multiples of the smallest of the set of frequencies.

Calculations are given aimed at comparing the predictions of all these models for all these operators. It is concluded that it will not be easy to demonstrate experimentally that a non-Condon operator is driving some nonradiative transition. It is also shown that the multiple-frequency derivative operator expression resembles more the W_p function than it does the SCC-model $W_{p,d/dz}$ function. Specifically, the cusp predicted at 0 K for the SCC $W_{p,d/dz}$ function near $p = S$ is significantly filled in in multiple coordinates by the L_p function.

The relationship between the W_p function in our notation to the same function under different names and under strikingly different forms, e.g., involving the I_p

13 Conclusions

Bessel functions of the second kind, is documented fully. The occurrence in the literature of the various forms, both exact and approximate, of the derivative-operator expressions is also documented.

The usefulness of the equal-force-constants W_p functions and of their unequal-force-constants counterparts in the investigation of the optical absorption bandshapes and of the thermal quenching of real phosphors is shown for a set of host-activator combinations. It is often, indeed generally, found that the anharmonicity-induced expansion of the lattice must be accounted for: we do so by allowing a temperature-dependent lowering of the difference in the zero-point energies between offset states.

Finally, we note that in the Einstein-Huang-Rhys-Pekar model every center in any given host should be described as having the same frequency, the longitudinal-optical frequency. This requirement is not forced with the SCC model here or indeed with the multiple-coordinate single-frequency model here, which can allow for localized vibrations of differing frequencies.

For the storage phenomena seen in oxysulfides with Eu activator, the additional concept of the dissociation of the charge-transfer is proposed and the consequences fully explored and confirmed. It is found that nonlinear efficiencies, fast and slow components of rise and decay that are temperature and excitation-intensity dependent, long-term storage of excitation energy and retrieval as line emissions, are all understandable through this extension to the SCC model. A measure of the energy required to free the hole from the CTS was obtained, in the neighborhood of 1000 cm^{-1}. A measure of the energy required to free the electron from the thus formed Eu^{2+} was also obtained, in the neighborhood of 10 000 cm^{-1}.

Much work remains to be done.

Within the framework of the SCC and the related multiple-coordinate models, there are many other host-activator systems which should be explored.

As extensions of the model, one would like a theoretical linkage between frequency and offset, replacing the equal S approximation adopted here for the DDESA model. One would like to have an experimental determination of, coupled to a theory of, the electronic factor, so that one can finally elucidate when a multiple-coordinate model is essential, when a non-Condon operator is essential. For small-offset cases, one would like a correlation between energy gap spanned and the offset parameter S and electronic integral A_{nr}. One would like also to treat the Jahn-Teller effect so as to obtain expressions for all these operators for the analog to the W_p distribution. We note that we have not found it necessary in treating the excited states of ruby to concern ourselves with the Jahn-Teller effect, although it is clearly theoretically called for.

The theory of energy transfer and of multiple-center models in general should be developed for the special case of two very close interacting centers such that the normal modes cannot be separated even conceptually into two disjoint sets. Our expectation here is that the obtained distribution will not differ much from the W_p distribution, but this expectation ought to be tested.

Another area of interest is the structure of the electronic states having CTS character. First, there are many possible CTS's. Second, in many hosts and specifically in oxysulfides, there is a required degeneracy of the CTS which has not been elucidated. There are three S^{2-} ions in the neighborhood of the Eu^{3+} ion. The CTS as pictured here, as a trapped exciton, must be split into at least two states because the known site symmetry will not support a triply-degenerate energy level. The existence of upper excited states of the CTS is an open question, particularly in view of the demonstrated ability to uncover impurity states with energy higher than the host band gap.

The more detailed treatment of the kinetic behavior of the CTS dissociation is needed, to determine the trap depth distribution, the spatial distribution of these traps, the chemical and/or crystallographic nature of them, and their capture crosssections. The loss process must be elucidated: the Eu^{3+} concentration dependence of the loss rate, the nature of the loss centers which collect the extra electron on the Eu^{3+} without forming the CTS or, at least, without forming a CTS which feeds efficiently emitting ^5D states.

Our hopes and our intentions in writing this book will be gratified if the approach taken here inspires similar approaches and the unfinished work outlined here is addressed.

14 References

1. Struck CW, Fonger WH (1970) J. Lum. 1/2:456
2. Fonger WH, Struck CW (1970) J. Chem. Phys. 52:6364
3. Fonger WH, Struck CW (1971) J. Elect. Soc. 118:273
4. Struck CW, Fonger WH (1971) Phys. Rev. B4:4
5. Struck CW, Fonger WH (1971) J. Appl. Phys. 42:4515
6. Fonger WH, Struck CW (1974) J. Lum. 8:452
7. Struck CW, Fonger WH (1974) J. Chem. Phys. 60:1988
8. Fonger WH, Struck CW (1974) J. Chem. Phys. 60:1994
9. Fonger WH, Struck CW (1975) Phys. Rev. B11:3251
10. Struck CW, Fonger WH (1976) J. Lum. 14:253
11. Struck CW, Fonger WH (1976) J. Chem. Phys. 64:1784
12. Struck CW, Fonger WH (1975) J. Lum. 10:1
13. Struck CW, Fonger WH (1975) J. Chem. Phys. 63:1533
14. Fonger WH, Struck CW (1978) J. Lum. 17:241
15. Fonger WH, Struck CW (1978) J. Chem. Phys. 69:4171
16. Struck CW, Fonger WH (1979) J. Lum. 18/19:101
17. Struck CW, Fonger WH (1979) Phys. Rev. B19:4400
18. Struck CW, Riseberg LA (1981) in Collins CB (ed) Proceedings of the International Conference on Lasers '81 STS Press, McLean, VA, p. 776
19. Alexander MN, Onorato PIK, Struck CW, Rozen JR, Tasker GW, Uhlmann DR (1986) J. Non-cryst. Solids 79:137
20. Bleijenberg KC, Kellendonk FA, Struck CW (1980) J. Chem. Phys. 73:3586
21. Fonger WH, Struck CW (1980) in: DiBartolo B (ed) *Radiationless Processes* Plenum, New York, p 471
22. Condon EU (1928) Phys. Rev. 32:858
23. Mott NF (1938) Proc. Roy. Soc. (London) A167:384; Gurney RW, Mott NF (1939) Trans. Faraday Soc. 35:69
24. Kisliuk P, Moore CA (1967) Phys. Rev. 160:307
25. Misu A (1964) J. Phys. Soc. Japan 19:2260
26. a. Partlow WD, Moos HW (1967) Phys. Rev. 157:252; b. Riseberg LA, Gandrud WB, Moos HW (1967) Phys. Rev. 159:262; c. Riseberg LA, Moos HW (1968) Phys. Rev. 174:429
27. Moos HW (1970) J. Lum. 1/2:106 and references therein
28. a. Chamberlain JR, Paxman DH, Page JL (1966) Proc. Phys. Soc. 89:143; b. Weber MJ (1968) Phys. Rev. 171:283; c. Fong FK, Naberhuis SL, Miller MM (1972) J. Chem. Phys. 56:4020
29. Kiel A (1966) in: Grivet P, Bloembergen N (eds) *Proc. Third Intern. Conf. Quantum Electronics, Paris, 1963* Columbia University Press, New York, p 765

30. Mahbub'ul Alam ASM, Di Bartolo B (1967) J. Chem. Phys. 47:3790
31. Sorokin PP, Stevenson MJ, Lankard JR, Petit GD (1962) Phys. Rev. 127:503
32. Manneback C (1951) Physica 17:1001
33. Hutchisson E (1930) Phys. Rev. 36:410; (1931) 37:45
34. Erdélyi A, Magnus W, Oberhettinger F, Tricomi FG (1953) *Higher Transcendental Functions, Bateman Manuscript Project*. Vol II, McGraw-Hill, New York
35. Ruamps J (1956) Compt. Rend. 243:2034
36. Wagner M (1959) Z. Naturforschung 14a:81
37. Koide S (1960) Z. Naturforschung 15a:123
38. Keil TH (1965) Phys. Rev. 140:A601
39. Fitchen DB (1968) in: Beall Fowler W (ed) *Physics of Color Centers*. Academic Press, New York, p 293
40. Huang K, Rhys A (1950) Proc. Roy. Soc. (London) 204A:406
41. Pekar SI (1950) J. Exper. Theoret. Phys. USSR 20:510
42. Pekar SI (1954) *Untersuchungen über die Elektronentheorie der Kristalle*, Akademie-Verlag, Berlin, Chap. 5, 6
43. Lax M (1952) J. Chem. Phys. 20:1752
44. Lax M (1956) in: Breckenridge RG, Russell BR, Hahn EE (eds) *Photoconductivity Conf Atlantic City, Nov., 1954* Wiley and Sons, New York, p 111
45. O'Rourke RC (1953) Phys. Rev. 91:265
46. a. Curie D (1958) Compt. Rend. 246:404; b. Curie D (1963) *Luminescence in crystals*, Wiley and Sons, New York, p 53, p 208
47. Lauer HV, Fong FK (1974) J. Chem. Phys. 60:274
48. Markham JJ (1959) Rev. Mod. Phys. 31:956
49. a. Perlin Yu E (1964) Sov. Phys. Usp. 6:642; b. Perlin Yu E (1963) Usp. Fiz. Nauk 80:553
50. Perlin Yu E (1957) Opt. i Spektr. 3:328
51. Miyakawa T, Dexter DL (1970) Phys. Rev. B1:2961
52. Fuchs BA, Levin VI (1961) *Functions of a complex variable and some of their applications, Vol II* Pergamon Press, London, p 233 ff
53. Mostoller M, Ganguly BN, Wood RF (1971) Phys. Rev. B4:2015
54. Förster T (1948) Ann. Phys. 2:55
55. Dexter DL (1953) J. Chem. Phys. 21:836
56. Wickersheim KA, Buchanan RA, Yates EC (1969) Proc. Conf. Rare Earth Res. 7th, Coronado, California 1968 p 835
57. Klick CC, Schulman JH (1957) in: Seitz F, Turnbull D (eds) *Solid State Physics* Vol 5, Academic Press, New York, p 97
58. Wortman DE, Leavitt RP, Morrison CA (1974) J. Phys. Chem. Solids 35:591
59. Buchanan RA (1971) private communication at Gordon Research Conference on Luminescence, 1971. Dr. Buchanan first informed us of the two broad emission bands in $Y_2O_2S:Yb^{3+}$ separated by the $10\,000\,cm^{-1}$ $^2F_{5/2}$ to $^2F_{7/2}$ energy difference. Our investigations were conducted during the few months after this Gordon Conference. One of us (WF) discussed this case with Dr. Nakazawa at the Paris, 1978 International Luminescence Conference
60. Nakazawa E (1979) J. Lum. 18/19:272
61. Gâcon JC, Souillat JC, Seriot J, Gaume-Mahn F, DiBartolo B (1979) J. Lum. 18/19:244
62. Fong FK, Lauer HV, Chilver CR, Miller MM (1975) J. Chem. Phys. 63:366
63. Lauer HV, Fong FK (1976) J. Chem. Phys. 65:3108

64. McClure DS (1962) J. Chem. Phys. 36:2757
65. Brossel J, Margerie J (1963) in: Low W (ed) *Paramagnetic Resonance* Academic, New York, Vol II, p 535
66. a. Bukke EE, Morgenshtern ZL (1963) Opt. Spektrosk. 14:687; b. Morgenshtern ZL, Neustruev VB (1973) in: Williams F (ed) *Luminescence of Crystals, Molecules, and Solutions* Plenum, New York, p. 524
67. Yamada N, Shionoya S, Kushida T (1972) J. Phys. Soc. Japan 32:1577
68. van Dijk JMF, Schuurmans MFH (1983) J. Chem. Phys. 78:5317
69. a. Englman R, Jortner J (1970) Mol. Phys. 18:145; b. Nitzan A, Mukamel S, Jortner J (1975) J. Chem. Phys. 63:200

Source Code

```
      options/g_float
      double precision function wp(S,wmav,ip)
c
c
c     finds one particular wp function, knowing its fundamental
c     arguments S,wmav,ip.
c
c
c     largest allowed ip is 70
c
c     eq. (4.45)
c
c
      IMPLICIT REAL*8(A-H,O-P,R-Z),LOGICAL*4(Q)
      dimension pfact1(47),pfact2(24),pfact(71)
      equivalence (pfact(1),pfact1(1)),(pfact(48),pfact2(1))
      data pfact1/1.d000,1.d000,2.d000,6.d000,24.d000,120.d000,720.d000,
     15040.d000,40320.d000,362880.d000,3.6288d006,3.99168d007,
     24.790016d008,6.2270208d009,8.71782912d010,1.307674368d012,
     32.0922789888d013,3.55687428096d014,6.402373705728000d+015,
     41.216451004088320d+017,2.432902008176640d+018,
     55.109094217170944d+019,1.124000727777608d+021,
     62.585201673888498d+022,6.204484017332395d+023,
     71.551121004333099d+025,4.032914611266057d+026,
     81.088886945041835d+028,3.048883446117139d+029,
     98.841761993739703d+030,2.652528598121911d+032,
     a8.222838654177925d+033,2.631308369336936d+035,
     b8.683317618811888d+036,2.952327990396042d+038,
     c1.033314796638615d+040,3.719933267899013d+041,
     d1.376375309122635d+043,5.230226174666011d+044,
     e2.039788208119744d+046,8.159152832478978d+047,
     f3.345252661316381d+049,1.405006117752880d+051,
     g6.041526306337384d+052,2.658271574788449d+054,
     h1.196222208654802d+056,5.502622159812090d+057/
      data pfact2/
     12.586232415111682d+059,1.241391559253608d+061,
     26.082818640342677d+062,3.041409320171339d+064,
     31.551118753287383d+066,8.065817517094390d+067,
     44.274883284060027d+069,2.308436973392415d+071,
     51.269640335365828d+073,7.109985878048638d+074,
     64.052691950487723d+076,
     72.350561331282880d+078,1.386831185456899d+080,
     88.320987112741393d+081,5.075802138772249d+083,
     93.146997326038795d+085,1.982608315404441d+087,
     a1.268869321858842d+089,8.247650592082473d+090,
     b5.443449390774432d+092,3.647111091818869d+094,
     c2.480035542436831d+096,1.711224524281413d+098,
     d1.197857166996989d+100/
      pr=dfloat(ip)
      ipr=ip
      if(ip.lt.0)then
         pr=-pr
         ipr=-ip
      end if
      boltz=wmav/(1.d000+wmav)
      boltzl=dlog(boltz)
      smav=S*wmav
      smplav=smav+S
      s2mpl=smav+smplav
```

Source Code

```
      if(s2mp1.le.235.d000)then
        bp=dexp(-s2mp1)*smp1av**pr/pfact(ipr+1)
      else
        bp=dexp(dlog(s2mp1)+pr*dlog(smp1av)-dlog(pfact(ipr+1)))
      end if
      bsum=1.d000
      blsum=0.d000
      blterm=1.d000
      wnum=smav*smp1av
      wdnom1=0.d000
      wdnom2=pr
      do 1 j=1,150
        wdnom1=wdnom1+1.d000
        wdnom2=wdnom2+1.d000
        bterm=wnum*blterm/(wdnom1*wdnom2)
        bsum=bsum+bterm
        if(bsum.eq.blsum)then
          wp=bsum*bp
          if(ip.lt.0)then
            if(pr*boltz1.le.200.d000)then
              wp=wp*(boltz**pr)
            else
              wp=0.d000
            end if
          end if
          return
        else
          blsum=bsum
          blterm=bterm
        end if
    1 continue
      END




      options/g_float
      subroutine wpdis(S,wmav,wpd,num)
c
c
c     finds the wp distribution for given s, wmav
c     there are n=num of these elements, in the first num elements of wpd.
c
c
c     wpd has two elements, the first being its value, the second its p number
c     given as a floating point number.
c
c
c     Eq. (4.45) to initiate, then Eq. (4.44) recursion formula to p=0,
c            then Eq. (4.89) for negative p values.
c
      IMPLICIT REAL*8(A-H,O-P,R-Z),LOGICAL*4(Q)
      real*8 wp
      dimension wpd(300,2)
      do 6 j=1,300
    6 wpd(j,1)=0.d000
      boltz=wmav/(wmav+1.d000)
      smav=S*wmav
      smp1av=smav+S
      s2mp1=smav+smp1av
      wa=dmin1(20.d000,3.d000/wmav)
      wb=dmin1(4.d000,1.5d000*s)
      pmax=dmin1(70.d000,S+(s2mp1*(1.d000+wa)*(8.d000-wb)))
      pmax=dmax1(30.d000,pmax)
      ipmx=idint(pmax)
      ipmxm1=ipmx-1
      pmxm1=dfloat(ipmx)
      wpd(300,2)=pmxm1
      wpd(299,2)=pmxm1-1.d000
      wpd(300,1)=wp(s,wmav,ipmx)
      wplim=wpd(300,1)
      wpd(299,1)=wp(s,wmav,ipmxm1)
      do 1 j=1,ipmxm1
      indx=299-j
```

```
              pmxm1=pmxm1-1.d000
              ipndx=ipmxm1-j
              wpd(indx,2)=ipndx
              wpd(indx,1)=wpd(indx+2,1)*boltz+pmxm1*wpd(indx+1,1)/smplav
              if(wpd(indx,1).lt.wplim)go to 3
            1 continue
              wr=1.d000
              indx0=indx
              do 2 j=1,ipmx
                 indx=indx0-j
                 wr=wr*boltz
                 wpd(indx,1)=wpd(indx0+j,1)*wr
                 wpd(indx,2)=-wpd(indx0+j,2)
                 if(wpd(indx,1).lt.wplim)go to 3
            2    continue
            3 do 4 j=indx,300
                 indxm1=indx-1
                 wpd(j-indxm1,1)=wpd(j,1)
                 wpd(j-indxm1,2)=wpd(j,2)
            4    continue
              num=300-indx+1
c             wpsum=0.d000
c             do 5 j=1,num
c           5 wpsum=wpsum+wpd(j,1)
c             write(6,*)' in wpdis, wpsum=',wpsum
              return
              END




              options/g_float
              subroutine wpdist(S,wmav,wpd,num)
c
c
c             finds the wp distribution for given s, wmav
c             the limits for p are so as to stop when exponents fall below -43
c             there are n=num of these elements, in the first num elements of wpd.
c
c
c             wpd has two elements, the first being its value, the second its p number
c             given as a floating point number.
c
c
c             Eq. (4.45) for positive p, Eq. (4.89) for negative p.
c
c
c
              IMPLICIT REAL*8(A-H,O-P,R-Z),LOGICAL*4(Q)
              dimension wpd(300,2)
              do 7 j=1,300
            7 wpd(j,1)=0.d000
              wl=0.d0
              do 1 j=1,70
              wpd(j+199,1)=wp(s,wmav,j-1)
              wpd(j+199,2)=wl
              wl=wl+1
              if(wpd(j+199,1).lt.1.d-040)go to 2
            1 continue
              j=70
            2 num1=j+199
              wplim=wpd(j+199,1)
              boltz=wmav/(wmav+1.d000)
              fac=1.d000
              jlast=num1-200
              do 3 j=1,jlast
              jr=200+j
              jl=200-j
              fac=fac*boltz
              wpd(jl,1)=wpd(jr,1)*fac
              wpd(jl,2)=-wpd(jr,2)
              if(wpd(jl,1).lt.wplim)go to 4
            3 continue
```

Source Code

```
      4 numf=j1
        num=num1-numf+1
        do 5 j=1,num
        jr=j+numf-1
        wpd(j,1)=wpd(jr,1)
      5 wpd(j,2)=wpd(jr,2)
c       wpsum=0.d000
c       do 6 j=1,num
c     6 wpsum=wpsum+wpd(j,1)
c       write(6,*)' wpsum in wpdist=',wpsum
        return
        END




        options/g_float
        subroutine wpzset(S,hom,t1,delt,t2,ip1,idelp,ip2,wparry,
     1       iparry,np,tem,nt)
c
c
c       Finds the wpz distributions for given s, at several values of t,
c          namely, those between t1 and t2 in intervals of delt.
c          There can be 20 temperatures.
c          They are given in degrees Kelvin.
c
c       Each distribution is found between two values of p between ip1 and ip2,
c          at intervals of idelp in p.
c          There can be 200 p values.
c          They are given as integers.
c
c       The array is in wparry(nt,np)
c
c       The wpz distributions depend upon the phonon frequency, hom, to be
c          specified in wavenumbers.
c
        IMPLICIT REAL*8(A-H,O-P,R-Z),LOGICAL*4(Q)
        real*8 wpz
        dimension wpd(300),wparry(20,200),tem(20),wmavar(20),iparry(200)
        do 6 j=1,20
        do 6 k=1,200
      6 wparry(j,k)=0.d000
        temrn=t1-delt
        do 1 j=1,20
        temrn=temrn+delt
        if(temrn.le.t2+0.001d000)then
          tem(j)=temrn
          r=dexp(-hom/(1.43868d000*temrn))
          wmavar(j)=r/(1.d000-r)
        else
          go to 2
        end if
      1 continue
      2 nt=j-1
        iprn=ip1-idelp
        do 3 k=1,200
        iprn=iprn+idelp
        if(iprn.gt.ip2)go to 4
      3   iparry(k)=iprn
      4 np=k-1
        do 5 j=1,nt
        do 5 k=1,np
      5   wparry(j,k)=wpz(s,wmavar(j),iparry(k))
        return
        end
```

```
      options/g_float
      double precision function wmavp(S,wmav,ip)
c
c
c     finds the wmavp for given s, wmav,ip
c
c     ip.le.68
c
c     Eq. (4.75)
c
      IMPLICIT REAL*8(A-H,O-P,R-Z),LOGICAL*4(Q)
      p=dfloat(ip)
      smmp1=s*wmav*(wmav+1.d000)
      wpp=wp(s,wmav,ip+1)
      wp0=wp(s,wmav,ip)
      wpm=wp(s,wmav,ip-1)
      wmavp=(smmp1*(wpp-wp0-wp0+wpm)+wmav*wp0)/wp0
      return
      END




      options/g_float
      double precision function wpz(S,wmav,ip)
c
c
c     finds the wpz for given s, wmav,ip
c
c     ip.le.69
c
c     eq. (4.23)
c
      IMPLICIT REAL*8(A-H,O-P,R-Z),LOGICAL*4(Q)
      real*8 wp
      ss=s+s
      p=dfloat(ip)
      wpz=wp(s,wmav,ip)*(p-s)**2/ss
      return
      END




      options/g_float
      subroutine wpzd(S,wmav,wpzdd,num)
C
C
C
C     finds the wpz distribution for given s, wmav
c     there are n=num of these elements, in the first num elements of wpzdd.
c
c     it first obtains the wp distribution named wpp using eq. (4.45)
c
c     it then uses Eq. (4.23)
c
c     wpzdd has two elements, the first being its value,
c     the second its p number
c     given as a floating point number.
c
c
      IMPLICIT REAL*8(A-H,O-P,R-Z),LOGICAL*4(Q)
      dimension wpzdd(300,2),wpp(300,2)
      ss=s+s
      call wpdist(s,wmav,wpp,num1)
      num=num1
      wpzsum=0.d000
      do 1 j=1,num
      wpzdd(j,1)=wpp(j,1)*(wpp(j,2)-s)**2/ss
      wpzdd(j,2)=wpp(j,2)
      wpzsum=wpzsum+wpzdd(j,1)
    1 continue
      sum=(2.d000*wmav+1.d000)/2.d000
      write(6,*)' wpzsum=',wpzsum,' vs (2m+1)/2=',sum
      return
      END
```

Source Code

```fortran
      options/g_float
      subroutine wpzds(wpdis,numwp,s,wpzdd)
c
c
c     finds the wpz distribution for given s, wmav
c
c     this subroutine is like wpzd except that it presupposes the wp
c        distribution as input under the name wpdis.
c
c     uses Eq. (4.23)
c
      IMPLICIT REAL*8(A-H,O-Z)
      dimension wpdis(300,2),wpzdd(300,2)
      do 1 j=1,300
    1 wpzdd(j,1)=0.d000
      ss=s+s
      do 2 j=1,numwp
      wpzdd(j,1)=wpdis(j,1)*(wpdis(j,2)-s)**2/ss
      wpzdd(j,2)=wpdis(j,2)
    2 continue
      return
      END




      options/g_float
      subroutine wpset(S,hom,t1,delt,t2,ip1,idelp,ip2,wparry,
     1    iparry,np,tem,nt)
c
c
c     Finds the wp distributions for given s, at several values of t,
c        namely, those between t1 and t2 in intervals of delt.
c        There can be 20 temperatures.
c        They are given in degrees Kelvin.
c
c     Each distribution is found between two values of p
c        between ip1 and ip2,
c        at intervals of idelp in p.
c        There can be 200 p values.
c        They are given as integers.
c
c     The array is in wparry(nt,np)
c
c     The wp distributions depend upon the phonon frequency, hom, to be
c        specified in wavenumbers.
c
      IMPLICIT REAL*8(A-H,O-P,R-Z),LOGICAL*4(Q)
      real*4 wp
      dimension wpd(300),wparry(20,200),tem(20),wmavar(20),iparry(200)
      do 6 j=1,20
      do 6 k=1,200
    6 wparry(j,k)=0.d000
      temrn=t1-delt
      do 1 j=1,20
        temrn=temrn+delt
        if(temrn.le.t2+0.001d000)then
          tem(j)=temrn
          r=dexp(-hom/(1.43868d000*temrn))
          wmavar(j)=r/(1.d000-r)
        else
          go to 2
        end if
    1   continue
    2 nt=j-1
      iprn=ip1-idelp
      do 3 k=1,200
        iprn=iprn+idelp
        if(iprn.gt.ip2)go to 4
    3   iparry(k)=iprn
    4 np=k-1
      do 5 j=1,nt
      do 5 k=1,np
    5 wparry(j,k)=wp(s,wmavar(j),iparry(k))
      return
      end
```

```
      options/g_float
      double precision function wpder(S,wmav,ip)
c
c
c     finds wpder for given s, wmav,ip
c     ip.le.68
c
c
c     Eq. (4.88)
c
      IMPLICIT REAL*8(A-H,O-P,R-Z),LOGICAL*4(Q)
      real*8 wp,wpz
      ss=s+s
      smmp1=s*wmav*(wmav+1.d000)*2.d000
      wpp=wp(s,wmav,ip+1)
      wp0=wp(s,wmav,ip)
      wpm=wp(s,wmav,ip-1)
      wpz0=wpz(s,wmav,ip)
      wpder=wpz0-smmp1*(wpp-wp0-wp0+wpm)
      return
      END




      options/g_float
      subroutine wpderd(S,wmav,wpd,num)
c
c
c     finds the wpder distribution for given s, wmav
c     there are n=num of these elements, in the first num elements of wpd.
c
c
c
c
c     wpd has two elements, the first being its value, the second its p number
c     given as a floating point number.
c
c     uses wpdist to get wp distribution, wpzd to get wpz distribution,
c           Eq. (4.88) to get derivative expression value.
c
c
      IMPLICIT REAL*8(A-H,O-P,R-Z),LOGICAL*4(Q)
      dimension wpzz(300,2),wpd(300,2),wp1(300,2)
      do 2 j=1,300
    2 wpd(j,1)=0.d000
      ss=s+s
      smmp1=s*wmav*(wmav+1.d000)*2.d000
      call wpdist(s,wmav,wp1,num0)
      call wpzd(s,wmav,wpzz,num1)
      num=num1-2
c     wpdsum=0.d000
      do 1 j=1,num
      wpd(j,1)=wpzz(j+1,1)-smmp1*
     1              (wp1(j,1)-wp1(j+1,1)-wp1(j+1,1)+wp1(j+2,1))
      wpd(j,2)=wpzz(j+1,2)
c   1 wpdsum=wpdsum+wpd(j,1)
    1 continue
c     write(6,*)' wpdsum=',wpdsum
      return
      END
```

Source Code

```fortran
      options/g_float
      subroutine wpdrs(wpzz,wp1,num1,s,wmav,wpd,num)
c
c     assumes the wp distribution wp1, the wpz distribution wpzz.
c
c     obtainss the wpder distribution using Eq. (4.88)
c
c
      IMPLICIT REAL*8(A-H,O-Z)
      dimension wpzz(300,2),wpd(300,2),wp1(300,2)
      ss=s+s
      smmp1=s*wmav*(wmav+1.d000)*2.d000
      num=num1-2
      klast=num1-1
      do 1 j=1,300
      wpd(j,1)=0.d000
  1   wpd(j,2)=0.d000
      do 2 k=2,klast
      kr=k-1
      wpd(kr,1)=wpzz(k,1)-smmp1*
  1               (wp1(k+1,1)-wp1(k,1)-wp1(k,1)+wp1(k-1,1))
      wpd(kr,2)=wpzz(k,2)
  2   continue
      return
      END




      options/g_float
      subroutine wpdset(S,hom,t1,delt,t2,ip1,idelp,ip2,wparry,
     1 iparry,np,tem,nt)
c
c
c     Finds the wpd distributions for given s, at several values of t,
c        namely, those between t1 and t2 in intervals of delt.
c        There can be 20 temperatures.
c        They are given in degrees Kelvin.
c
c     Each distribution is found between two values of p between ip1 and ip2,
c        at intervals of idelp in p.
c        There can be 200 p values.
c        They are given as integers.
c
c     The array is in wparry(nt,np)
c
c     The wpd distributions depend upon the phonon frequency, hom, to be
c        specified in wavenumbers.
c
      IMPLICIT REAL*8(A-H,O-P,R-Z),LOGICAL*4(Q)
      real*8 wpder
      dimension wpd(300),wparry(20,200),tem(20),wmavar(20),iparry(200)
      do 6 j=1,20
      do 6 k=1,200
  6   wparry(j,k)=0.d000
      temrn=t1-delt
      do 1 j=1,20
        temrn=temrn+delt
        if(temrn.le.t2+0.001d000)then
          tem(j)=temrn
          r=dexp(-hom/(1.43868d000*temrn))
          wmavar(j)=r/(1.d000-r)
        else
          go to 2
        end if
  1     continue
  2   nt=j-1
      iprn=ip1-idelp
      do 3 k=1,200
        iprn=iprn+idelp
        if(iprn.gt.ip2)go to 4
  3     iparry(k)=iprn
  4   np=k-1
      do 5 j=1,nt
      do 5 k=1,np
  5   wparry(j,k)=wpder(s,wmavar(j),iparry(k))
      return
      end
```

```
      options/g_float
      double precision function wlp(S,wmav,ip)
c
c
c     finds the wlp for given s, wmav,ip
c
c     ip.le.69
c
c     Eq. (5.7)
c
      IMPLICIT REAL*8(A-H,O-P,R-Z),LOGICAL*4(Q)
      wmavp1=wmav+1.d000
      wlp=(wp(s,wmav,ip-1)*wmavp1+wp(s,wmav,ip+1)*wmav)/2.d000
      return
      END




      options/g_float
      double precision function wmzp(S,wmav,gamma,ip)
c
c
c     finds the wmpz for given s, wmav,ip, and gamma
c
c     ip.le.69
c
c     Eq. (5.6) without the A squared electronic factor
c
      IMPLICIT REAL*8(A-H,O-P,R-Z),LOGICAL*4(Q)
      real*8 wpz,wlp
      wmzp=wpz(s,wmav,ip)*(1.d000-gamma)+gamma*wlp(s,wmav,ip)
      return
      END




      options/g_float
      subroutine wmzpd(S,wmav,gamma,wmzpdd,num)
c
c
c     finds the wmzp distribution for given s, wmav
c     there are n=num of these elements, in the first num elements of wmzpdd.
c
c     it first obtains the wmp distribution named wmpp using wmzp
c
c
c     wmzpdd has two elements, the first being its value,
c     the second its p number
c     given as a floating point number.
c
c
      IMPLICIT REAL*8(A-H,O-P,R-Z),LOGICAL*4(Q)
      real*8 wmzp
      dimension wmzpdd(300,2),wpp(300,2)
      wmzpsm=0.d000
      s2mp1=s*(2.d000*wmav+1.d000)
      pmax=dmin1(70.d000,S+(s2mp1*(1.d000+wa)*(8.d000-wb)))
      pmax=dmax1(15.d000,pmax)
      jlast=idint(pmax-1)
      jfirst=idint(s-s2mpq*3.d000)
      jfirst=min(jjfirst,-10)
      num=jlast-jfirst+1
      ww=dfloat(jfirst-1)
      jr=-1
      do 1 j=jfirst,jlast
      jr=jr+1
      wmzpdd(jr,1)=wmzp(s,wmav,gamma,j)
      ww=ww+1.d000
      wmzpdd(jr,2)=ww
      wmzpsm=wmzpsm+wmzpdd(jr,1)
    1 continue
      sum=(2.d000*wmav+1.d000)/2.d000
      write(6,*)' wmzpsm=',wmzpsm,' vs (2m+1)/2=',sum
      return
      END
```

Source Code

```fortran
      options/g_float
      double precision function wmder(S,wmav,gamma,ip)
c
c     finds wmder for given s, wmav,gamma,ip
c     ip.le.68
c
c     Eq. (5.6)
c
      IMPLICIT REAL*8(A-H,O-P,R-Z),LOGICAL*4(Q)
      wmder=(1.d000-gamma)*wpder(s,wmav,ip)+gamma*wlp(s,wmav,ip)
      return
      END




      options/g_float
      subroutine wmdpd(S,wmav,gamma,wmdpdd,num)
c
c
c     finds the wmdp distribution for given s, wmav
c     there are n=num of these elements, in the first num elements of wmdpdd.
c
c     it first obtains the wmdp distribution using wmdp
c
c
c     wmzpdd has two elements, the first being its value,
c     the second its p number
c     given as a floating point number.
c
c
      IMPLICIT REAL*8(A-H,O-P,R-Z),LOGICAL*4(Q)
      real*8 wmdp
      dimension wmdpdd(300,2)
      wmzpsm=0.d000
      s2mp1=s*(2.d000*wmav+1.d000)
      pmax=dmin1(70.d000,S+(s2mp1*(1.d000+wa)*(8.d000-wb)))
      pmax=dmax1(15.d000,pmax)
      jlast=idint(pmax-1)
      jfirst=idint(s-s2mpq*3.d000)
      jfirst=min(jjfirst,-10)
      num=jlast-jfirst+1
      ww=dfloat(jfirst-1)
      jr=-1
      do 1 j=jfirst,jlast
      jr=jr+1
      wmdpdd(jr,1)=wmdp(s,wmav,gamma,j)
      ww=ww+1.d000
      wmdpdd(jr,2)=ww
      wmdpsm=wmdpsm+wmdpdd(jr,1)
    1 continue
      sum=(2.d000*wmav+1.d000)/2.d000
      write(6,*)' wmdpsm=',wmdpsm,' vs (2m+1)/2=',sum
      return
      END




      options/g_float
      subroutine debo(S,sgnrts,wkt,homd,ndis,zder,wdebo,ndebo,wnorm)
C
C
c     zder=true operator is z
c          false           d/dz
c
C     finds the discretized-Debye equal-S**0.5,A distributions from its
c        arguments  S,wkt,homd and the number of partitions desired ndis.
c
c     results are put into wdebo(1200,2)
c
c     wnorm is the calculated exact normalization sum
```

```
      c
            IMPLICIT REAL*8(A-H,O-Z)
            logical*4 zder
            dimension wdebo(1200,2),sk(25),wmavk(25),homk(25),gbar(25),
           1xth1(1200,2),wpd1(300,2),xph1(1200,2),
           2xps1(1200,2),wpzd1(300,2),wpdd1(300,2),xps12b(1200,2),wpd(300,2),
           3xps12a(1200,2),xth12a(1200,2),xth12b(1200,2),xth12(1200,2),
           4xps12(1200,2),xph12a(1200,2),xph12b(1200,2),xph12(1200,2),
           5wpd2(300,2),wpzd2(300,2),wpdd2(300,2),xwpdcv(1200,2),
           6xwpcv(1200,2),xwpzcv(1200,2),xd1(1200,2),xch1(1200,2),
           7xchk(1200,2),xd12(1200,2),gammak(25),ch1(300,2),chk(300,2)
            call gtdbpr(S,sgnrts,wkt,homd,ndis,sk,wmavk,homk,gbar,gammak)
            write(6,*)' sk'
            write(6,*) sk(1),sk(2),sk(3)
            write(6,*)' gbar'
            write(6,*) gbar(1),gbar(2),gbar(3)
            write(6,*)' homk'
            write(6,*) homk(1),homk(2),homk(3)
            write(6,*)' wmavk'
            write(6,*) wmavk(1),wmavk(2),wmavk(3)
            write(6,*)' gammak'
            write(6,*) gammak(1),gammak(2),gammak(3)
            wnorm=0.d000
            do 40 j=1,ndis
         40 wnorm=wnorm+gbar(j)*(2.d000*wmavk(j)+1.d000)
            wnorm=wnorm/2.d000
            k=0
            i300=300
            i1200=1200
          1 k=k+1
            if(k.gt.ndis)go to 2
            if(k.eq.1)go to 11
            go to 12
         11 call wpdis(sk(1),wmavk(1),wpd1,num1)
            call wpzds(wpd1,num1,sk(1),wpzd1)
            do 110 j=1,1200
            xth1(j,1)=0.d000
            xph1(j,1)=0.d000
            xch1(j,1)=0.d000
            xd1(j,1)=0.d000
        110 xps1(j,1)=0.d000
            call chisd(wpd1,num1,sk(1),wmavk(1),zder,ch1,nch1)
            call cnvwx(ch1,nch1,xch1)
            do 111 j=1,num1
            xth1(j,1)=xch1(j,1)*gammak(1)
            xth1(j,2)=xch1(j,2)
            xph1(j,1)=0.d000
            xph1(j,2)=wpd1(j,2)
            xd1(j,1)=wpd1(j,1)
            xd1(j,2)=wpd1(j,2)
        111 continue
            nth1=num1
            nph1=num1
            nd1=num1
            if(zder)then
              do 112 j=1,num1
                xps1(j,1)=gbar(1)*wpzd1(j,1)
                xps1(j,2)=wpzd1(j,2)
        112     continue
              nps1=num1
            else
              call wpdrs(wpzd1,wpd1,num1,sk(1),wmavk(1),wpdd1,nmd1)
              do 113 j=1,nmd1
                xps1(j,1)=gbar(1)*wpdd1(j,1)
                xps1(j,2)=wpdd1(j,2)
        113     continue
              nps1=nmd1
            end if
            go to 1
         12 if(k.gt.2)then
              do 220 j=1,1200
              xth1(j,1)=0.d000
              xph1(j,1)=0.d000
              xd1(j,1)=0.d000
        220   xps1(j,1)=0.d000
              do 120 j=1,ns12
                xps1(j,1)=xps12(j,1)
```

```
120       xps1(j,2)=xps12(j,2)
          do 1201 j=1,nph12
          xph1(j,1)=xph12(j,1)
1201      xph1(j,2)=xph12(j,2)
          do 1202 j=1,nth12
          xth1(j,1)=xth12(j,1)
1202      xth1(j,2)=xth12(j,2)
          do 1203 j=1,nd12
          xd1(j,1)=xd12(j,1)
1203      xd1(j,2)=xd12(j,2)
          nps1=ns12
          nph1=nph12
          nth1=nth12
          nd1=nd12
        end if
        call wpdis(sk(k),wmavk(k),wpd2,num2)
        call cnvwx(wpd2,num2,xwpcv)
        call chisd(wpd2,num2,sk(k),wmavk(k),zder,chk,nchk)
        call cnvwx(chk,nchk,xchk)
        call cmbww(xd1,nd1,homk(1),i1200,xwpcv,num2,k,i1200,
     1             xd12,nd12)
        call wpzds(wpd2,sk(k),wpzd2)
        call cnvwx(wpzd2,num2,xwpzcv)
        if(.not.zder)then
           call wpdrs(wpzd2,wpd2,num2,sk(k),wmavk(k),wpdd2,nmd2)
           call cnvwx(wpdd2,nmd2,xwpdcv)
        end if
        call cmbww(xps1,nps1,homk(1),i1200,xwpcv,num2,k,i1200,
     1             xps12a,ns12a)
        call cmbww(xth1,nth1,homk(1),i1200,xwpcv,num2,k,i1200,
     1             xth12a,nt12a)
        call cmbww(xph1,nph1,homk(1),i1200,xwpcv,num2,k,i1200,
     1             xph12a,np12a)
        if(zder)then
           call cmbww(xd1,nd1,homk(1),i1200,xwpzcv,num2,k,i1200,
     1                xps12b,ns12b)
           do 121 j=1,ns12b
121        xps12b(j,1)=xps12b(j,1)*gbar(k)
           call addxx(xps12a,ns12a,xps12b,ns12b,
     1                xps12,ns12)
           call cmbww(xd1,nd1,homk(1),i1200,xchk,nchk,k,i1200,
     1                xth12b,nt12b)
           do 122 j=1,nt12b
122        xth12b(j,1)=xth12b(j,1)*gammak(k)
           call addxx(xth12a,nt12a,xth12b,nt12b,
     1                xth12,nth12)
           call cmbww(xth1,nth1,homk(1),i1200,xchk,nchk,k,i1200,
     1                xph12b,np12b)
           do 123 j=1,np12b
123        xph12b(j,1)=xph12b(j,1)*gammak(k)*2.d000
           call addxx(xph12a,np12a,xph12b,np12b,
     1                xph12,nph12)
        else
           call cmbww(xd1,nd1,homk(1),i1200,xwpdcv,nmd2,k,i1200,
     1                xps12b,ns12b)
           do 124 j=1,ns12b
124        xps12b(j,1)=xps12b(j,1)*gbar(k)
           call addxx(xps12a,ns12a,xps12b,ns12b,
     1                xps12,ns12)
           call cmbww(xd1,nd1,homk(1),i1200,xchk,nchk,k,i1200,
     1                xth12b,nt12b)
           do 125 j=1,nt12b
125        xth12b(j,1)=xth12b(j,1)*gammak(k)
           call addxx(xth12a,nt12a,xth12b,nt12b,
     1                xth12,nth12)
           call cmbww(xth1,nth1,homk(1),i1200,xchk,nchk,k,i1200,
     1                xph12b,np12b)
           do 126 j=1,np12b
126        xph12b(j,1)=xph12b(j,1)*gammak(k)*2.d000
           call addxx(xph12a,np12a,xph12b,np12b,
     1                xph12,nph12)
        end if
        go to 1
    2 call addxx(xps12,ns12,xph12,nph12,wdebo,ndebo)
      return
      END
```

```fortran
      options/g_float
      subroutine addxx(a,na,b,nb,ab,nab)
c
c     adds the distribution a and b, assuming the same frequency,
c          respectively, finding the sum at the same p values.
c
c     these distributions a and b have dimensions (1200,2), with the
c          first element its value and the second element its index
c          as a floating point number. This is the format of the
c          unequal-force-constants distributions. The first na, nb, nab
c          elements are non-zero in these distributions.
c
      IMPLICIT REAL*8(A-H,O-Z)
      logical*4 alow,ahigh
      dimension a(1200,2),b(1200,2),ab(1200,2)
      do 1 j=1,1200
    1 ab(j,1)=0.d000
      nabl=idint(dmin1(a(1,2),b(1,2)))
      nabh=idint(dmax1(a(na,2),b(nb,2)))
      nab=nabh-nabl+1
      wnrn=dfloat(nabl-1)
      alow=.true.
      if(a(1,2).ne.wnrn+1.d000)alow=.false.
      ahigh=.true.
      if(a(na,2).ne.qfloat(nabh))ahigh=.false.
      jratb=idint(a(1,2)-b(1,2))
      if(alow.and.ahigh)then
         do 2 j=1,nab
           ab(j,2)=a(j,2)
           ab(j,1)=a(j,1)
           jro=j+jratb
           if(jro.ge.1.and.jro.le.nb)ab(j,1)=ab(j,1)+b(jro,1)
    2    continue
         return
      end if
      if(.not.alow.and..not.ahigh)then
         do 3 j=1,nab
           ab(j,2)=b(j,2)
           ab(j,1)=b(j,1)
           jro=j-jratb
           if(jro.ge.1.and.jro.le.na)ab(j,1)=ab(j,1)+a(jro,1)
    3    continue
         return
      end if
      if(alow.and..not.ahigh)then
         do 4 j=1,nab
           if(j.le.na)then
             ab(j,2)=a(j,2)
             ab(j,1)=a(j,1)
           end if
           jro=j+jratb
           if(jro.ge.1.and.jro.le.nb)ab(j,1)=ab(j,1)+b(jro,1)
           if(j.gt.na)ab(j,2)=b(jro,2)
    4    continue
         return
      end if
      if(.not.alow.and.ahigh)then
         do 5 j=1,nab
           if(j.lt.nb)then
             ab(j,2)=b(j,2)
             ab(j,1)=b(j,1)
           end if
           jro=j-jratb
           if(jro.ge.1.and.jro.le.na)ab(j,1)=ab(j,1)+a(jro,1)
           if(j.gt.nb)ab(j,2)=a(jro,2)
    5    continue
          return
         end if
         end
```

Source Code

```fortran
      options/g_float
      subroutine debc(S,wkt,homd,ndis,wdeb,ndeb)
c
c
c     finds the discretized-Debye equal-S**0.5,A distributions from its
c        arguments  S,wkt,homd and the number of partitions desired ndis.
c
c     wdeb has dimensions (1200,2).
c
c     wdeb has two elements, the first being its value, the second its p number
c     given as a floating point number.
c
c     Eq. (5.36) and (5.37), as calculated using (5.38) and (5.39)
c
      IMPLICIT REAL*8(A-H,O-Z)
      dimension wdeb(1200,2),sk(25),wmavk(25),wpd1(300,2),wpd2(300,2),
     1   homk(25),gbar(25),wpdk(300,2),ab(1200,2),gammak(25)
      sgnrts=1.d000
      call gtdbpr(S,sgnrts,wkt,homd,ndis,sk,wmavk,homk,gbar,gammak)
      write(6,*)' sk'
      write(6,*) sk(1),sk(2),sk(3)
      write(6,*)' gbar'
      write(6,*) gbar(1),gbar(2),gbar(3)
      write(6,*)' homk'
      write(6,*) homk(1),homk(2),homk(3)
      write(6,*)' wmavk'
      write(6,*) wmavk(1),wmavk(2),wmavk(3)
      call wpdis(sk(1),wmavk(1),wpd1,num1)
      call wpdis(sk(2),wmavk(2),wpd2,num2)
      idim1=300
      ifac=2
      call cmbww(wpd1,num1,homk(1),idim1,wpd2,num2,ifac,idim1,ab,numab)
      idimab=1200
      if(ndis.gt.2)then
        do 1 k=3,ndis
          call wpdis(sk(k),wmavk(k),wpdk,numk)
          ifac=k
          call cmbww(ab,numab,homk(1),idimab,wpdk,numk,ifac,idim1,
     1                        wdeb,ndeb)
          if(k.ne.ndis)then
            do 2 l=1,ndeb
              ab(l,1)=wdeb(l,1)
              ab(l,2)=wdeb(l,2)
 2          continue
          end if
 1      continue
      end if
      sum=0.d000
      do 3 j=1,ndeb
 3    sum=sum+wdeb(j,1)
      write(6,*)' sum of wdeb=',sum
      return
      END




      options/g_float
      subroutine gtdbpr(S,sgnrts,wkt,homd,ndis,sk,wmavk,homk,gbar
     1             ,gammak)
c
c
c     finds the discretized-Debye equal-S**0.5,A parameters for each of the
c        separate groups.
c     the fundamental required parameters are S,wkt,homd,
c               and the number of partitions desired ndis.
c
c     Eqs.(5.57), (5.59)-(5.61)
c
c
      IMPLICIT REAL*8(A-H,O-Z)
      dimension sk(25),wmavk(25),gbar(25),homk(25),
```

```fortran
    1 gammak(25)
      wndis=dfloat(ndis)
      gbar(1)=27.d000/(8.d000*wndis**3)
      gbar(ndis)=(12.d000*wndis**2-6.d000*wndis+1.d000)/
     1                                    (8.d000*wndis**3)
      wk=1.d000
      do 1 k=2,ndis-1
      wk=wk+1.d000
    1 gbar(k)=(12.d000*wk**2+1.d000)/(4.d000*wndis**3)
      smgbar=0.d000
      do 10 j=1,ndis
   10 smgbar=smgbar+gbar(j)
      write(6,*)' sum of gbar=',smgbar
      wk=0.d000
      do 2 k=1,ndis
      wk=wk+1.d000
      sk(k)=(S*gbar(k)*8.d000*wndis**4)/
     1       (6.d000*wndis**4+wndis**2+1.d000)
      gammak(k)=sgnrts*dsqrt(gbar(k)*sk(k)/2.d000)
      homk(k)=homd*wk/wndis
      r=dexp(-homk(k)/wkt)
    2 wmavk(k)=r/(1.d000-r)
  100 format('    ')
      write(6,100)
      write(6,100)
      write(6,100)
      write(6,100)
      write(6,*)' check of sk and homk calculation'
      ss=0.d000
      do 20 j=1,ndis
   20 ss=ss+homk(j)*sk(j)
      ss=ss/homd
      write(6,*) ' (sum of homk*sk)/homd=',ss
      write(6,*) '       should be', s
      return
      END




      options/g_float
      subroutine cnvwx(a,numa,b)
c
c     comverts the distribution a from w to x format
c
c     the distribution a has dimensions (300,2), b dimensions (1200,2).
c
      IMPLICIT REAL*8(A-H,O-Z)
      dimension a(300,2),b(1200,2)
      do 1 j=1,1200
    1 b(j,1)=0.d000
      do 2 j=1,numa
      b(j,1)=a(j,1)
    2 b(j,2)=a(j,2)
      return
      end




      options/g_float
      subroutine cnvwu(a,numa,idima,b,nblo,nbhi)
c
c
c
c     comverts the distribution a from its equal-force-constants format,
c           into the unequal-force-constants format, so that it can be
c           combined with another unequal-force-constants distribution.
c
c
c
```

Source Code

```
c
c       the distribution a has dimensions (idima,2), b dimensions (-50:75,2).
c
c
        IMPLICIT REAL*8(A-H,O-Z)
        dimension a(idima,2),b(-50:75,2)
        logical*4 swtch
        ww=-51.d000
        do 1 j=-50,75
        b(j,1)=0.d000
        ww=ww+1.d000
    1   b(j,2)=ww
        swtch=.true.
        do 2 j=1,numa
        if(a(j,2).lt.-50.d000)go to 2
        if(a(j,2).gt.75.d000)go to 3
        iww=idint(a(j,2))
        if(swtch)then
          swtch=.false.
          nblo=iww
        end if
        b(iww,1)=a(j,1)
    2   continue
    3   nbhi=iww
        write(6,*)' b'
        do 5 j=nblo,nbhi
        write(6,*)b(j,1),b(j,2)
    5   continue
        return
        end




        options/g_float
        subroutine cmbuu(a,nalo,nahi,homa,b,nblo,nbhi,homb,ab,nablo,nabhi,
    1              homab)
c
c
c
c       combines the distribution a and b, with frequencies homa and homb,
c           respectively, finding the full combined
c           distribution which has frequency homab.
c
c
c       these distributions a and b have dimensions (-50:75,2), with the
c           first element its value and the second element its index
c           as a floating point number.  This is the format of the
c           unequal-force-constants distributions.  The lowest and
c           highest nonzero elements have the lo and hi indices as passed
c           in the argument list.
c
c       pm*homab=na*homa+nb*homb, with pm generally nonintegral.
c
c       cmbuum(m) draws intensity from nearby pm entries.
c
        IMPLICIT REAL*8(A-H,O-Z)
        real*8 cmbuum
        logical*4 sw
        dimension a(-50:75,2),b(-50:75,2),ab(-50:75,2)
        wnab=-51.d000
        sw=.false.
        do 1 j=-50,75
        wnab=wnab+1.d000
        ab(j,1)=0.d000
    1   ab(j,2)=wnab
        wnablo=dmin1(dfloat(nalo)*homa/homab,dfloat(nblo)*homb/homab)
        nablo=idint(wnablo)
        write(6,*)'          '
        sum=0.d000
        do 2 nab=nablo,75
          ab(nab,1)=cmbuum(a,nalo,nahi,homa,b,nblo,nbhi,homb,homab,nab)
```

```
      1              *homab/homb
        write(6,*)' in cmbuu,nab=',nab,'    ab=',ab(nab,1)
        if(.not.sw)then
          if(ab(nab,1).ne.0.d000)sw=.true.
        end if
        if(sw.and.ab(nab,1).lt.1.d-060)go to 3
        sum=sum+ab(nab,1)
      2 continue
        nabhi=75
        go to 4
      3 nabhi=nab
      4 write(6,*)'  sum=',sum
        return
        end




        options/g_float
        subroutine cmbww(a,numa,homa,idima,b,numb,ifac,idimb,ab,numab)
c
c
c
c
c       combines the distribution a and b, with frequencies homa and homb,
c           respectively, where homb is some multiple of homa,
c           namely ifac*homa
c       The combined distribution has frequency homa.
c
c
c
c
c       the distribution a has dimensions (idima,2), b dimensions (idimb,2),
c           ab dimensions (1200,2). Idima can be either 300 or 1200.
c           The first element its value and the second element its index
c           as a floating point number.
c           There are num (numa,numb,numab) of these elements.
c           This is the format of the
c           equal-force-constants distributions.
c
c           nab=na+nb*ifac, with nab always integral.
c
c           cmbww(nab) draws intensity from only those combinations of
c                na and nb satisfying this equation.
c
        IMPLICIT REAL*8(A-H,O-Z)
        dimension a(idima,2),b(idimb,2),ab(1200,2)
        ialo=idint(a(1,2))
        iahi=idint(a(numa,2))
        iblo=idint(b(1,2))
        ibhi=idint(b(numb,2))
        iablo=ialo+iblo*ifac
        iabhi=iahi+ibhi*ifac
    100 numab=iabhi-iablo+1
        if(numab.gt.1200)then
          wlow=a(1,1)*b(1,1)
          whigh=a(numa,1)*b(numb,1)
          if(wlow.lt.whigh*1.d-010)then
            iablo=iablo+(numab-1200)
            go to 100
          else if (whigh.lt.wlow*1.d-010)then
            iabhi=iabhi-(numab-1200)
            go to 100
          else
            idel=1+(numab-1200)/2
            iablo=iablo+idel
            iabhi=iabhi-idel
            go to 100
          end if
        end if
        wnabrn=dfloat(iablo-1)
        do 1 j=1,1200
      1 ab(j,1)=0.d000
        do 2 j=1,numab
        wnabrn=wnabrn+1.d000
        ab(j,1)=0.d000
```

Source Code

```fortran
    2 ab(j,2)=wnabrn
      do 3 k=1,numb
      nb=idint(b(k,2))
      do 3 j=1,numa
      na=idint(a(j,2))
      nab=na+ifac*nb
      if(nab.lt.iablo.or.nab.gt.iabhi)go to 3
      indx=nab-iablo+1
      ax=a(j,1)*b(k,1)
      ay=ab(indx,1)+ax
      ab(indx,1)=ay
    3 continue
c     write(6,*)' ab'
c     do 5 k=1,numab
c     write(6,*)k,ab(k,1),ab(k,2)
c   5 sum=sum+ab(k,1)
      return
      end




      options/g_float
      subroutine gta(theta,auv,a)
c
c
c
c     finds the a matrix, given its fundamental variables auv and theta
c
c     with double precision, about 65-70 rows of A exhibit orthonormality
c         to full precision.
c
c
c     Eq. (3.36) for A(0,0);
c     Eqs. (3.32)-(3.33), the Manneback recursion formulas.
c     Eqs. (3.28)-(3.30) to convert from auv,theta to su,sv,k,k+ variables.
c
c
      IMPLICIT REAL*8(A-H,O-P,R-Z),LOGICAL*4(Q)
      dimension a(0:100,0:100),sqrtn(0:100),summl(0:100,0:100),
     1las(0:100,0:100)
      common as
      data pi/3.141592653589793d000/
      thrad=theta*(pi/180.d000)
      wn=1.d000
      sqrtn(0)=0.d000
      sqrtn(1)=1.d000
      do 1 j=2,100
      wn=wn+1.d000
    1 sqrtn(j)=dsqrt(wn)
      wkp=dsin(thrad+thrad)
      wk=dcos(thrad+thrad)
      sqrtsu=auv*dcos(thrad)/sqrtn(2)
      sqrtsv=auv*dsin(thrad)/sqrtn(2)
      a(0,0)=dsqrt(wkp)*dexp(-(auv*wkp)**2/8.d000)
      a(1,0)=a(0,0)*wkp*sqrtsv
      a(0,1)=-a(0,0)*wkp*sqrtsu
      a(1,1)=a(0,0)*wkp-a(1,0)*wkp*sqrtsu
      do 2 n=2,100
      a(n,0)=(a(n-1,0)*wkp*sqrtsv+a(n-2,0)*wk*sqrtn(n-1))/sqrtn(n)
      m=n
      a(0,m)=-(a(0,m-1)*wkp*sqrtsu+a(0,m-2)*wk*sqrtn(m-1))/sqrtn(m)
      a(n,1)=(wkp*a(n-1,0)*sqrtn(n)-wkp*sqrtsu*a(n,0))/sqrtn(1)
    2 a(1,m)=(wkp*a(0,m-1)*sqrtn(m)+wkp*sqrtsv*a(0,m))/sqrtn(1)
      do 3 n=2,100
      do 3 m=2,100
      as(n,m)=(wkp*sqrtn(n)*a(n-1,m-1)-wkp*sqrtsu*a(n,m-1)
     1       -wk*sqrtn(m-1)*a(n,m-2))/sqrtn(m)
      a(n,m)=(wkp*sqrtn(m)*a(n-1,m-1)+wkp*sqrtsv*a(n-1,m)
     1       +wk*sqrtn(n-1)*a(n-2,m))/sqrtn(n)
    3 continue
c     check of orthonormality of this matrix
c              rows
c     do 6 m=0,100
c     do 5 l=m,100
c     sum=0.d000
```

```
c       sum1=0.d000
c       do 4 j=0,100
c       sum=sum+a(j,m)*a(j,l)
c       if(sum.eq.sum1)then
c          summ1(m,l)=sum
c          go to 5
c       else
c          sum1=sum
c       end if
c    4  continue
c       summ1(m,l)=sum
c    5  continue
c    6  continue
c       write(6,*)' check of orthonormality'
c       do 7 m=0,100
c       do 7 l=m,100
c       write(6,100)m,l,summ1(m,l)
c    7  continue
c 100   format(1x,I4,1x,I4,1x,1pe23.15)
        return
        end




        options/g_float
        double precision function upum(a,thrad,pu,m)
c
c
c       finds the u(pu,m) matrix element, given its fundamental a matrix
c
c
c       pu must be an integer, but it is given as a floating point number
c
c       Eq. (4.31)
c
        IMPLICIT REAL*8(A-H,O-P,R-Z),LOGICAL*4(Q)
        dimension a(0:100,0:100)
        tansq=(dtan(thrad))**2
        wm=dfloat(m)
        pm=pu+wm*tansq
        jpm=idint(pm)
      1 wpm=dfloat(jpm)
        if(wpm.le.pm)then
           jpm=jpm+1
           go to 1
        end if
        wtpmm1=wpm-pm
        wtpm=1.d000-wtpmm1
        if(jpm.ge.1)then
           wa=a(jpm,m)**2
           wb=a(jpm-1,m)**2
           upum=wa*wtpm+wb*wtpmm1
        else if(jpm.eq.0)then
           upum=a(jpm,m)**2*wtpm
        else
           upum=0.d000
        end if
        return
        end




        options/g_float
        double precision function vpvn(a,thrad,pv,n)
c
c
c       finds the v(n,pv) matrix element, given its fundamental a matrix
c
c       pv is an integer in floating point form
c
c       Eq. (4.30)
```

Source Code

```fortran
c
      IMPLICIT REAL*8(A-H,O-P,R-Z),LOGICAL*4(Q)
      dimension a(0:100,0:100)
      tansq=(dtan(thrad))**2
      wn=dfloat(n)
      pn=pv+wn/tansq
      jpn=idint(pn)
    1 wpn=dfloat(jpn)
      if(wpn.le.pn)then
         jpn=jpn+1
         go to 1
      end if
      wtpnm1=wpn-pn
      wtpn=1.d000-wtpnm1
      if(jpn.ge.1)then
         vpvn=a(n,jpn)**2*wtpn+a(n,jpn-1)**2*wtpnm1
      else if(jpn.eq.0)then
         vpvn=a(n,jpn)**2*wtpn
      else
         vpvn=0.d000
      end if
      return
      end




      options/g_float
      subroutine upu(a,theta,homv,wkt,pu,upurn,upurnl)
c
c
c
c     finds the u vector element, from its a matrix and its other needed
c              fundamental variables homv,wkt,pu
c
c
c          pu must satisfy -50.le.pu.le.75
c
c
c
c     this matrix element is put into upurn, and its next-to-last
c              computed value is in upurnl, so that one can
c              assess its computational accuracy.
c
c          this accuracy will fall off for large abs(p) values
c              because the a matrix is obtained only to dimension
c              100 x 100
c
c     Eq. (4.25)
c
c
      IMPLICIT REAL*8(A-H,O-P,R-Z),LOGICAL*4(Q)
      dimension a(0:100,0:100)
      data pi/3.141592653589793d000/
      thrad=theta*(pi/180.d000)
      m=0
      upurun=upum(a,thrad,pu,m)
      upurnl=upurun
      if(wkt.eq.0.d00)go to 2
      boltz=dexp(-homv/wkt)
      wnorm=1.d000-boltz
      fac=1.d000
      do 1 m=1,100
         fac=fac*boltz
         uu=upum(a,thrad,pu,m)
         upurun=upurun+uu*fac
         if(upurun.eq.upurnl.and.upurun.ne.0.d000)go to 2
         upurnl=upurun
    1 continue
    2 upurnl=upurnl*wnorm
      upurn=upurun*wnorm
      return
      end
```

```
      options/g_float
      subroutine vpv(a,theta,homu,wkt,pv,vpvrn,vpvrnl)
c
c
c     finds the v vector element, from its a matrix and its other needed
c                fundamental variables homu,wkt,pv
c
c
c         pv must satisfy -50.le.pv.le.75
c
c
c     this matrix element is put into vpvrn, and its next-to-last
c                computed value is in vpvrnl, so that one can
c                assess its computational accuracy.
c
c           this accuracy will fall off for large abs(p) values
c                because the a matrix is obtained only to dimension
c                100 x 100
c
c     Eq. (4.24)
c
c
      IMPLICIT REAL*8(A-H,O-P,R-Z),LOGICAL*4(Q)
      dimension a(0:100,0:100)
      data pi/3.141592653589793d000/
      thrad=theta*(pi/180.d000)
      boltz=dexp(-homu/wkt)
      wnorm=1.d000-boltz
      n=0
      vpvrun=vpvn(a,thrad,pv,n)
      vpvrnl=vpvrun
      if(wkt.eq.0.d000)go to 2
      fac=1.d000
      do 1 n=1,100
        fac=fac*boltz
        vpvrun=vpvrun+vpvn(a,thrad,pv,n)*fac
        if(vpvrun.eq.vpvrnl.and.vpvrun.ne.0.d000)go to 2
        vpvrnl=vpvrun
    1 continue
    2 vpvrnl=vpvrnl*wnorm
      vpvrn=vpvrun*wnorm
      return
      end




      options/g_float
      subroutine upudis(a,theta,homv,wkt,upuds,numlo,numhi)
c
c
c     gets the upu distribution from the a matrix, plus the other required
c         parameters homu,wkt.
c
c
c     the distribution is in  upuds(j,1), j=1,num
c     the corresponding wp is in upuds(j,2)
c
      IMPLICIT REAL*8(A-H,O-P,R-Z),LOGICAL*4(Q)
      dimension upuds(-50:75,2),a(0:100,0:100)
      do 1 m=-50,75
      upuds(m,1)=0.d000
    1 upuds(m,2)=0.d000
      wm=-1.d000
      do 2 m=0,75
      wm=wm+1.d000
      call upu(a,theta,homv,wkt,wm,upurn,upurnl)
      if(upurn.eq.0.d000.or.wm.eq.0.d000
     1        .or.dabs((upurn-upurnl)/upurn).lt.1.d-003)then
        upuds(m,1)=upurn
        upuds(m,2)=wm
```

Source Code

```
          else
            numhi=m-1
            go to 3
          end if
     2  continue
        numhi=75
     3  wm=0.d000
        do 4 m=-1,-50,-1
        wm=wm-1.d000
        call upu(a,theta,homv,wkt,wm,upurn,upurnl)
        if(upurn.eq.0.d000.or.dabs((upurn-upurnl)/upurn).lt.1.d-003)then
          upuds(m,1)=upurn
          upuds(m,2)=wm
          if(upurn.ne.0.d000.and.upurn.lt.1.d-040)then
            numlo=m
            return
          end if
        else
          numlo=m+1
          return
        end if
     4  continue
        numlo=-50
        return
        end



        options/g_float
        subroutine vpvdis(a,theta,homu,wkt,vpvds,numlo,numhi)
c
c       gets the vpv distribution from the a matrix, plus the other required
c           parameters homu,wkt.
c
c
c
c       the distribution is in  vpvds(j,1), j=1,num
c       the corresponding wp is in vpvds(j,2)
c
c
c
c
c
c
c
        IMPLICIT REAL*8(A-H,O-P,R-Z),LOGICAL*4(Q)
        dimension vpvds(-50:75,0:75),a(0:100,0:100)
        do 1 n=-50,75
        vpvds(n,1)=0.d000
     1  vpvds(n,2)=0.d000
        wn=-1.d000
        do 2 n=0,75
        wn=wn+1.d000
        call vpv(a,theta,homu,wkt,wn,vpvrn,vpvrnl)
        if(vpvrn.eq.0.d000.or.wn.eq.0.d000
     1    .or.dabs((vpvrn-vpvrnl)/vpvrn).lt.1.d-003)then
          vpvds(n,1)=vpvrn
          vpvds(n,2)=wn
        else
          numhi=n-1
          go to 3
        end if
     2  continue
        numhi=75
     3  wn=0.d000
        do 4 n=-1,-50,-1
        wn=wn-1.d000
        call vpv(a,theta,homu,wkt,wn,vpvrn,vpvrnl)
        if(vpvrn.eq.0.d000.or.dabs((vpvrn-vpvrnl)/vpvrn).lt.1.d-003)then
          vpvds(n,1)=vpvrn
          vpvds(n,2)=wn
          if(vpvrn.ne.0.d000.and.vpvrn.lt.1.d-040)then
            numlo=n
            return
          end if
```

```
          else
            numlo=n+1
            return
          end if
    4   continue
        numlo=-50
        return
        end
```

```
        options/g_float
        subroutine chisd(wpds,npds,s,wmav,zder,chsd,num)
c
c
c       finds the chis distribution for given s, wmav
c       there are n=num of these elements, in the first num elements of chsd.
c
c       chisd uses wp distribution as given through calling sequence.
c
c
c       chsd has two elements, the first being its value, the second its p number
c       given as a floating point number.
c
c
c
        IMPLICIT REAL*8(A-H,O-P,R-Z),LOGICAL*4(Q)
        logical*4 zder
        dimension wpds(300,2),chsd(300,2)
        do 1 j=1,300
    1   chsd(j,1)=0.d000
        num=npds-2
        jlast=npds-1
        wmavp1=wmav+1.d000
        sign=-1.d000
        if(zder)sign=-sign
        do 2 j=2,jlast
        chsd(j-1,1)=wmav*(wpds(j,1)-wpds(j+1,1))
    1              +sign*wmavp1*(wpds(j-1,1)-wpds(j,1))
        chsd(j-1,2)=wpds(j,2)
    2   continue
        return
        END
```

```
        options/g_float
        double precision function wmdp(S,wmav,gamma,ip)
c
c       finds the wmpd for given s, wmav,ip, and gamma
c
c       ip.le.69
c
c       Eq. (5.6) without the A squared electronic factor.
c
        IMPLICIT REAL*8(A-H,O-P,R-Z),LOGICAL*4(Q)
        real*8 wpder,wlp
        wmdp=wpder(s,wmav,ip)*(1.d000-gamma)+gamma*wlp(s,wmav,ip)
        return
        END
```

```
        options/g_float
        IMPLICIT REAL*8(A-H,O-z)
        logical*4 zder
        dimension wpda(300,2),wpdb(300,2),ab(1200,2),wdebo(1200,2)
```

Source Code

```
   10 format(/,/,/)
      sa=3.d000
      homa=100.d000
      ifac=2
      wkt=200.d000
      boltza=dexp(-homa/wkt)
      wmava=boltza/(1.d000-boltza)
      call wpdis(sa,wmava,wpda,numa)
      sb=5.d000
      homb=200.d000
      ifac=2
      wkt=200.d000
      boltzb=dexp(-homb/wkt)
      wmavb=boltzb/(1.d000-boltzb)
      call wpdis(sb,wmavb,wpdb,numb)
      write(6,10)
      write(6,*)' wpda'
      do 1 j=1,numa
      write(6,*)j,wpda(j,1),wpda(j,2)
    1 continue
      write(6,10)
      write(6,10)
      write(6,*)' wpdb'
      do 2 j=1,numb
      write(6,*)j,wpdb(j,1),wpdb(j,2)
    2 continue
      idima=300
      idimb=300
      call cmbww(wpda,numa,homa,idima,wpdb,numb,ifac,idimb,ab,numab)
      write(6,10)
      write(6,10)
      write(6,*)' combined wpda+wpdb=wua'
      do 3 j=1,numab
      write(6,*)j,ab(j,1),ab(j,2)
    3 continue
      sa=6.25d000
      homa=400.d000
      wkt=200.d000
      ndis=3
      zder=.false.
      sgnrts=+1.d000
      call debo(sa,sgnrts,wkt,homa,ndis,zder,wdebo,ndebo,wnorm)
      write(6,10)
      write(6,*)' test of debo for derivative operator: indx,wdeb,n'
      write(6,*)' s=',sa
      write(6,*)' homa=',homa
      write(6,*)' kt=',wkt
      write(6,*)' sgnrts=',sgnrts
      write(6,*)' ndis=',ndis
      sum=0.d000
      do 4  j=1,ndebo
      sum=sum+wdebo(j,1)
      if(dabs(wdebo(j,1)).lt.1.d-040)go to 4
      write(6,*) j,wdebo(j,1),wdebo(j,2)
    4 continue
      write(6,*)' sum in debo=',sum
      write(6,*)' normalization factor should be',wnorm
      sa=6.25d000
      homa=400.d000
      wkt=50.d000
      ndis=3
      zder=.false.
      sgnrts=+1.d000
      call debo(sa,sgnrts,wkt,homa,ndis,zder,wdebo,ndebo,wnorm)
      write(6,10)
      write(6,*)' test of debo for derivative operator: indx,wdeb,n'
      write(6,*)' s=',sa
      write(6,*)' homa=',homa
      write(6,*)' kt=',wkt
      write(6,*)' ndis=',ndis
      write(6,*)' sgnrts=',sgnrts
      sum=0.d000
      do 5 j=1,ndebo
      sum=sum+wdebo(j,1)
      if(wdebo(j,1).lt.1.d-040)go to 5
      write(6,*) j,wdebo(j,1),wdebo(j,2)
```

```
    5 continue
      write(6,*)' sum in debo=',sum
      write(6,*)' normalization factor should be',wnorm
      sa=6.25d000
      homa=400.d000
      wkt=200.d000
      ndis=3
      call debc(sa,wkt,homa,ndis,wdebo,ndebo)
      write(6,10)
      write(6,*)' test of debc: indx,wdeb,n'
      write(6,*)' s=',sa
      write(6,*)' homa=',homa
      write(6,*)' kt=',wkt
      write(6,*)' ndis=',ndis
      sum=0.d000
      do 6 j=1,ndebo
      sum=sum+wdebo(j,1)
      if(wdebo(j,1).lt.1.d-040)go to 6
      write(6,*) j,wdebo(j,1),wdebo(j,2)
    6 continue
      write(6,*)' sum in debc=',sum
      write(6,*)' normalization factor should be unity'
      END
```

wpda

1	1.753355400761211E-044	-57.0000000000000
2	2.183950210240399E-043	-56.0000000000000
3	2.673554527104853E-042	-55.0000000000000
4	3.215725350191892E-041	-54.0000000000000
5	3.799076445890522E-040	-53.0000000000000
6	4.407042016910789E-039	-52.0000000000000
7	5.018135470001542E-038	-51.0000000000000
8	5.606790910837636E-037	-50.0000000000000
9	6.144809273933999E-036	-49.0000000000000
10	6.603345538998110E-035	-48.0000000000000
11	6.955279911832709E-034	-47.0000000000000
12	7.177730331030023E-033	-46.0000000000000
13	7.254404235679048E-032	-45.0000000000000
14	7.177470267634344E-031	-44.0000000000000
15	6.948665433263352E-030	-43.0000000000000
16	6.579440612003210E-029	-42.0000000000000
17	6.090076429068653E-028	-41.0000000000000
18	5.507851533822478E-027	-40.0000000000000
19	4.864488979953241E-026	-39.0000000000000
20	4.193215833893363E-025	-38.0000000000000
21	3.525824383028740E-024	-37.0000000000000
22	2.890109746009242E-023	-36.0000000000000
23	2.307981816843255E-022	-35.0000000000000
24	1.794426233933603E-021	-34.0000000000000
25	1.357345649347246E-020	-33.0000000000000
26	9.981779808721061E-020	-32.0000000000000
27	7.130874204497012E-019	-31.0000000000000
28	4.944719827883039E-018	-30.0000000000000
29	3.325313169796816E-017	-29.0000000000000
30	2.166819319732414E-016	-28.0000000000000
31	1.366776142523939E-015	-27.0000000000000
32	8.337158913687426E-015	-26.0000000000000
33	4.912703304427183E-014	-25.0000000000000
34	2.793269120899304E-013	-24.0000000000000
35	1.530639259160260E-012	-23.0000000000000
36	8.073212101200580E-012	-22.0000000000000
37	4.093020319911116E-011	-21.0000000000000
38	1.991765305195160E-010	-20.0000000000000
39	9.288828431102083E-010	-19.0000000000000
40	4.144764951813793E-009	-18.0000000000000
41	1.766425201918821E-008	-17.0000000000000
42	7.176889289461994E-008	-16.0000000000000
43	2.774328008101860E-007	-15.0000000000000
44	1.018209695667356E-006	-14.0000000000000
45	3.539902702077417E-006	-13.0000000000000
46	1.162985475914795E-005	-12.0000000000000

```
 47    3.601444951077348E-005   -11.0000000000000
 48    1.048399669006072E-004   -10.0000000000000
 49    2.860841774540390E-004   -9.00000000000000
 50    7.296185568240838E-004   -8.00000000000000
 51    1.733857274737393E-003   -7.00000000000000
 52    3.827447854587438E-003   -6.00000000000000
 53    7.824541040750583E-003   -5.00000000000000
 54    1.477030503468892E-002   -4.00000000000000
 55    2.567623531499413E-002   -3.00000000000000
 56    4.100883608575932E-002   -2.00000000000000
 57    6.006849148565736E-002   -1.00000000000000
 58    8.060137638311822E-002    0.000000000000000E+000
 59    9.903619961127284E-002    1.00000000000000
 60    0.111473573938175         2.00000000000000
 61    0.115072903178638         3.00000000000000
 62    0.109138612499634         4.00000000000000
 63    9.532242397422053E-002    5.00000000000000
 64    7.687634520489149E-002    6.00000000000000
 65    5.741746728479540E-002    7.00000000000000
 66    3.983582343244756E-002    8.00000000000000
 67    2.575247696488205E-002    9.00000000000000
 68    1.555963068792870E-002   10.0000000000000
 69    8.812445240223387E-003   11.0000000000000
 70    4.691818273978813E-003   12.0000000000000
 71    2.354536664077928E-003   13.0000000000000
 72    1.116602514502175E-003   14.0000000000000
 73    5.016102710261581E-004   15.0000000000000
 74    2.139400544953603E-004   16.0000000000000
 75    8.681571541109755E-005   17.0000000000000
 76    3.358537826462022E-005   18.0000000000000
 77    1.240962104071209E-005   19.0000000000000
 78    4.387155036616410E-006   20.0000000000000
 79    1.486400903734785E-006   21.0000000000000
 80    4.833766454441354E-007   22.0000000000000
 81    1.510982346073450E-007   23.0000000000000
 82    4.546179331491126E-008   24.0000000000000
 83    1.318261474192129E-008   25.0000000000000
 84    3.688470754721878E-009   26.0000000000000
 85    9.969488922742609E-010   27.0000000000000
 86    2.605826196921222E-010   28.0000000000000
 87    6.593295491578148E-011   29.0000000000000
 88    1.616437501935705E-011   30.0000000000000
 89    3.843326183454361E-012   31.0000000000000
 90    8.869919857166981E-013   32.0000000000000
 91    1.988609027669734E-013   33.0000000000000
 92    4.334428090044252E-014   34.0000000000000
 93    9.191487824930949E-015   35.0000000000000
 94    1.897645167264662E-015   36.0000000000000
 95    3.816880753942438E-016   37.0000000000000
 96    7.484148104685628E-017   38.0000000000000
 97    1.431461332166593E-017   39.0000000000000
 98    2.672217865695115E-018   40.0000000000000
 99    4.871465396604303E-019   41.0000000000000
100    8.677069803207772E-020   42.0000000000000
101    1.510889706942300E-020   43.0000000000000
102    2.573060536517013E-021   44.0000000000000
103    4.287731628907098E-022   45.0000000000000
104    6.994557126606724E-023   46.0000000000000
105    1.117467593255154E-023   47.0000000000000
106    1.749168264480796E-024   48.0000000000000
107    2.683633559414007E-025   49.0000000000000
108    4.037164151406378E-026   50.0000000000000
109    5.957330162831733E-027   51.0000000000000
110    8.625886127064391E-028   52.0000000000000
111    1.225975533201841E-028   53.0000000000000
112    1.710921014828216E-029   54.0000000000000
113    2.345240029536296E-030   55.0000000000000
114    3.158553419621063E-031   56.0000000000000
115    4.180831741903980E-032   57.0000000000000
116    5.440469520825155E-033   58.0000000000000
117    6.961950730233737E-034   59.0000000000000
118    8.763229051796667E-035   60.0000000000000
119    1.085300826670284E-035   61.0000000000000
120    1.322815998690694E-036   62.0000000000000
121    1.587154852202028E-037   63.0000000000000
122    1.875053797519411E-038   64.0000000000000
```

```
       123   2.181642945023062E-039      65.0000000000000
       124   2.500501540708793E-040      66.0000000000000
       125   2.823834740467185E-041      67.0000000000000
       126   3.142770116624203E-042      68.0000000000000
       127   3.447760144724686E-043      69.0000000000000
       128   3.729065324241758E-044      70.0000000000000


 wpdb
         1   7.902535783621105E-043     -47.0000000000000
         2   1.289303139197071E-041     -46.0000000000000
         3   2.059636582440293E-040     -45.0000000000000
         4   3.220179394499526E-039     -44.0000000000000
         5   4.925181376345941E-038     -43.0000000000000
         6   7.365584259646981E-037     -42.0000000000000
         7   1.076504596541358E-035     -41.0000000000000
         8   1.536807128770496E-034     -40.0000000000000
         9   2.141796639400441E-033     -39.0000000000000
        10   2.912338740594245E-032     -38.0000000000000
        11   3.861426308424467E-031     -37.0000000000000
        12   4.989079381466343E-030     -36.0000000000000
        13   6.277268447838821E-029     -35.0000000000000
        14   7.685898652630330E-028     -34.0000000000000
        15   9.151081040872132E-027     -33.0000000000000
        16   1.058685432019743E-025     -32.0000000000000
        17   1.189111978836215E-024     -31.0000000000000
        18   1.295580346966228E-023     -30.0000000000000
        19   1.368026715341175E-022     -29.0000000000000
        20   1.398597683718678E-021     -28.0000000000000
        21   1.382970413356465E-020     -27.0000000000000
        22   1.321237609195904E-019     -26.0000000000000
        23   1.218127492417392E-018     -25.0000000000000
        24   1.082458129325132E-017     -24.0000000000000
        25   9.258968424436758E-017     -23.0000000000000
        26   7.612620535534675E-016     -22.0000000000000
        27   6.007160970776121E-015     -21.0000000000000
        28   4.542170606106443E-014     -20.0000000000000
        29   3.285183250768295E-013     -19.0000000000000
        30   2.268519857725606E-012     -18.0000000000000
        31   1.492564860987442E-011     -17.0000000000000
        32   9.336447697341555E-011     -16.0000000000000
        33   5.539368688565959E-010     -15.0000000000000
        34   3.109249928789788E-009     -14.0000000000000
        35   1.646494595250219E-008     -13.0000000000000
        36   8.200950291710340E-008     -12.0000000000000
        37   3.829534160854497E-007     -11.0000000000000
        38   1.670573112753598E-006     -10.0000000000000
        39   6.782006157604936E-006     -9.00000000000000
        40   2.551720482950821E-005     -8.00000000000000
        41   8.858860309329301E-005     -7.00000000000000
        42   2.824712158664061E-004     -6.00000000000000
        43   8.232469787402430E-004     -5.00000000000000
        44   2.182406697055553E-003     -4.00000000000000
        45   5.237809118529717E-003     -3.00000000000000
        46   1.133241580449745E-002     -2.00000000000000
        47   2.202675500760044E-002     -1.00000000000000
        48   3.837433452780335E-002     0.000000000000000E+000
        49   5.987492787707956E-002      1.00000000000000
        50   8.373585611584000E-002      2.00000000000000
        51   0.105204208446838           3.00000000000000
        52   0.119155368279178           4.00000000000000
        53   0.122180684836491           5.00000000000000
        54   0.113957021813410           6.00000000000000
        55   9.714919961096299E-002      7.00000000000000
        56   7.606571554350224E-002      8.00000000000000
        57   5.495516509240582E-002      9.00000000000000
        58   3.679682152579090E-002      10.0000000000000
        59   2.292900710501932E-002      11.0000000000000
        60   1.334743954164935E-002      12.0000000000000
        61   7.284312588090038E-003      13.0000000000000
```

Source Code

```
 62   3.739197284901625E-003    14.0000000000000
 63   1.810829247545016E-003    15.0000000000000
 64   8.296470610751825E-004    16.0000000000000
 65   3.605283369879831E-004    17.0000000000000
 66   1.489509438456846E-004    18.0000000000000
 67   5.863470656828485E-005    19.0000000000000
 68   2.203703089696237E-005    20.0000000000000
 69   7.922338407833006E-006    21.0000000000000
 70   2.729058115056340E-006    22.0000000000000
 71   9.022682741116103E-007    23.0000000000000
 72   2.867336558813531E-007    24.0000000000000
 73   8.771114747161647E-008    25.0000000000000
 74   2.586053212106068E-008    26.0000000000000
 75   7.358069752306486E-009    27.0000000000000
 76   2.022731780179830E-009    28.0000000000000
 77   5.378170345430065E-010    29.0000000000000
 78   1.384518644617722E-010    30.0000000000000
 79   3.454233510835806E-011    31.0000000000000
 80   8.359693561455899E-012    32.0000000000000
 81   1.964220793523819E-012    33.0000000000000
 82   4.484427820753105E-013    34.0000000000000
 83   9.955832202055007E-014    35.0000000000000
 84   2.150907642043964E-014    36.0000000000000
 85   4.525260467051189E-015    37.0000000000000
 86   9.277526417396878E-016    38.0000000000000
 87   1.854654540223663E-016    39.0000000000000
 88   3.617417560826773E-017    40.0000000000000
 89   6.887944618521837E-018    41.0000000000000
 90   1.281077613246181E-018    42.0000000000000
 91   2.328546689927644E-019    43.0000000000000
 92   4.138445787460754E-020    44.0000000000000
 93   7.195190265088133E-021    45.0000000000000
 94   1.224337657917299E-021    46.0000000000000
 95   2.039891745183925E-022    47.0000000000000
 96   3.329242470794345E-023    48.0000000000000
 97   5.324689646105594E-024    49.0000000000000
 98   8.348819902884350E-025    50.0000000000000
 99   1.283815789894720E-025    51.0000000000000
100   1.936805338658298E-026    52.0000000000000
101   2.867661733343139E-027    53.0000000000000
102   4.168464797218322E-028    54.0000000000000
103   5.950768119204454E-029    55.0000000000000
104   8.345571378712043E-030    56.0000000000000
105   1.150159578802152E-030    57.0000000000000
106   1.558143603049982E-031    58.0000000000000
107   2.075526935543162E-032    59.0000000000000
108   2.719198527156992E-033    60.0000000000000
109   3.504785274051245E-034    61.0000000000000
110   4.445314849866013E-035    62.0000000000000
111   5.549750000749419E-036    63.0000000000000
112   6.821500890927398E-037    64.0000000000000
113   8.257046919461946E-038    65.0000000000000
114   9.844813264637509E-039    66.0000000000000
115   1.156444648041983E-039    67.0000000000000
116   1.338661555566860E-040    68.0000000000000
117   1.527342986457469E-041    69.0000000000000
118   1.717951570726394E-042    70.0000000000000


combined wpda+wpdb=wua
  1   1.385595379592080E-086   -151.000000000000
  2   1.725874468607159E-085   -150.000000000000
  3   2.338846694223787E-084   -149.000000000000
  4   2.822815851210198E-083   -148.000000000000
  5   3.383048729551547E-082   -147.000000000000
  6   3.942266640208118E-081   -146.000000000000
  7   4.516127256672686E-080   -145.000000000000
  8   5.072252858284234E-079   -144.000000000000
  9   5.590667351327222E-078   -143.000000000000
 10   6.043402650756295E-077   -142.000000000000
 11   6.405722008052658E-076   -141.000000000000
 12   6.655014138225698E-075   -140.000000000000
 13   6.774372276647843E-074   -139.000000000000
 14   6.753939295754163E-073   -138.000000000000
```

15	6.592344038985550E-072	-137.000000000000
16	6.297043123290325E-071	-136.000000000000
17	5.883851345254432E-070	-135.000000000000
18	5.375550201922399E-069	-134.000000000000
19	4.799791712604900E-068	-133.000000000000
20	4.186547572320162E-067	-132.000000000000
21	3.565444799933491E-066	-131.000000000000
22	2.963325370547040E-065	-130.000000000000
23	2.402320654224870E-064	-129.000000000000
24	1.898633702439052E-063	-128.000000000000
25	1.462101316820123E-062	-127.000000000000
26	1.096485547656105E-061	-126.000000000000
27	8.003447713836687E-061	-125.000000000000
28	5.682735176009304E-060	-124.000000000000
29	3.922840031687892E-059	-123.000000000000
30	2.631267781372519E-058	-122.000000000000
31	1.714011226966023E-057	-121.000000000000
32	1.083721193891208E-056	-120.000000000000
33	6.647539122861710E-056	-119.000000000000
34	3.954093327752545E-055	-118.000000000000
35	2.279846709413179E-054	-117.000000000000
36	1.273806652497202E-053	-116.000000000000
37	6.895330636885575E-053	-115.000000000000
38	3.616074667754089E-052	-114.000000000000
39	1.837412109857835E-051	-113.000000000000
40	9.049344113571805E-051	-112.000000000000
41	4.322572833793188E-050	-111.000000000000
42	2.004448927999350E-049	-110.000000000000
43	9.035350863788186E-049	-109.000000000000
44	3.965794172166479E-048	-108.000000000000
45	1.698457468009961E-047	-107.000000000000
46	7.114922656632628E-047	-106.000000000000
47	2.923042790447760E-046	-105.000000000000
48	1.181008886827757E-045	-104.000000000000
49	4.705398397376378E-045	-103.000000000000
50	1.853225403584274E-044	-102.000000000000
51	7.230042395840712E-044	-101.000000000000
52	2.798475978893827E-043	-100.000000000000
53	1.075853529327962E-042	-99.0000000000000
54	4.110942256391478E-042	-98.0000000000000
55	1.561917311125808E-041	-97.0000000000000
56	5.901806520480370E-041	-96.0000000000000
57	2.217942528255537E-040	-95.0000000000000
58	8.289978532517697E-040	-94.0000000000000
59	3.081625499114673E-039	-93.0000000000000
60	1.139223727911674E-038	-92.0000000000000
61	4.188105569911999E-038	-91.0000000000000
62	1.531021015156578E-037	-90.0000000000000
63	5.565118179327605E-037	-89.0000000000000
64	2.011275419952357E-036	-88.0000000000000
65	7.226804262336383E-036	-87.0000000000000
66	2.581497875523087E-035	-86.0000000000000
67	9.166859683161450E-035	-85.0000000000000
68	3.235674042574268E-034	-84.0000000000000
69	1.135208874031373E-033	-83.0000000000000
70	3.958443858740020E-033	-82.0000000000000
71	1.371769557586390E-032	-81.0000000000000
72	4.724060162144449E-032	-80.0000000000000
73	1.616576457212117E-031	-79.0000000000000
74	5.496572569530085E-031	-78.0000000000000
75	1.856821679378868E-030	-77.0000000000000
76	6.231575554216244E-030	-76.0000000000000
77	2.077500620996607E-029	-75.0000000000000
78	6.879637023382881E-029	-74.0000000000000
79	2.262745022530412E-028	-73.0000000000000
80	7.391205984486320E-028	-72.0000000000000
81	2.397549743150853E-027	-71.0000000000000
82	7.722444010326113E-027	-70.0000000000000
83	2.469664964838646E-026	-69.0000000000000
84	7.841115436745840E-026	-68.0000000000000
85	2.471346590185445E-025	-67.0000000000000
86	7.731503740167607E-025	-66.0000000000000
87	2.400635886155760E-024	-65.0000000000000
88	7.397365935198075E-024	-64.0000000000000
89	2.261891715624597E-023	-63.0000000000000
90	6.862221439665069E-023	-62.0000000000000

Source Code

```
 91   2.065423134936727E-022   -61.0000000000000
 92   6.166755690448954E-022   -60.0000000000000
 93   1.826243324225372E-021   -59.0000000000000
 94   5.363697479925302E-021   -58.0000000000000
 95   1.562146981086301E-020   -57.0000000000000
 96   4.511057995613470E-020   -56.0000000000000
 97   1.291455232449090E-019   -55.0000000000000
 98   3.664956581212247E-019   -54.0000000000000
 99   1.030835748679297E-018   -53.0000000000000
100   2.873309247222172E-018   -52.0000000000000
101   7.935733211324114E-018   -51.0000000000000
102   2.171408010244637E-017   -50.0000000000000
103   5.885477692752623E-017   -49.0000000000000
104   1.579946522222669E-016   -48.0000000000000
105   4.200066163069143E-016   -47.0000000000000
106   1.105487381357653E-015   -46.0000000000000
107   2.880476191137775E-015   -45.0000000000000
108   7.428730326786911E-015   -44.0000000000000
109   1.895962012225837E-014   -43.0000000000000
110   4.787758631198444E-014   -42.0000000000000
111   1.196032923755455E-013   -41.0000000000000
112   2.955158467515424E-013   -40.0000000000000
113   7.220402005061098E-013   -39.0000000000000
114   1.744210224169560E-012   -38.0000000000000
115   4.164900393665183E-012   -37.0000000000000
116   9.828512801501747E-012   -36.0000000000000
117   2.291685177378109E-011   -35.0000000000000
118   5.278494682156253E-011   -34.0000000000000
119   1.200755682803340E-010   -33.0000000000000
120   2.697040538210345E-010   -32.0000000000000
121   5.980049115641966E-010   -31.0000000000000
122   1.308579215808071E-009   -30.0000000000000
123   2.825297301870975E-009   -29.0000000000000
124   6.017074564208821E-009   -28.0000000000000
125   1.263715605604823E-008   -27.0000000000000
126   2.616610418703849E-008   -26.0000000000000
127   5.339924243630633E-008   -25.0000000000000
128   1.073780690626743E-007   -24.0000000000000
129   2.126945862634296E-007   -23.0000000000000
130   4.148885238862035E-007   -22.0000000000000
131   7.967323389781404E-007   -21.0000000000000
132   1.505808566277993E-006   -20.0000000000000
133   2.800091986450392E-006   -19.0000000000000
134   5.121391745189874E-006   -18.0000000000000
135   9.210522105400985E-006   -17.0000000000000
136   1.628277330315438E-005   -16.0000000000000
137   2.828708455290410E-005   -15.0000000000000
138   4.827614850267203E-005   -14.0000000000000
139   8.091562794212417E-005   -13.0000000000000
140   1.331560593089714E-004   -12.0000000000000
141   2.150775180446473E-004   -11.0000000000000
142   3.408912620244138E-004   -10.0000000000000
143   5.300417472225053E-004   -9.00000000000000
144   8.082963586757513E-004   -8.00000000000000
145   1.208642733175132E-003   -7.00000000000000
146   1.771744150169056E-003   -6.00000000000000
147   2.545647927634784E-003   -5.00000000000000
148   3.584421194919988E-003   -4.00000000000000
149   4.945425726315412E-003   -3.00000000000000
150   6.685057672609940E-003   -2.00000000000000
151   8.852976255283069E-003   -1.00000000000000
152   1.148511651103658E-002   0.000000000000000E+000
153   1.459609026108837E-002   1.00000000000000
154   1.817187079365631E-002   2.00000000000000
155   2.216386042579647E-002   3.00000000000000
156   2.648548929145983E-002   4.00000000000000
157   3.101234050448808E-002   5.00000000000000
158   3.558643254666233E-002   6.00000000000000
159   4.002475036568364E-002   7.00000000000000
160   4.413148586222285E-002   8.00000000000000
161   4.771283755448626E-002   9.00000000000000
162   5.059274910750747E-002   10.0000000000000
163   5.262773347693748E-002   11.0000000000000
164   5.371898835326541E-002   12.0000000000000
165   5.382035290823448E-002   13.0000000000000
166   5.294122520924653E-002   14.0000000000000
```

167	5.114424865295557E-002	15.0000000000000
168	4.853826312922791E-002	16.0000000000000
169	4.526758704651115E-002	17.0000000000000
170	4.149906713726535E-002	18.0000000000000
171	3.740846403690251E-002	19.0000000000000
172	3.316764087864917E-002	20.0000000000000
173	2.893373538681945E-002	21.0000000000000
174	2.484109427517180E-002	22.0000000000000
175	2.099631007280917E-002	23.0000000000000
176	1.747629523327097E-002	24.0000000000000
177	1.432900781762877E-002	25.0000000000000
178	1.157623490904653E-002	26.0000000000000
179	9.217748495601603E-003	27.0000000000000
180	7.236159649056443E-003	28.0000000000000
181	5.601884397532376E-003	29.0000000000000
182	4.277768189732517E-003	30.0000000000000
183	3.223066160667310E-003	31.0000000000000
184	2.396620030082718E-003	32.0000000000000
185	1.759193461147325E-003	33.0000000000000
186	1.275017896574829E-003	34.0000000000000
187	9.126586809620450E-004	35.0000000000000
188	6.453398472124628E-004	36.0000000000000
189	4.508712410971560E-004	37.0000000000000
190	3.113106541733000E-004	38.0000000000000
191	2.124730123870561E-004	39.0000000000000
192	1.433740035359019E-004	40.0000000000000
193	9.567093400443842E-005	41.0000000000000
194	6.314171415732336E-005	42.0000000000000
195	4.122503114936229E-005	43.0000000000000
196	2.663135077894575E-005	44.0000000000000
197	1.702511807973313E-005	45.0000000000000
198	1.077275724363874E-005	46.0000000000000
199	6.748021483323068E-006	47.0000000000000
200	4.185139638577776E-006	48.0000000000000
201	2.570375213507885E-006	49.0000000000000
202	1.563520151980586E-006	50.0000000000000
203	9.420985763860170E-007	51.0000000000000
204	5.623916967270673E-007	52.0000000000000
205	3.326543765650318E-007	53.0000000000000
206	1.949933700915961E-007	54.0000000000000
207	1.132863563016044E-007	55.0000000000000
208	6.524149493584531E-008	56.0000000000000
209	3.724900086548103E-008	57.0000000000000
210	2.108648786233543E-008	58.0000000000000
211	1.183711474512954E-008	59.0000000000000
212	6.590087793645414E-009	60.0000000000000
213	3.639072729159546E-009	61.0000000000000
214	1.993396389704685E-009	62.0000000000000
215	1.083298121110283E-009	63.0000000000000
216	5.841179117977651E-010	64.0000000000000
217	3.125337736124397E-010	65.0000000000000
218	1.659517639917776E-010	66.0000000000000
219	8.745785073098311E-011	67.0000000000000
220	4.574990876082658E-011	68.0000000000000
221	2.375733197933423E-011	69.0000000000000
222	1.224790008511451E-011	70.0000000000000
223	6.269338712455964E-012	71.0000000000000
224	3.186520041154408E-012	72.0000000000000
225	1.608363496221760E-012	73.0000000000000
226	8.062344575022805E-013	74.0000000000000
227	4.014063455354455E-013	75.0000000000000
228	1.985126462315299E-013	76.0000000000000
229	9.752315691913670E-014	77.0000000000000
230	4.759669094728594E-014	78.0000000000000
231	2.307960038519624E-014	79.0000000000000
232	1.111974161820509E-014	80.0000000000000
233	5.323627720721609E-015	81.0000000000000
234	2.532784547518843E-015	82.0000000000000
235	1.197558606832448E-015	83.0000000000000
236	5.627726781181697E-016	84.0000000000000
237	2.628669974735505E-016	85.0000000000000
238	1.220490754340359E-016	86.0000000000000
239	5.633214064186803E-017	87.0000000000000
240	2.584810744192668E-017	88.0000000000000
241	1.179176561714054E-017	89.0000000000000
242	5.348510445217651E-018	90.0000000000000

243	2.412218232145477E-018	91.0000000000000
244	1.081820969445392E-018	92.0000000000000
245	4.824741790927706E-019	93.0000000000000
246	2.139917313443491E-019	94.0000000000000
247	9.439479332698987E-020	95.0000000000000
248	4.141433237035845E-020	96.0000000000000
249	1.807290129109110E-020	97.0000000000000
250	7.845168235375814E-021	98.0000000000000
251	3.387630154660705E-021	99.0000000000000
252	1.455227293938255E-021	100.000000000000
253	6.219108908484305E-022	101.000000000000
254	2.644289346502747E-022	102.000000000000
255	1.118649136749728E-022	103.000000000000
256	4.708723287280514E-023	104.000000000000
257	1.972226144360714E-023	105.000000000000
258	8.220042756053852E-024	106.000000000000
259	3.409372748501702E-024	107.000000000000
260	1.407268061985365E-024	108.000000000000
261	5.780956306656434E-025	109.000000000000
262	2.363531048269713E-025	110.000000000000
263	9.617886786225181E-026	111.000000000000
264	3.895582358952256E-026	112.000000000000
265	1.570568444768324E-026	113.000000000000
266	6.303041610041427E-027	114.000000000000
267	2.518077634043049E-027	115.000000000000
268	1.001452202988933E-027	116.000000000000
269	3.965060116181996E-028	117.000000000000
270	1.562945294460939E-028	118.000000000000
271	6.133777084233762E-029	119.000000000000
272	2.396723437803733E-029	120.000000000000
273	9.324570814020962E-030	121.000000000000
274	3.612233451095587E-030	122.000000000000
275	1.393392934332390E-030	123.000000000000
276	5.352255202352414E-031	124.000000000000
277	2.047292033695786E-031	125.000000000000
278	7.798600953615254E-032	126.000000000000
279	2.958431049070473E-032	127.000000000000
280	1.117706516522014E-032	128.000000000000
281	4.205612130329862E-033	129.000000000000
282	1.576082982275345E-033	130.000000000000
283	5.882886992204379E-034	131.000000000000
284	2.187139273082511E-034	132.000000000000
285	8.099340138943215E-035	133.000000000000
286	2.987601575102728E-035	134.000000000000
287	1.097761419839921E-035	135.000000000000
288	4.018072002469567E-036	136.000000000000
289	1.465088932195238E-036	137.000000000000
290	5.321801453130658E-037	138.000000000000
291	1.925810159536203E-037	139.000000000000
292	6.942884562689751E-038	140.000000000000
293	2.493729816885471E-038	141.000000000000
294	8.923860669734789E-039	142.000000000000
295	3.181707374525840E-039	143.000000000000
296	1.130264191748009E-039	144.000000000000
297	4.000512680117016E-040	145.000000000000
298	1.410785567433367E-040	146.000000000000
299	4.956668515129706E-041	147.000000000000
300	1.734762192775813E-041	148.000000000000
301	6.046296395409911E-042	149.000000000000
302	2.097625891579243E-042	150.000000000000
303	7.238137104486302E-043	151.000000000000
304	2.481498972974406E-043	152.000000000000
305	8.440450840866013E-044	153.000000000000
306	2.843245241103248E-044	154.000000000000
307	9.466301137309474E-045	155.000000000000
308	3.108255734833936E-045	156.000000000000
309	1.004296838172387E-045	157.000000000000
310	3.186326596389785E-046	158.000000000000
311	9.907207488860211E-047	159.000000000000
312	3.013689844227081E-047	160.000000000000
313	8.955786293477207E-048	161.000000000000
314	2.596925466653637E-048	162.000000000000
315	7.341329891675212E-049	163.000000000000
316	2.021921910143022E-049	164.000000000000
317	5.422961112042586E-050	165.000000000000
318	1.416050641572150E-050	166.000000000000

```
319  3.599527068096590E-051   167.000000000000
320  8.907279520365752E-052   168.000000000000
321  2.146002100148717E-052   169.000000000000
322  5.034864746502436E-053   170.000000000000
323  1.150621127754245E-053   171.000000000000
324  2.562115842232029E-054   172.000000000000
325  5.560804448654720E-055   173.000000000000
326  1.176822111821937E-055   174.000000000000
327  2.429339284255990E-056   175.000000000000
328  4.893782964624275E-057   176.000000000000
329  9.624063264374989E-058   177.000000000000
330  1.848454794249650E-058   178.000000000000
331  3.468767005960772E-059   179.000000000000
332  6.362614277537293E-060   180.000000000000
333  1.141212619596840E-060   181.000000000000
334  2.002358537946913E-061   182.000000000000
335  3.438208736235195E-062   183.000000000000
336  5.779698976282679E-063   184.000000000000
337  9.515347766178916E-064   185.000000000000
338  1.534796162412530E-064   186.000000000000
339  2.426272676407299E-065   187.000000000000
340  3.760483887662097E-066   188.000000000000
341  5.716253889701179E-067   189.000000000000
342  8.524864002106028E-068   190.000000000000
343  1.247705891426114E-068   191.000000000000
344  1.792762005774854E-069   192.000000000000
345  2.529599594916159E-070   193.000000000000
346  3.506138628169926E-071   194.000000000000
347  4.775076897403873E-072   195.000000000000
348  6.391871649173381E-073   196.000000000000
349  8.411839207567210E-074   197.000000000000
350  1.088641968833092E-074   198.000000000000
351  1.385868130529453E-075   199.000000000000
352  1.735852192088750E-076   200.000000000000
353  2.139717092587529E-077   201.000000000000
354  2.596393661440762E-078   202.000000000000
355  3.101655610260506E-079   203.000000000000
356  3.649547478695239E-080   204.000000000000
357  4.225407182328862E-081   205.000000000000
358  4.825668902909448E-082   206.000000000000
359  5.377802555460645E-083   207.000000000000
360  5.968683035188762E-084   208.000000000000
361  5.923084956117634E-085   209.000000000000
362  6.406353631122458E-086   210.000000000000
sum of gbar=   1.00000000000000


check of sk and homk calculation
(sum of homk*sk)/homd)=   6.25000000000000
     should be   6.25000000000000
sk
  1.02066532258065         3.70463709677419         3.44002016129032
gbar
  0.125000000000000        0.453703703703704        0.421296296296296
homk
  133.333333333333         266.666666666667         400.000000000000
wmavk
  1.05514833980972         0.357952354287769        0.156517642749666
gammak
  0.252569955975152        0.916735395761664        0.851254296064402


test of debo for derivative operator: indx,wdeb,n
s=   6.25000000000000
homa=    400.000000000000
kt=    200.000000000000
sqnrts=    1.00000000000000
ndis=         3
     175  2.228494236619836E-040   -97.0000000000000
     176  7.427612171943742E-040   -96.0000000000000
     177  2.465314811426753E-039   -95.0000000000000
```

178	8.148164034674319E-039	-94.0000000000000
179	2.681572851760524E-038	-93.0000000000000
180	8.786979568086425E-038	-92.0000000000000
181	2.866731314150350E-037	-91.0000000000000
182	9.311257467682524E-037	-90.0000000000000
183	3.010784564497169E-036	-89.0000000000000
184	9.691175247810980E-036	-88.0000000000000
185	3.105084503363426E-035	-87.0000000000000
186	9.902497560491227E-035	-86.0000000000000
187	3.143140962584555E-034	-85.0000000000000
188	9.928954202515173E-034	-84.0000000000000
189	3.121301289944176E-033	-83.0000000000000
190	9.764087041190524E-033	-82.0000000000000
191	3.039217537883207E-032	-81.0000000000000
192	9.412299021782732E-032	-80.0000000000000
193	2.900027708956303E-031	-79.0000000000000
194	8.888917732596597E-031	-78.0000000000000
195	2.710209022722842E-030	-77.0000000000000
196	8.219202955832987E-030	-76.0000000000000
197	2.479102342797155E-029	-75.0000000000000
198	7.436366161870510E-029	-74.0000000000000
199	2.218151572312123E-028	-73.0000000000000
200	6.578807875163863E-028	-72.0000000000000
201	1.939942476234842E-027	-71.0000000000000
202	5.686892026027872E-027	-70.0000000000000
203	1.657154612324493E-026	-69.0000000000000
204	4.799643465986255E-026	-68.0000000000000
205	1.381552149734865E-025	-67.0000000000000
206	3.951762149940679E-025	-66.0000000000000
207	1.123131691473242E-024	-65.0000000000000
208	3.171292632852779E-024	-64.0000000000000
209	8.895189683093170E-024	-63.0000000000000
210	2.478180272764284E-023	-62.0000000000000
211	6.856663058155130E-023	-61.0000000000000
212	1.883806088155137E-022	-60.0000000000000
213	5.138566499055271E-022	-59.0000000000000
214	1.391446895843292E-021	-58.0000000000000
215	3.739762728740520E-021	-57.0000000000000
216	9.974827181468040E-021	-56.0000000000000
217	2.639847961386506E-020	-55.0000000000000
218	6.930904440087415E-020	-54.0000000000000
219	1.804930320169684E-019	-53.0000000000000
220	4.661323366963595E-019	-52.0000000000000
221	1.193579948206079E-018	-51.0000000000000
222	3.029694688194775E-018	-50.0000000000000
223	7.621826834814810E-018	-49.0000000000000
224	1.899923176619444E-017	-48.0000000000000
225	4.691693848040053E-017	-47.0000000000000
226	1.147455508085274E-016	-46.0000000000000
227	2.778731364035326E-016	-45.0000000000000
228	6.661161347191842E-016	-44.0000000000000
229	1.580267928856114E-015	-43.0000000000000
230	3.709118141125081E-015	-42.0000000000000
231	8.610918900001754E-015	-41.0000000000000
232	1.976724645553479E-014	-40.0000000000000
233	4.485808868462635E-014	-39.0000000000000
234	1.006044354133445E-013	-38.0000000000000
235	2.229290842924694E-013	-37.0000000000000
236	4.879709319269808E-013	-36.0000000000000
237	1.054929317298527E-012	-35.0000000000000
238	2.252214215639172E-012	-34.0000000000000
239	4.748443154276739E-012	-33.0000000000000
240	9.887976130943834E-012	-32.0000000000000
241	2.034330092580411E-011	-31.0000000000000
242	4.137599061222335E-011	-30.0000000000000
243	8.326983726188970E-011	-29.0000000000000
244	1.660419199497786E-010	-28.0000000000000
245	3.286438232436813E-010	-27.0000000000000
246	6.471703722855680E-010	-26.0000000000000
247	1.271472094463749E-009	-25.0000000000000
248	2.499917560589868E-009	-24.0000000000000
249	4.933938178447791E-009	-23.0000000000000
250	9.799842665713484E-009	-22.0000000000000
251	1.961875107743487E-008	-21.0000000000000
252	3.959259465599318E-008	-20.0000000000000
253	8.043843224782016E-008	-19.0000000000000

254	1.640725193289291E-007	-18.0000000000000
255	3.347817894578484E-007	-17.0000000000000
256	6.806341688847521E-007	-16.0000000000000
257	1.373166739972364E-006	-15.0000000000000
258	2.739205146038642E-006	-14.0000000000000
259	5.385681297941206E-006	-13.0000000000000
260	1.040712083615241E-005	-12.0000000000000
261	1.972492118163259E-005	-11.0000000000000
262	3.659604723166636E-005	-10.0000000000000
263	6.635168609630354E-005	-9.00000000000000
264	1.174641032380132E-004	-8.00000000000000
265	2.027013202551350E-004	-7.00000000000000
266	3.407270782921270E-004	-6.00000000000000
267	5.576804513751331E-004	-5.00000000000000
268	8.869666120130683E-004	-4.00000000000000
269	1.372278304376002E-003	-3.00000000000000
270	2.062070771954459E-003	-2.00000000000000
271	3.006453644785981E-003	-1.00000000000000
272	4.263285454855759E-003	0.000000000000000E+000
273	5.855772106802558E-003	1.00000000000000
274	7.822811684128189E-003	2.00000000000000
275	1.013984137437971E-002	3.00000000000000
276	1.276558081677431E-002	4.00000000000000
277	1.563268922395776E-002	5.00000000000000
278	1.860306814094844E-002	6.00000000000000
279	2.155580070267142E-002	7.00000000000000
280	2.433001654700120E-002	8.00000000000000
281	2.676818066804045E-002	9.00000000000000
282	2.875614901406725E-002	10.0000000000000
283	3.018849601679559E-002	11.0000000000000
284	3.102318997863694E-002	12.0000000000000
285	3.126987606297555E-002	13.0000000000000
286	3.097702221735155E-002	14.0000000000000
287	3.024601022856756E-002	15.0000000000000
288	2.920036103970466E-002	16.0000000000000
289	2.797473328781010E-002	17.0000000000000
290	2.670358866097040E-002	18.0000000000000
291	2.549922816604524E-002	19.0000000000000
292	2.444595768769046E-002	20.0000000000000
293	2.359359409568555E-002	21.0000000000000
294	2.295469382893656E-002	22.0000000000000
295	2.251001370356483E-002	23.0000000000000
296	2.221454373556004E-002	24.0000000000000
297	2.200635922782561E-002	25.0000000000000
298	2.181672947906975E-002	26.0000000000000
299	2.157874329135445E-002	27.0000000000000
300	2.123479513100049E-002	28.0000000000000
301	2.074186003599869E-002	29.0000000000000
302	2.007419057065302E-002	30.0000000000000
303	1.922386535833720E-002	31.0000000000000
304	1.819937009206284E-002	32.0000000000000
305	1.702275486289232E-002	33.0000000000000
306	1.572598843041796E-002	34.0000000000000
307	1.434700576940252E-002	35.0000000000000
308	1.292592161162735E-002	36.0000000000000
309	1.150175076211772E-002	37.0000000000000
310	1.010983349636389E-002	38.0000000000000
311	8.780056177966138E-003	39.0000000000000
312	7.535852009601219E-003	40.0000000000000
313	6.393892608882601E-003	41.0000000000000
314	5.364338313893820E-003	42.0000000000000
315	4.451494838909594E-003	43.0000000000000
316	3.654725435801977E-003	44.0000000000000
317	2.969484209064835E-003	45.0000000000000
318	2.388360977072350E-003	46.0000000000000
319	1.902056578071151E-003	47.0000000000000
320	1.500235581455535E-003	48.0000000000000
321	1.172228168026015E-003	49.0000000000000
322	9.075730591755249E-004	50.0000000000000
323	6.964082364261490E-004	51.0000000000000
324	5.297259778359431E-004	52.0000000000000
325	3.995141211603357E-004	53.0000000000000
326	2.988073387214568E-004	54.0000000000000
327	2.216715296163467E-004	55.0000000000000
328	1.631420960378079E-004	56.0000000000000
329	1.191336262745557E-004	57.0000000000000

```
330    8.633494231971238E-005    58.0000000000000
331    6.210000036169951E-005    59.0000000000000
332    4.434201987296003E-005    60.0000000000000
333    3.143559892581085E-005    61.0000000000000
334    2.212949385356237E-005    62.0000000000000
335    1.547118051587566E-005    63.0000000000000
336    1.074321366405252E-005    64.0000000000000
337    7.410682911698887E-006    65.0000000000000
338    5.078646689809849E-006    66.0000000000000
339    3.458232128874312E-006    67.0000000000000
340    2.340051904860678E-006    68.0000000000000
341    1.573653809703556E-006    69.0000000000000
342    1.051842718314276E-006    70.0000000000000
343    6.988666703709642E-007    71.0000000000000
344    4.616175747645304E-007    72.0000000000000
345    3.031489060054172E-007    73.0000000000000
346    1.979498634791567E-007    74.0000000000000
347    1.285341562263360E-007    75.0000000000000
348    8.300102475348387E-008    76.0000000000000
349    5.330723923705361E-008    77.0000000000000
350    3.405351495128908E-008    78.0000000000000
351    2.163938237245671E-008    79.0000000000000
352    1.367943260512781E-008    80.0000000000000
353    8.603275084733284E-009    81.0000000000000
354    5.383482673384946E-009    82.0000000000000
355    3.351946079606509E-009    83.0000000000000
356    2.076799134316904E-009    84.0000000000000
357    1.280514497468496E-009    85.0000000000000
358    7.857692691471326E-010    86.0000000000000
359    4.799028998245645E-010    87.0000000000000
360    2.917333318636023E-010    88.0000000000000
361    1.765301704095570E-010    89.0000000000000
362    1.063352483656993E-010    90.0000000000000
363    6.376546912858022E-011    91.0000000000000
364    3.806868190089598E-011    92.0000000000000
365    2.262806815636950E-011    93.0000000000000
366    1.339205721257392E-011    94.0000000000000
367    7.892045349299982E-012    95.0000000000000
368    4.631226920652030E-012    96.0000000000000
369    2.706375240707959E-012    97.0000000000000
370    1.575020405375789E-012    98.0000000000000
371    9.128740670252004E-013    99.0000000000000
372    5.269651577359574E-013    100.000000000000
373    3.029832872385630E-013    101.000000000000
374    1.735162506691885E-013    102.000000000000
375    9.898397813005780E-014    103.000000000000
376    5.624851743972831E-014    104.000000000000
377    3.184171550727792E-014    105.000000000000
378    1.795719251202793E-014    106.000000000000
379    1.008913800873143E-014    107.000000000000
380    5.647549168440490E-015    108.000000000000
381    3.149725186648806E-015    109.000000000000
382    1.750281944750758E-015    110.000000000000
383    9.691294080267514E-016    111.000000000000
384    5.346989552996877E-016    112.000000000000
385    2.939718840463867E-016    113.000000000000
386    1.610593546602092E-016    114.000000000000
387    8.793551889929181E-017    115.000000000000
388    4.784705008983543E-017    116.000000000000
389    2.594612112152941E-017    117.000000000000
390    1.402264270146556E-017    118.000000000000
391    7.553372676793100E-018    119.000000000000
392    4.055260836110256E-018    120.000000000000
393    2.170083722671291E-018    121.000000000000
394    1.157515320752029E-018    122.000000000000
395    6.154348570788891E-019    123.000000000000
396    3.261781136334503E-019    124.000000000000
397    1.723284628502934E-019    125.000000000000
398    9.076126912079997E-020    126.000000000000
399    4.765375509444572E-020    127.000000000000
400    2.494354614017347E-020    128.000000000000
401    1.301652462854765E-020    129.000000000000
402    6.772028376362584E-021    130.000000000000
403    3.512693693796961E-021    131.000000000000
404    1.816646020171874E-021    132.000000000000
405    9.367407613683161E-022    133.000000000000
```

```
406   4.816127638377425E-022    134.000000000000
407   2.468972434519515E-022    135.000000000000
408   1.262072301856163E-022    136.000000000000
409   6.432975109833649E-023    137.000000000000
410   3.269707417244729E-023    138.000000000000
411   1.657238040584834E-023    139.000000000000
412   8.376244013779197E-024    140.000000000000
413   4.221934343378321E-024    141.000000000000
414   2.122171698047752E-024    142.000000000000
415   1.063813441029693E-024    143.000000000000
416   5.318329346840727E-025    144.000000000000
417   2.651664291664347E-025    145.000000000000
418   1.318572570976703E-025    146.000000000000
419   6.539434909645407E-026    147.000000000000
420   3.234710896291918E-026    148.000000000000
421   1.595871966881541E-026    149.000000000000
422   7.853008821364394E-027    150.000000000000
423   3.854407382650708E-027    151.000000000000
424   1.886994114429742E-027    152.000000000000
425   9.214731390218739E-028    153.000000000000
426   4.488504975781921E-028    154.000000000000
427   2.180897353320478E-028    155.000000000000
428   1.057038259027426E-028    156.000000000000
429   5.110642899166140E-029    157.000000000000
430   2.464886893570636E-029    158.000000000000
431   1.185938775631146E-029    159.000000000000
432   5.692177639567035E-030    160.000000000000
433   2.725539734237013E-030    161.000000000000
434   1.301941258745068E-030    162.000000000000
435   6.204429788456034E-031    163.000000000000
436   2.949787251714984E-031    164.000000000000
437   1.399150679099623E-031    165.000000000000
438   6.621094801934208E-032    166.000000000000
439   3.126029838313462E-032    167.000000000000
440   1.472518681268391E-032    168.000000000000
441   6.920527445828601E-033    169.000000000000
442   3.245147889000423E-033    170.000000000000
443   1.518283450124546E-033    171.000000000000
444   7.087619006906395E-034    172.000000000000
445   3.301284948506394E-034    173.000000000000
446   1.534287434804709E-034    174.000000000000
447   7.115040378579809E-035    175.000000000000
448   3.292309983225672E-035    176.000000000000
449   1.520136172592845E-035    177.000000000000
450   7.003712987284948E-036    178.000000000000
451   3.219910798033955E-036    179.000000000000
452   1.477183707967120E-036    180.000000000000
453   6.762476937333929E-037    181.000000000000
454   3.089321391253503E-037    182.000000000000
455   1.408353631890757E-037    183.000000000000
456   6.407032699570597E-038    184.000000000000
457   2.909733701111643E-038    185.000000000000
458   1.317825569001794E-038    186.000000000000
459   5.958321426624796E-039    187.000000000000
460   2.688481095699274E-039    188.000000000000
461   1.210631808555115E-039    189.000000000000
462   5.440566543625841E-040    190.000000000000
463   2.440102045030637E-040    191.000000000000
464   1.092215858727096E-040    192.000000000000
sum in debo=   0.860238154561498
normalization factor should be   0.860238154561498
sum of gbar=   1.00000000000000




check of sk and homk calculation
(sum of homk*sk)/homd)=   6.25000000000000
      should be    6.25000000000000
sk
 1.02066532258065          3.70463709677419          3.44002016129032
gbar
0.125000000000000         0.453703703703704         0.421296296296296
homk
  133.333333333333          266.666666666667          400.000000000000
```

Source Code

```
wmavk
 7.467191348086226E-002    4.851372176099109E-003    3.355752008412450E-004
gammak
 0.252569955975152         0.916735395761664         0.851254296064402


test of debo for derivative operator: indx,wdeb,n
s=    6.25000000000000
homa=    400.000000000000
kt=    50.0000000000000
ndis=         3
sqnrts=    1.00000000000000
        108   1.447614726961549E-039   -32.0000000000000
        109   2.238419915289568E-038   -31.0000000000000
        110   3.380074284603419E-037   -30.0000000000000
        111   4.973015316917952E-036   -29.0000000000000
        112   7.109115570041523E-035   -28.0000000000000
        113   9.840987406518270E-034   -27.0000000000000
        114   1.313689330853731E-032   -26.0000000000000
        115   1.682429846880698E-031   -25.0000000000000
        116   2.054155989214676E-030   -24.0000000000000
        117   2.375224875219750E-029   -23.0000000000000
        118   2.588294388646334E-028   -22.0000000000000
        119   2.669565887026420E-027   -21.0000000000000
        120   2.728597441950322E-026   -20.0000000000000
        121   3.159896175766011E-025   -19.0000000000000
        122   4.837826882063998E-024   -18.0000000000000
        123   9.652336022208480E-023   -17.0000000000000
        124   2.077386692966241E-021   -16.0000000000000
        125   4.264793137964755E-020   -15.0000000000000
        126   8.337370525294722E-019   -14.0000000000000
        127   1.514052124522727E-017   -13.0000000000000
        128   2.539810312654327E-016   -12.0000000000000
        129   4.109645864380201E-015   -11.0000000000000
        130   6.157726921614267E-014   -10.0000000000000
        131   8.612714209373683E-013   -9.00000000000000
        132   1.192147029236623E-011   -8.00000000000000
        133   1.464304758681507E-010   -7.00000000000000
        134   1.721730507956665E-009   -6.00000000000000
        135   2.045439164943837E-008   -5.00000000000000
        136   1.861749696348411E-007   -4.00000000000000
        137   1.914940693884670E-006   -3.00000000000000
        138   1.763182740403196E-005   -2.00000000000000
        139   7.815102492306781E-005   -1.00000000000000
        140   1.110620423613837E-003    0.000000000000000E+000
        141   1.124742993442761E-003    1.00000000000000
        142   3.652031903097536E-003    2.00000000000000
        143   5.708357756146738E-003    3.00000000000000
        144   7.987222179575763E-003    4.00000000000000
        145   1.262931103996058E-002    5.00000000000000
        146   1.529948758023310E-002    6.00000000000000
        147   1.872672369083405E-002    7.00000000000000
        148   2.194212921017597E-002    8.00000000000000
        149   2.281432385615380E-002    9.00000000000000
        150   2.347505450554518E-002    10.0000000000000
        151   2.254806105097660E-002    11.0000000000000
        152   2.005509405896856E-002    12.0000000000000
        153   1.720611678309364E-002    13.0000000000000
        154   1.363608690351503E-002    14.0000000000000
        155   1.003869470784527E-002    15.0000000000000
        156   7.037448665000544E-003    16.0000000000000
        157   4.705967735977047E-003    17.0000000000000
        158   3.394575361989009E-003    18.0000000000000
        159   3.190998228326584E-003    19.0000000000000
        160   3.965626753602040E-003    20.0000000000000
        161   5.583822837831220E-003    21.0000000000000
        162   7.791539483451439E-003    22.0000000000000
        163   1.029041657798322E-002    23.0000000000000
        164   1.280796882799563E-002    24.0000000000000
        165   1.509740913655974E-002    25.0000000000000
        166   1.696589408384558E-002    26.0000000000000
        167   1.829116523819251E-002    27.0000000000000
        168   1.901677595378220E-002    28.0000000000000
        169   1.914519220420770E-002    29.0000000000000
        170   1.872771524218665E-002    30.0000000000000
```

171	1.784919901079376E-002	31.0000000000000
172	1.661302741417938E-002	32.0000000000000
173	1.512879150942800E-002	33.0000000000000
174	1.350171599148284E-002	34.0000000000000
175	1.182497665055852E-002	35.0000000000000
176	1.017548891929000E-002	36.0000000000000
177	8.612047133952475E-003	37.0000000000000
178	7.175449730515656E-003	38.0000000000000
179	5.890260219115942E-003	39.0000000000000
180	4.767345290910352E-003	40.0000000000000
181	3.806767410193846E-003	41.0000000000000
182	3.000736087479838E-003	42.0000000000000
183	2.336265560848121E-003	43.0000000000000
184	1.797421127364833E-003	44.0000000000000
185	1.367111091784168E-003	45.0000000000000
186	1.028399843960481E-003	46.0000000000000
187	7.653998755288593E-004	47.0000000000000
188	5.638131730674740E-004	48.0000000000000
189	4.111915014972367E-004	49.0000000000000
190	2.969935584364371E-004	50.0000000000000
191	2.125041744031072E-004	51.0000000000000
192	1.506680826804924E-004	52.0000000000000
193	1.058807930922426E-004	53.0000000000000
194	7.376626548666570E-005	54.0000000000000
195	5.096122528858996E-005	55.0000000000000
196	3.491844126459083E-005	56.0000000000000
197	2.373501309048577E-005	57.0000000000000
198	1.600759161645122E-005	58.0000000000000
199	1.071377639518716E-005	59.0000000000000
200	7.117273513808084E-006	60.0000000000000
201	4.693644067776905E-006	61.0000000000000
202	3.073263229537306E-006	62.0000000000000
203	1.998237833433715E-006	63.0000000000000
204	1.290370093813341E-006	64.0000000000000
205	8.276760156953943E-007	65.0000000000000
206	5.274025921770658E-007	66.0000000000000
207	3.338985238041295E-007	67.0000000000000
208	2.100532897779228E-007	68.0000000000000
209	1.313220374917396E-007	69.0000000000000
210	8.159947366360477E-008	70.0000000000000
211	5.039940682814888E-008	71.0000000000000
212	3.094545207994719E-008	72.0000000000000
213	1.889056725694879E-008	73.0000000000000
214	1.146599762892127E-008	74.0000000000000
215	6.920506326524228E-009	75.0000000000000
216	4.153959023667368E-009	76.0000000000000
217	2.479834648124531E-009	77.0000000000000
218	1.472502564766186E-009	78.0000000000000
219	8.697565392042556E-010	79.0000000000000
220	5.110719400290297E-010	80.0000000000000
221	2.987737228107375E-010	81.0000000000000
222	1.737845335365334E-010	82.0000000000000
223	1.005818432940744E-010	83.0000000000000
224	5.792929883284203E-011	84.0000000000000
225	3.320295723996844E-011	85.0000000000000
226	1.894016854199660E-011	86.0000000000000
227	1.075344109432263E-011	87.0000000000000
228	6.077078342004381E-012	88.0000000000000
229	3.418632657119022E-012	89.0000000000000
230	1.914458860316715E-012	90.0000000000000
231	1.067334966661451E-012	91.0000000000000
232	5.924353800172073E-013	92.0000000000000
233	3.274090019495127E-013	93.0000000000000
234	1.801660092852178E-013	94.0000000000000
235	9.872118824466768E-014	95.0000000000000
236	5.386730730473009E-014	96.0000000000000
237	2.927110449205195E-014	97.0000000000000
238	1.584064874002051E-014	98.0000000000000
239	8.537827837372920E-015	99.0000000000000
240	4.583344340670164E-015	100.000000000000
241	2.450743569154326E-015	101.000000000000
242	1.305307197333809E-015	102.000000000000
243	6.925415331942385E-016	103.000000000000
244	3.660289922345663E-016	104.000000000000
245	1.927257118002619E-016	105.000000000000
246	1.010964470242039E-016	106.000000000000

Source Code

```
247   5.283495804916606E-017    107.000000000000
248   2.751141291962688E-017    108.000000000000
249   1.427338372152348E-017    109.000000000000
250   7.378696184368587E-018    110.000000000000
251   3.800902693515069E-018    111.000000000000
252   1.951029266577570E-018    112.000000000000
253   9.979893162062193E-019    113.000000000000
254   5.087306926925745E-019    114.000000000000
255   2.584428385828341E-019    115.000000000000
256   1.308488378468692E-019    116.000000000000
257   6.602646694280199E-020    117.000000000000
258   3.320647838816277E-020    118.000000000000
259   1.664553066828742E-020    119.000000000000
260   8.316790732748101E-021    120.000000000000
261   4.142001022701251E-021    121.000000000000
262   2.056239449277980E-021    122.000000000000
263   1.017557023129078E-021    123.000000000000
264   5.019699410622748E-022    124.000000000000
265   2.468554026075686E-022    125.000000000000
266   1.210222997848848E-022    126.000000000000
267   5.915039783768535E-023    127.000000000000
268   2.882245058220664E-023    128.000000000000
269   1.400220289116788E-023    129.000000000000
270   6.782114734225339E-024    130.000000000000
271   3.275277403224174E-024    131.000000000000
272   1.577088048083427E-024    132.000000000000
273   7.571801408037941E-025    133.000000000000
274   3.624835202928215E-025    134.000000000000
275   1.730347312660337E-025    135.000000000000
276   8.236529994004025E-026    136.000000000000
277   3.909589255944012E-026    137.000000000000
278   1.850561540025941E-026    138.000000000000
279   8.735163422824791E-027    139.000000000000
280   4.111904713312351E-027    140.000000000000
281   1.930317781671902E-027    141.000000000000
282   9.037274137048602E-028    142.000000000000
283   4.219664791986365E-028    143.000000000000
284   1.964984539917401E-028    144.000000000000
285   9.126194478674994E-029    145.000000000000
286   4.227447989263156E-029    146.000000000000
287   1.953139434152391E-029    147.000000000000
288   9.000424037802828E-030    148.000000000000
289   4.136905914573942E-030    149.000000000000
290   1.896615749568622E-030    150.000000000000
291   8.673254019878813E-031    151.000000000000
292   3.956322076361820E-031    152.000000000000
293   1.800179520353781E-031    153.000000000000
294   8.170755837990318E-032    154.000000000000
295   3.699460688132261E-032    155.000000000000
296   1.670905129197160E-032    156.000000000000
297   7.528521262842571E-033    157.000000000000
298   3.383913912798857E-033    158.000000000000
299   1.517357061246160E-033    159.000000000000
300   6.787692529658452E-034    160.000000000000
301   3.029209802079185E-034    161.000000000000
302   1.348702641113448E-034    162.000000000000
303   5.990864210177515E-035    163.000000000000
304   2.654946780032890E-035    164.000000000000
305   1.173874896610958E-035    165.000000000000
306   5.178380431477831E-036    166.000000000000
307   2.279180074441116E-036    167.000000000000
308   1.000880435770538E-036    168.000000000000
309   4.385416837582030E-037    169.000000000000
310   1.917215114137995E-037    170.000000000000
311   8.363117175784090E-038    171.000000000000
312   3.640063044556933E-038    172.000000000000
313   1.580880176796159E-038    173.000000000000
314   6.850845784897819E-039    174.000000000000
315   2.962445792928975E-039    175.000000000000
316   1.278272186305451E-039    176.000000000000
317   5.503875814992278E-040    177.000000000000
318   2.364786074541142E-040    178.000000000000
319   1.013908018631552E-040    179.000000000000
sum in debo=    0.511676451298693
normalization factor should be  0.511676451298692
sum of gbar=    1.00000000000000
```

```
check of sk and homk calculation
(sum of homk*sk)/homd)=    6.25000000000000
        should be    6.25000000000000
sk
  1.02066532258065          3.70463709677419          3.44002016129032
gbar
0.125000000000000          0.453703703703704         0.421296296296296
homk
  133.333333333333          266.666666666667          400.000000000000
wmavk
  1.05514833980972          0.357952354287769         0.156517642749666
sum of wdeb=    1.00000000000000



test of debc: indx,wdeb,n
s=    6.25000000000000
homa=    400.000000000000
kt=    200.000000000000
ndis=           3
        178   1.878964229896919E-040   -94.0000000000000
        179   6.359260563854359E-040   -93.0000000000000
        180   2.143802253617919E-039   -92.0000000000000
        181   7.198377899299807E-039   -91.0000000000000
        182   2.407342162506640E-038   -90.0000000000000
        183   8.018171079543081E-038   -89.0000000000000
        184   2.659670195592647E-037   -88.0000000000000
        185   8.785681803080978E-037   -87.0000000000000
        186   2.890002286277365E-036   -86.0000000000000
        187   9.466192106565216E-036   -85.0000000000000
        188   3.087349369764589E-035   -84.0000000000000
        189   1.002554604423112E-034   -83.0000000000000
        190   3.241306221310554E-034   -82.0000000000000
        191   1.043276281947342E-033   -81.0000000000000
        192   3.342895574213563E-033   -80.0000000000000
        193   1.066269332958878E-032   -79.0000000000000
        194   3.385380871153565E-032   -78.0000000000000
        195   1.069842802473394E-031   -77.0000000000000
        196   3.364952486124382E-031   -76.0000000000000
        197   1.053315669131772E-030   -75.0000000000000
        198   3.281196738695941E-030   -74.0000000000000
        199   1.017121742240467E-029   -73.0000000000000
        200   3.137275732777230E-029   -72.0000000000000
        201   9.628152729722986E-029   -71.0000000000000
        202   2.939777725830675E-028   -70.0000000000000
        203   8.929707288956689E-028   -69.0000000000000
        204   2.698238486392037E-027   -68.0000000000000
        205   8.109838162454488E-027   -67.0000000000000
        206   2.424377858630183E-026   -66.0000000000000
        207   7.207942953770955E-026   -65.0000000000000
        208   2.131135268713561E-025   -64.0000000000000
        209   6.265621014544654E-025   -63.0000000000000
        210   1.831616136364221E-024   -62.0000000000000
        211   5.323347413421705E-024   -61.0000000000000
        212   1.538072764342876E-023   -60.0000000000000
        213   4.417446297384261E-023   -59.0000000000000
        214   1.261035887623046E-022   -58.0000000000000
        215   3.577697576045145E-022   -57.0000000000000
        216   1.008687887298918E-021   -56.0000000000000
        217   2.825808058426654E-021   -55.0000000000000
        218   7.865303707336852E-021   -54.0000000000000
        219   2.174844455331644E-020   -53.0000000000000
        220   5.973545376833629E-020   -52.0000000000000
        221   1.629585793909423E-019   -51.0000000000000
        222   4.414805766673894E-019   -50.0000000000000
        223   1.187631725313317E-018   -49.0000000000000
        224   3.171995248576198E-018   -48.0000000000000
        225   8.410206966112850E-018   -47.0000000000000
        226   2.213324931633807E-017   -46.0000000000000
        227   5.780785883999875E-017   -45.0000000000000
        228   1.498195702220838E-016   -44.0000000000000
        229   3.852340466424588E-016   -43.0000000000000
        230   9.826240242569298E-016   -42.0000000000000
        231   2.485915587520804E-015   -41.0000000000000
        232   6.236612419025664E-015   -40.0000000000000
```

Source Code

```
233  1.551311079644245E-014  -39.0000000000000
234  3.825236868309566E-014  -38.0000000000000
235  9.348579149467817E-014  -37.0000000000000
236  2.263998494882903E-013  -36.0000000000000
237  5.432028246733123E-013  -35.0000000000000
238  1.290957376052582E-012  -34.0000000000000
239  3.038301159626335E-012  -33.0000000000000
240  7.079783380066294E-012  -32.0000000000000
241  1.632963740154220E-011  -31.0000000000000
242  3.727293510777523E-011  -30.0000000000000
243  8.417046233805516E-011  -29.0000000000000
244  1.880004603723590E-010  -28.0000000000000
245  4.152125783957894E-010  -27.0000000000000
246  9.065015708399055E-010  -26.0000000000000
247  1.955789594994885E-009  -25.0000000000000
248  4.168644695282272E-009  -24.0000000000000
249  8.774945618292866E-009  -23.0000000000000
250  1.823573341941462E-008  -22.0000000000000
251  3.740050169611921E-008  -21.0000000000000
252  7.567397629820893E-008  -20.0000000000000
253  1.509960171236210E-007  -19.0000000000000
254  2.970032481249748E-007  -18.0000000000000
255  5.756451633590259E-007  -17.0000000000000
256  1.098915006024364E-006  -16.0000000000000
257  2.065360456670601E-006  -15.0000000000000
258  3.819938303193094E-006  -14.0000000000000
259  6.949407916877906E-006  -13.0000000000000
260  1.242966077758687E-005  -12.0000000000000
261  2.184705986921940E-005  -11.0000000000000
262  3.771704941294765E-005  -10.0000000000000
263  6.392597129008576E-005  -9.00000000000000
264  1.063237804805746E-004  -8.00000000000000
265  1.734490807835338E-004  -7.00000000000000
266  2.774103640821206E-004  -6.00000000000000
267  4.348415193092345E-004  -5.00000000000000
268  6.676851530288751E-004  -4.00000000000000
269  1.004118317125466E-003  -3.00000000000000
270  1.478457777970433E-003  -2.00000000000000
271  2.130717333089028E-003  -1.00000000000000
272  3.006164835649893E-003   0.000000000000000E+000
273  4.150070681522735E-003   1.00000000000000
274  5.608777805931060E-003   2.00000000000000
275  7.419486575203905E-003   3.00000000000000
276  9.609268700368889E-003   4.00000000000000
277  1.218931435784231E-002   5.00000000000000
278  1.514609267890479E-002   6.00000000000000
279  1.844503929587353E-002   7.00000000000000
280  2.202255214250811E-002   8.00000000000000
281  2.578957747041053E-002   9.00000000000000
282  2.963700113358593E-002  10.0000000000000
283  3.343637593111216E-002  11.0000000000000
284  3.705229657116690E-002  12.0000000000000
285  4.034905005517194E-002  13.0000000000000
286  4.319878309874076E-002  14.0000000000000
287  4.549259145280139E-002  15.0000000000000
288  4.714531887289665E-002  16.0000000000000
289  4.810154082593598E-002  17.0000000000000
290  4.833870169934697E-002  18.0000000000000
291  4.786619760237338E-002  19.0000000000000
292  4.672396034457111E-002  20.0000000000000
293  4.497800356966498E-002  21.0000000000000
294  4.271453039274812E-002  22.0000000000000
295  4.003376997681578E-002  23.0000000000000
296  3.704303748300718E-002  24.0000000000000
297  3.385037594525681E-002  25.0000000000000
298  3.055903111497707E-002  26.0000000000000
299  2.726284508289898E-002  27.0000000000000
300  2.404303239656763E-002  28.0000000000000
301  2.096619864273790E-002  29.0000000000000
302  1.808353084506021E-002  30.0000000000000
303  1.543106263347507E-002  31.0000000000000
304  1.303073512710464E-002  32.0000000000000
305  1.089204485756097E-002  33.0000000000000
306  9.014054089079666E-003  34.0000000000000
307  7.387541451119206E-003  35.0000000000000
308  5.997133263273508E-003  36.0000000000000
```

309	4.823283946927400E-003	37.0000000000000
310	3.844016187146411E-003	38.0000000000000
311	3.036375117213986E-003	39.0000000000000
312	2.377578907448643E-003	40.0000000000000
313	1.845874695365216E-003	41.0000000000000
314	1.421126936624104E-003	42.0000000000000
315	1.085175076382446E-003	43.0000000000000
316	8.220028993929134E-004	44.0000000000000
317	6.177622141060019E-004	45.0000000000000
318	4.606905330139096E-004	46.0000000000000
319	3.409576583842389E-004	47.0000000000000
320	2.504701345129750E-004	48.0000000000000
321	1.826563882682899E-004	49.0000000000000
322	1.322495890737221E-004	50.0000000000000
323	9.508009669123793E-005	51.0000000000000
324	6.788506003074912E-005	52.0000000000000
325	4.813931382960901E-005	53.0000000000000
326	3.390914547071277E-005	54.0000000000000
327	2.372868452562833E-005	55.0000000000000
328	1.649747441114811E-005	56.0000000000000
329	1.139708898300856E-005	57.0000000000000
330	7.824334579081517E-006	58.0000000000000
331	5.338520319154872E-006	59.0000000000000
332	3.620396680496011E-006	60.0000000000000
333	2.440583893531775E-006	61.0000000000000
334	1.635584726309902E-006	62.0000000000000
335	1.089763762366179E-006	63.0000000000000
336	7.219529759444509E-007	64.0000000000000
337	4.755967628866737E-007	65.0000000000000
338	3.115713476522187E-007	66.0000000000000
339	2.030014060544397E-007	67.0000000000000
340	1.315518153503646E-007	68.0000000000000
341	8.479756680695700E-008	69.0000000000000
342	5.437387909292341E-008	70.0000000000000
343	3.468553899136365E-008	71.0000000000000
344	2.201343530033044E-008	72.0000000000000
345	1.390073371375245E-008	73.0000000000000
346	8.734271986275354E-009	74.0000000000000
347	5.461131573094902E-009	75.0000000000000
348	3.398072855676328E-009	76.0000000000000
349	2.104279254453699E-009	77.0000000000000
350	1.296942120286219E-009	78.0000000000000
351	7.956272188939002E-010	79.0000000000000
352	4.858421370549827E-010	80.0000000000000
353	2.953257781350585E-010	81.0000000000000
354	1.787111873127322E-010	82.0000000000000
355	1.076637166272279E-010	83.0000000000000
356	6.457683626777852E-011	84.0000000000000
357	3.856523258922248E-011	85.0000000000000
358	2.293234580922288E-011	86.0000000000000
359	1.357861330236715E-011	87.0000000000000
360	8.006401989209959E-012	88.0000000000000
361	4.701263330945488E-012	89.0000000000000
362	2.749202539397458E-012	90.0000000000000
363	1.601154392977000E-012	91.0000000000000
364	9.287801959596158E-013	92.0000000000000
365	5.366170878726508E-013	93.0000000000000
366	3.088204454411819E-013	94.0000000000000
367	1.770334776974075E-013	95.0000000000000
368	1.010950808438328E-013	96.0000000000000
369	5.751048469776534E-014	97.0000000000000
370	3.259291405381021E-014	98.0000000000000
371	1.840242125158917E-014	99.0000000000000
372	1.035186249312090E-014	100.000000000000
373	5.801891470273056E-015	101.000000000000
374	3.239991099118713E-015	102.000000000000
375	1.802836672776581E-015	103.000000000000
376	9.995908170401710E-016	104.000000000000
377	5.522759365885026E-016	105.000000000000
378	3.040691689208123E-016	106.000000000000
379	1.668342486513038E-016	107.000000000000
380	9.122386914146748E-017	108.000000000000
381	4.971138633673759E-017	109.000000000000
382	2.699857114545824E-017	110.000000000000
383	1.461424493772322E-017	111.000000000000
384	7.884526681009102E-018	112.000000000000

```
385   4.239857312183353E-018   113.000000000000
386   2.272561953452078E-018   114.000000000000
387   1.214175301547632E-018   115.000000000000
388   6.466366539374129E-019   116.000000000000
389   3.432925603492942E-019   117.000000000000
390   1.816792401626987E-019   118.000000000000
391   9.585053843875644E-020   119.000000000000
392   5.041310643918004E-020   120.000000000000
393   2.643401554555335E-020   121.000000000000
394   1.381861611623614E-020   122.000000000000
395   7.202088747067034E-021   123.000000000000
396   3.742445596795080E-021   124.000000000000
397   1.938947602130158E-021   125.000000000000
398   1.001614208308396E-021   126.000000000000
399   5.159039160419840E-022   127.000000000000
400   2.649604661205517E-022   128.000000000000
401   1.356897527953557E-022   129.000000000000
402   6.929093205180147E-023   130.000000000000
403   3.528406814458135E-023   131.000000000000
404   1.791691253107728E-023   132.000000000000
405   9.072755425069519E-024   133.000000000000
406   4.581586005548721E-024   134.000000000000
407   2.307289346716083E-024   135.000000000000
408   1.158795215551947E-024   136.000000000000
409   5.804147650583655E-025   137.000000000000
410   2.899385229357303E-025   138.000000000000
411   1.444500073383903E-025   139.000000000000
412   7.177641036091938E-026   140.000000000000
413   3.557186616204783E-026   141.000000000000
414   1.758330160802185E-026   142.000000000000
415   8.669041138436536E-027   143.000000000000
416   4.263109180227493E-027   144.000000000000
417   2.091097983142704E-027   145.000000000000
418   1.023110461045830E-027   146.000000000000
419   4.993194182502185E-028   147.000000000000
420   2.430802351964863E-028   148.000000000000
421   1.180438797004472E-028   149.000000000000
422   5.718303643228838E-029   150.000000000000
423   2.763300032569241E-029   151.000000000000
424   1.332088410096038E-029   152.000000000000
425   6.406033866898510E-030   153.000000000000
426   3.073289616501821E-030   154.000000000000
427   1.470898282505800E-030   155.000000000000
428   7.023173808528261E-031   156.000000000000
429   3.345510979618447E-031   157.000000000000
430   1.589923874760032E-031   158.000000000000
431   7.538440461242663E-032   159.000000000000
432   3.566025072342169E-032   160.000000000000
433   1.683027838510719E-032   161.000000000000
434   7.925167909475160E-033   162.000000000000
435   3.723420416690080E-033   163.000000000000
436   1.745412976185613E-033   164.000000000000
437   8.163621541489422E-034   165.000000000000
438   3.809798516136395E-034   166.000000000000
439   1.774032248224761E-034   167.000000000000
440   8.242655947811103E-035   168.000000000000
441   3.821419639653395E-035   169.000000000000
442   1.767827374623513E-035   170.000000000000
443   8.160525851419921E-036   171.000000000000
444   3.758937800418097E-036   172.000000000000
445   1.727771522993487E-036   173.000000000000
446   7.924777405170207E-037   174.000000000000
447   3.627210587674946E-037   175.000000000000
448   1.656718394591958E-037   176.000000000000
449   7.551271276421607E-038   177.000000000000
450   3.434725820537129E-038   178.000000000000
451   1.559085074898112E-038   179.000000000000
452   7.062498123431120E-039   180.000000000000
453   3.192733497031490E-039   181.000000000000
454   1.440415310162296E-039   182.000000000000
455   6.485424629392252E-040   183.000000000000
456   2.914201446971625E-040   184.000000000000
457   1.306880786543645E-040   185.000000000000
sum in debc=    1.00000000000000
normalization factor should be unity
```

Subject Index

a_{uv}: see Manneback
A matrix: see overlap integrals
$Al_2O_3:Cr^{3+}$ phosphor 6, 8
  anharmonicity and temperature-dependent placement of offset states 166
  $^2E \to {}^4A_2$, $^4T_2 \to {}^4A_2$ emissions in competition 14
  energy diagram 165
  $^4T_2 \to {}^2E$ feeding 6
  phonon energies, participating differently in different aspects of luminescence 169
  $^4T_2 \to {}^4A_2$ quenching 8
  thermal equilibrium between $^4T_2$ and ^2E states, temperature for establishing it 171, 172
  transition rates, assessment 166
  spectra, fitting with W_p functions 167
  W_1/W_0 ratio and derived S_0 168

$BaBr_2:Sm^{2+}$ phosphor 13
$BaCl_2:Sm^{2+}$ phosphor 13
  energy diagram 159
$BaClF:Sm^{2+}$ phosphor 13
  $f^5d \to {}^5D$ crossover skipping 161
  energy diagram 159
  quenching and fit by W_p functions 162–165
  two-step quenchings 160
Bandshapes: see Optical bandshapes
Brossel, J., and Marjorie, J., ref. 65 on ruby 168

Cauchy's integral theorem 97
Combining distributions
  equations for single-frequency model 61, 82
  equations for arbitrary frequencies 76
  code for Eq. (6.6), cmbuu 219
  code for combining equal-force constants distributions, cmbww, used in 220

Crossover types
  fast bottom 6, 7
  outside 6–8
  small offset multiphonon emission 8–10
  tunnelling 14, 15
Curie, D.
  W_p in I_p Bessel function form 90

D_p^K
  definition 68
  recursion algebra 68, 83
  with DDESA parameters, code, debc 217
DDESA parameters 72–74
  equations 73–74, 84
  code for obtaining parameters, gtdbpr 218
Derivative-operator integral, $A_{nm,d/dz}$ 31, 78
Dexter, D. L. (see also Miyakawa and Dexter)
  energy transfer rate 108

Energy-Gap Law
  exponential, per Riseberg, Moos, et al. 194–195
  requiring S,A vs. Energy gap correlation 196–197
  for energy transfer, exponential, per Yamada, Shionoya, Kushida 195
Energy Transfer 75–76
  model equations 49, 84–85
  relationship to Förster and Dexter work 108

Feeding fractions
  fitting in oxysulfides:Eu^{3+} 133, 145–147
  fitting in $LaOCl:Eu^{3+}$ 145–146
Fitchen, D. B.
  Laguerre polynomial expression for A_{nm} 88

Fong, F. K.
  derivative-operator expression 91
  as approximate form of $W_{p,d/dz}$ 91–92
  Sm^{2+} luminescence 162
Förster, T., energy transfer rate theory 108
Franck-Condon Principle 14–18
  in optical transitions 14, 16
  in nonradiative transitions 16–18

Gâcon, J. C., ref. 61 158–162

Harmonic oscillator wavefunctions
  derivative operator expression 20
  explicit expressions 20, 77
  generating function 20
  z operator expression 20, 23
Hermite Polynomials
  derivative operator expression 20
  explicit expressions 20
  generating function 19
  recursion formula 20, 77
Huang, K. and Rhys, A. (see also Single-frequency multiple coordinate model)
  W_p expression in I_p Bessel functions 88
  W_p series expression 88
  multicoordinate single-frequency derivative expression 95
  relation of Eq. (8.95) to Eq. (5.6) 95

Inverse transitions, rate relations 48

$KCl:Sm^{2+}$ phosphor 13
  energy diagram 159
  two-step quenchings 160
Keil, T. H.
  derivation of W_p in I_p Bessel function form 90
  Laguerre Polynominal Expression for A_{nm} 88
  moments for S = 0 94
Koide, S.
  derivation of W_p in I_p Bessel function form 90
  Laguerre Polynominal Expression for A_{nm} 88

L_p
  equation 62
  code, wlp 212

Lax, M.
  arbitrary-force-constants moments 92–93
  arbitrary-force-constants moments expression 93
  equal-force-constants moments 92
  W_p as single-configurational-coordinate result 88
  W_p derivation 89
$La_2O_2S:Tm^{3+}$
  excitation and emission spectra 148–149
  two-photon absorption spectra 147–149
  uncovering higher-than-bandgap impurity states 147–149

$<m>$ equation 4
$<m>_p$
  code, wmavp 208
  definition 5, 45, 80
  derivation 45
  expression 45, 80
$M_{p,d/dz}$ Single-frequency multiple coordinate derivative operator distribution
  equation 61
  code for single value, wmder 213
  code for single value, wmdp 226
  code for distribution, wmdpd 213
$M_{p,z}$ Single-frequency multiple coordinate z operator distribution
  equation 61
  code for single value, wmzp 212
  code for distribution, wmzpd 212
Manneback
  a_{uv} definition 25
  θ definition 25
  recursion formulas for A_{nm} 24–31
Markham, J. J., Arbitrary force constants moments 93–94
McClure, D. S., ref. 64, on ruby 167
Miyakawa, T. and Dexter, D. L.
  $M_{p,d/dz}$ approximate expression at T = 0 K 96
  relation to Eq. (5.6) 96
  energy-transfer expression 109
Moments: see Optical Bandshapes
Mostoller, M., Ganguly, B. N., and Wood, R. W., D_p^K at 0 K 107
Multiple-frequency model 66–74
  discretized-Debye-Equal-S-and-A parameters 73, 74, 84

Subject Index

Condon operator recursion algebra 68, 83
Condon operator recursion algebra, code, debc 217
z and d/dz operator recursion algebra 69, 83–84
z and d/dz operator recursion algebra, code, debo 213
supporting code, chis 218
supporting code, cnvwu 219

Nakazawa, E. 154–158
Non-radiative rate model functions
one- vs. N_{Av} dimensional single-frequency model comparisons 121–128

Optical bandshapes
in oxysulfides:Eu^{3+} 130–131, 138–141
in LaOCl:Eu^{3+} 136
in KCl:Sm^{2+} 162
in $BaCl_2Sm^{2+}$ 162
in Y_2O_2S:Yb^{3+} 153, 156
in La_2O_2S:Tm^{3+} 150, 153
in ruby 167–169
descriptions in equal or arbitrary force constants with matching width and position of absorption band 137
descriptions with related first and second moments 110–112
moments, arbitrary force constants, expressions 52, 81–82
moments, equal force constants, expressions 50–51, 81
moments, equal force constants, recursion formula 49, 81, 110
nonradiative rates, calculated for such related fits 112–113
one- vs. N_{Av} dimensional single-frequency bandshapes 113–119
one vs. multi-freqeuncy bandshapes 113–119
O'Rourke, R. C.
moments 94
W_p as single-configurational-coordinate result 88
W_p derivation 89
Overlap integrals A_{nm}, $<u_n \mid v_m>$: see also derivative operator integral and z-operator integral
code gta 221
definition 24

explicit expressions, A_{00} 29
explicit expressions, A_{10} 29
explicit expressions, A_{01} 29
explicit expressions, general 87
explicit expressions, Laguerre polynomial 88
generating function derivation 25–27
generating function expression 27, 87
Manneback recursion formulas for arbitrary force constants 28, 28, 78
Manneback recursion formula derivation for arbitrary force constants 27–28
Manneback recursion formulas for equal force constants 28, 78
orthonormality 29–30
recursion formulas, additional, equal force constants 31, 78
Oxysulfides:Eu^{3+}
absorption, broad band, into CTS 129
absorption and emission, narrow lines 129
CTS feeding, bypassing upper crossovers 130
CTS feeding fractions 133
CTS $\to {}^5D$ feeding 6
energy levels 130
phosphors 6
quenching 141, 143–145
Oxysulfides:Eu^{3+} storage and loss processes 173–193
B_∞/G temperature behavior 182–189
B_0/G and f_{feed} 178–181
binding energy of electron at Eu^{3+} to form Eu^{3+} 182
CTS dissociation, energy barrier 180
energy level diagram for storage and loss 182
f_{feed} and f_{ionize}: breakup of the CTS 175–180
model: word description 177–178
model: equation for rates 177
phosphorescence, excitation energy dependence 190–193
quantum efficiency, nonlinear with excitation intensity 184
quantum efficiency, equation 184
quantum efficiency, derivation of equation 186–189
rise curves and rise behavior 176
rise time, excitation energy dependence 189–190

storage and glow curves 173, 178
Oxysulfides:Tb^{3+} 13

Pekar, S. I.
  W_p expression in Bessel function form 88
  W_p series expression 88
Perlin, Yu. E.
  $M_{p,d/dz}$ approximate form 95
  $M_{p,d/dz}$, (8.41) relation to (5.6) 95
  multiple-frequency Condon-operator approximate form 96
  relation of (8.46) to (5.32,35) 104–106
  multiple-frequency derivative-operator approximate form 105–106
  relation of (8.78–79) to (5.50–52) with DDESA parameters (5.55–61) 107–108

Quenchings
  fitting to W_p functions in oxysulfides: Eu^{3+} 141–145
  fitting to W_p functions in LaOCl:Eu^{3+} 141–142
  fitting to W_p functions in La_2O_2S:Tm^{3+} 150
  fitting to W_p functions in ruby 169–172
  fitting to W_p functions in Y_2O_2S:Yb^{3+} 155–159
  "quenching efficiencies" 135
  quenching expressions 135
  two-step quenching with fast second step 10–15, 132
  two-step quenching, efficiency expression 11, 12
  two-step quenching, lifetime expression 12

Residues, theorem 98
Riseberg, L. A. and coworkers on exponential energy gab law 194–195
Ruamps, J.
  Laguerre polynomial expression for A_m 88
Ruby: see Al_2O_3:Cr^{3+}

Saddle point approximation
  statement 98
  Perlin's use 99–101, 104–107
Semiclassical model
  equations 136
  relating W_p fits to semiclassical-model fits 138–139
  use for LaOCl:Eu^{3+} 138

use for oxysulfides:Eu^{3+} 138
Single-coordinate, arbitrary force constants model: see also overlap integrals
  curve equations for arbitrary force constants 2
  distribution functions, Condon operator 37
  distribution functions, z operator 37
  distribution functions, d/dz operator 39
  energy balance equation 33
  nonradiative rate model energy balance 5
  nonradiative rate model equation 5
  S_u, S_v definition 32
  $S_u, S_v \rightarrow \theta, a_{uv}$ transformation 28
Single-coordinate, equal force constants model: see also W_p $W_{p,z}$, and $W_{p,d/dz}$
  curve equations for equal force constants 2
  crossover expression 3
  distribution functions, Condon operator, W_p, definition 34
  distribution functions, z operator, definition 36
  distribution functions, z operator, expression 36, 80
  distribution functions, z operator, $5-W_p$ expression 44, 48, 80
  distribution functions, d/dz operator, definition 45
  distribution functions, d/dz operator, expression 48, 80
  energy balance equation 3, 34
  $<m>_p$ 5
  S, definition 34
Single-frequency multiple-coordinate model
  reproductive property of W_p distributions 60–61, 82
  z- and d/dz distributions 61–66, 82
SrF_2:Sm^{2+} 13

Test programs for codes 226, 227
  output 228–248
Theta, θ: see Manneback
Two-step quenching with fast second step 10–13
  efficiency equation 11, 12
  lifetime 12
Two-step quenching with slow second step 13–14
  efficiency equation 13

Subject Index

U_{pU} distribution
  apportioning thermal weights 37–38
  definition 37
  code for individual member, upum 222
  code for distribution, upu 223
  code for a set of distributions at several temperatures, upudis 224
$U_{pU, d/dz}$ distribution
  definition 39
$U_{pU, z}$ distribution
  definition 37

V_{pV} distribution
  apportioning thermal weights 37–38
  definition 37
  code for individual member, vpvm 223
  code for distribution, vpv 224
  code for a set of distributions at several temperatures, vpvdis 225
$V_{pV, d/dz}$ distribution
  definition 39
$V_{pV, z}$ distribution
  definition 37
Vibrationally-enhanced radiative transitions
  moments 120–121

W_p: see also single-coordinate, equal force constants model
  definition
  code for single value, wp 204
  code for distribution wpdis 205
  code for distribution wpdist 206
  code for a set of distributions at several temperatures, wpset 209
W_p expressions
  definition 3, 34, 79
  series expansion 40, 79
  I_p Bessel function form 42, 79
  0 K Poisson distribution 42, 79
  Gaussian 42, 79
  B_p^* 42
  W_p^* 42, 79
W_p recursion formula
  derivation 39–40
  r-form 40
  $<m>$-form 40, 79
$W_{p, d/dz}$
  code for a single value, (4.88) wpder 210
  code for a distribution, wpderd 210
  code for a distribution, wpdrs 211
  code for set of distributions at several temperatures, wpdset 207
  expression 47, 48
$W_{p, z}$
  code for a single value, (4.23) wpz 208
  code for a distribution, wpzd 208
  code for a distribution, wpzds 209
  code for set of distributions at several temperatures, wpzset 207
  expression 36, 46
Wagner, M.
  derivation of W_p in I_p Bessel function form 90
  Laguerre polynomial expression for A_{nm} 88
Wavefunctions u_n, v_m 25

X function, Eq. (5.48), code chisd 226

$Y_2O_2S:Yb^{3+}$
  emission in sequence of CTS → $^2F_{7/2}$ and $^2F_{7/2}$ → $^2F_{5/2}$ 154
  energy-transfer depopulation of CTS 154
  energy loss and storage by CTS breakup into Yb^{2+} + hole 158
Yamada, N. and coworkers on energy transfer exponential energy gap law 195

z-operator integral, $A_{nm, z}$ 31, 78

Source of Illustrations

A substantial number of the illustrations in this book have been previously published elsewhere. Permission for their reproduction was granted by the following publishers:

Elsevier Science Publishers, Inc.:

Figures: 1–13, 18, 22, 42, 44, 46, 69, 70
Tables: 1–4, 42

The American Institute of Physics:

Figures: 25–38, 45, 47, 48
Tables: 19, 22, 26, 27, 32

Plenum Press:

Figures: 40, 41

The American Physical Society:

Figures: 50–56, 60–68
Tables: 35, 37, 38, 41

The Electrochemical Society:

Figures: 57–59